Springer Topics in Signal Processing

Volume 22

The aim of the Springer Topics in Signal Processing series is to publish very high quality theoretical works, new developments, and advances in the field of signal processing research. Important applications of signal processing are covered as well. Within the scope of the series are textbooks, monographs, and edited books.

Topics include but are not limited to:

* Audio & Acoustic Signal Processing
* Biomedical Signal & Image Processing
* Design & Implementation of Signal Processing
* Graph Theory & Signal Processing
* Industrial Signal Processing
* Machine Learning for Signal Processing
* Multimedia Signal Processing
* Quantum Signal Processing
* Remote Sensing & Signal Processing
* Sensor Array & Multichannel Signal Processing
* Signal Processing for Big Data
* Signal Processing for Communication & Networking
* Signal Processing for Cyber Security
* Signal Processing for Education
* Signal Processing for Smart Systems
* Signal Processing Implementation
* Signal Processing Theory & Methods
* Spoken language processing

** Indexing: The books of this series are indexed in Scopus and zbMATH **

Jacob Benesty • Gongping Huang • Jingdong Chen • Ningning Pan

Microphone Arrays

Springer

Jacob Benesty
INRS-EMT, University of Quebec
Montreal, QC, Canada

Gongping Huang
Wuhan University
Wuhan, China

Jingdong Chen
Northwestern Polytechnical University
Xi'an, China

Ningning Pan
Northwestern Polytechnical University
Xi'an, China

ISSN 1866-2609 ISSN 1866-2617 (electronic)
Springer Topics in Signal Processing
ISBN 978-3-031-36973-5 ISBN 978-3-031-36974-2 (eBook)
https://doi.org/10.1007/978-3-031-36974-2

This Springer imprint is published by the registered company Springer Nature Switzerland AG
The registered company address is: Gewerbestrasse 11, 6330 Cham, Switzerland

Preface

Microphone arrays refer to systems that comprise a number of sensors arranged to sample the spatial sound field from the one-, two-, or three-dimensional perspective. They have been used in a wide spectrum of applications to achieve such functions as sound source localization, noise reduction, signal enhancement, source separation, dereverberation, spatial sound acquisition, to name but a few. Generally, a microphone array system is composed of two important components: hardware and algorithm. The former is composed of an array of microphones (arranged into a particular geometry such as a line, circle, cuboid, sphere, etc.) to observe the sound field, a multichannel pre-amplifier to amplify the microphones' outputs, an automatic gain controller to ensure that small and large signals are sensed properly, a multichannel analog-to-digital (A/D) converter to digitize the amplified microphone signals, as well as digital processors to handle the storage, processing, and transmission of the digitized sound signals. The algorithm component, in contrast, consists of a number of processing techniques that take the digitized array signals as their inputs to accomplish multiple functions. While both components are important from the performance perspective, most studies and publications are devoted to the latter as algorithms can be flexibly integrated into the array processors to achieve different objectives/functions and also can be optimized to improve performance.

The reason behind this endeavor is that we believe there is still a need to clarify the advantages of using microphone arrays as compared to using a single sensor for sound acquisition from a signal enhancement and spatial filtering perspective, and summarize the most useful ideas, concepts, results, and new algorithms. The material presented in this book includes analysis about the advantages of using microphone arrays, dimensionality reduction to remove the redundancy while preserving the variability of the array signals using the principal component analysis (PCA), beamforming with low-rank approximation, fixed/adaptive and robust distortionless beamforming, differential beamforming, a new form of binaural beamforming that takes advantage of both beamforming and human binaural hearing properties to potentially improve speech intelligibility, and beamforming with large arrays, which consist of a large number of microphones.

Montreal, QC, Canada	Jacob Benesty
Wuhan, China	Gongping Huang
Xi'an, China	Jingdong Chen
Xi'an, China	Ningning Pan

Contents

Chapter 1
Introduction

In this chapter, we discuss the important applications of microphone arrays and their associated beamforming in today's products and technologies. Then, we present the organization of this work.

1.1 Sound Signal Acquisition and Processing

Sound signal acquisition refers to the process of sensing sound waves, which radiate from certain sources of interest, and converting them into electrical signals that can be stored, processed, and transmitted. It is an integral part of such systems as voice communication, human-machine speech interface, music and sound recording, and public address (PA), to name but a few. Generally, the acquired signal consists of not only the sound signals of interest but also signal components from various distortion sources, e.g., noise generated by sensors, preamplifiers, analog-to-digital (A/D) converters, and the associated circuits, noise from different ambient sound sources, interference from competing sources, signal components due to the coupling between loudspeakers and sensors, and the multipath effect. Consequently, signal processing tools are needed either to estimate some important parameters related to the sources, signals of interest, and the propagation environment or to enhance the signals of interest before they are stored and transmitted. This acquisition and processing step has been an important area of research in the field of signal and information processing over more than a century, and it is still an active research topic as the resulting technologies are needed in a wide spectrum of applications, whose types and number are still increasing.

Chronologically, research in sound signal acquisition and processing can be divided into three periods, i.e., waveform transformation, digital sound signal processing, and intelligent sound processing and analysis. Before pulse code modulation (PCM) and solid-state A/D converters were widely used, the signals picked up

J. Benesty et al., *Microphone Arrays*, Springer Topics in Signal Processing 22,
https://doi.org/10.1007/978-3-031-36974-2_1

by sensors were primarily processed with low-pass, high-pass, or band-pass filters, implemented with analog circuits to reject unwanted frequencies and suppress out-of-band or even in-band noise [1]. While a large number of efforts were devoted to developing high-quality microphones and filters via resistor-capacitor (RC), resistor-inductor (RL), and RLC circuits, the means to process sound signals in this period were rather limited. After the 1960s, as digital processors, A/D converters, and computers started to gain popularity, sound signals were generally converted to the digital domain, in which more complicated and effective transformations, filters, and estimation techniques were developed to either enhance signals or estimate the signal parameters. Many technologies were developed in this period such as echo cancellation based on adaptive filters, speech coding based on linear predictive coding (LPC), noise reduction based on statistical models and optimal filtering, speech recognition and enhancement based on statistical Markov models, etc., most of which are playing a vital role in our daily lives today. As data-driven and machine learning methods are gaining more and more popularity today, research in sound signal acquisition and processing is moving to the third period in which technologies are developed not only to process the signals but also to identify, recognize, and interpret the embedded information.

While the fundamental theory and methods developed in the aforementioned three periods may differ significantly, most sound systems built in these periods share a common structure: they use a single microphone and produce a single output. There are strong reasons to adopt such a single-input/single-output (SISO) structure: it is cost-effective; it can be conveniently integrated into various devices, large or small; and it is simple from the system design perspective. However, the signal captured by a single sensor does not contain any spatial information, and, as a result, it is difficult if not impossible to accomplish such tasks as direction-of-arrival (DOA) estimation, source localization, and spatial sound recording and preproduction. This leads to the development of multichannel systems in which multiple microphones are used to sample the sound field.

Depending on how the multiple microphones are arranged and their outputs converted to the digital domain, these systems fall into two basic categories: microphone arrays and microphone sensor networks (which are also called acoustic sensor networks). A microphone array consists of a set of sensors of same frequency responses, which are positioned in a way that the spatial information is well captured. As the level of the sensed signals is generally too small to match the requirements of the subsequent devices in the circuit chain, signal amplification is needed. In a microphone array, the microphones' outputs are amplified with the same gain and then converted to the digital domain synchronously with the same clock for subsequent spatiotemporal processing. In contrast, in a sensor network, microphones may have very different frequency responses and they may be arbitrarily placed in the sound field. Signals from different sensors may be amplified with different gains and converted to the digital domain with different clocks, and consequently the subsequent processors have to deal with the clock-shift and clock-skew problem. Moreover, unlike what is assumed to be true in microphone arrays, the signal components at different sensors due to the same source in a sensor network may not be coherent. Consequently, the signal processing theory and methods needed in sensor networks are rather different from those needed in microphone arrays. In this book, we focus only on microphone arrays and leave the topic on sensor networks for future discussion.

1.2 Microphone Arrays and Their Applications

When we talk about a microphone array, we generally mean such a system that consists of two components: hardware array and signal processing unit.

The hardware array is composed of an array of microphones, a multichannel preamplifier, a multichannel A/D converter, some memory to store the sensed signals, and a processor with sufficient computational power to run processing algorithms. The microphones in the array should have the same characteristics and are arranged into a particular geometry to sample the sound field. The selection of the sensors should be made according to the application requirements in terms of sensitivity, signal-to-noise ratio (SNR), dynamic range, frequency response, etc. The geometry is also an important factor that plays a critical role on the array performance, and its determination may be restricted by the device in which the array is integrated.

The signal processing unit consists of a number of processing algorithms that take the digital microphone signals as their inputs to accomplish certain tasks. Typical problems that can be solved with microphone arrays include, but not limited to, the following:

- DOA estimation, which is to estimate the incidence angles of the sources located in the far field with respect to the array;
- source localization, which is to estimate the positions of the sources located in the near field with respect to the array;
- source number estimation, which is to determine the number of active sources in the sound field;
- multichannel echo cancellation, which is to mitigate or eliminate the detrimental effects due to the coupling between loudspeakers and microphones;
- acoustic channel identification, which is to estimate the channel impulse response between a source and a sensor;
- noise reduction, which is to reduce the impact of background noise on speech, thereby improving the quality and/or intelligibility of the noisy speech;
- dereverberation, which is to mitigate the detrimental effect on speech caused by late reflections;
- source separation, which is to simultaneously recover the multiple source signals from the observed mixed signals;
- signal extraction in cocktail party environments, which is to extract the source signals of interest in complicated acoustic environments, where multiple sources, noise, reverberation, and echo may coexist;

- spatial sound recording, which is to record the sound signals and meanwhile preserve the sound realism;
- distant sound acquisition, which is to pick up signals from sources that are far away from the sensors;
- environment parameter estimation, which is to estimate some important parameters related to the acoustic environment or sound field, such as sound level, reverberation time, coherent-to-diffusion ratio (CDR), noise level, noise covariance matrices, SNR, etc.

Each of these tasks is a rich topic of research. The interested reader is encourage to read [2–23] and many other books and papers for the descriptions of these problems, solutions, and applications. Since they have the potential to solve many acoustic tasks, microphone arrays have been used in a wide spectrum of applications such as smart speakers, smart glasses, smart televisions, hearing aids, audio bridging, teleconferencing, hands-free human-machine interfaces, computer games, acoustic surveillance (security and monitoring), acoustic scene analysis, concert recordings, etc. In the following, we present two examples to illustrate how microphone arrays can be deployed and solve complicated acoustic problems.

1.2.1 Airborne Dipping and Reconfigurable Microphone Arrays

Earthquakes are quite common in the planet we live and large ones happen almost every year, which often cause great damage and destruction. A critical issue after a damaging earthquake is to respond to the immediate effects, search and rescue quake survivors speedily. This is extremely challenging as the area hit by the disaster can be vast and survivors are often trapped in the rubble or buried under falling debris.

To help quickly find quake survivors, our team developed an airborne dipping and reconfigurable microphone array system, which can be carried by either a drone or a helicopter. The system is illustrated in Fig. 1.1, which consists of a microphone array with reconfigurable topology and a releasing apparatus. The array consists of four arms, in which the sensors are mounted. If the four arms are in their original positions, i.e., they are parallel with system's axis as shown in Fig. 1.1a, this gives a cylindrical topology. When the four arms are fully extended, i.e., they are perpendicular to the axis as illustrated in Fig. 1.1c, a concentric circular array is formed. The arms can be configured to any angle between 0° and 90°, forming an array of conical frustum-shaped topology as illustrated in Fig. 1.1b. The functions for this array to achieve include (1) recording the sound signals and transmitting them to a searching team or rescuing center; (2) detecting the presence of survivors based on their voice signals such as screaming, crying, calling for help, etc.; and (3) if any survivors are detected, estimating their positions and passing the information to the rescuing team/center. The topology of the array can be controlled through two different modes: remote and automatic. In the former, the angle parameters

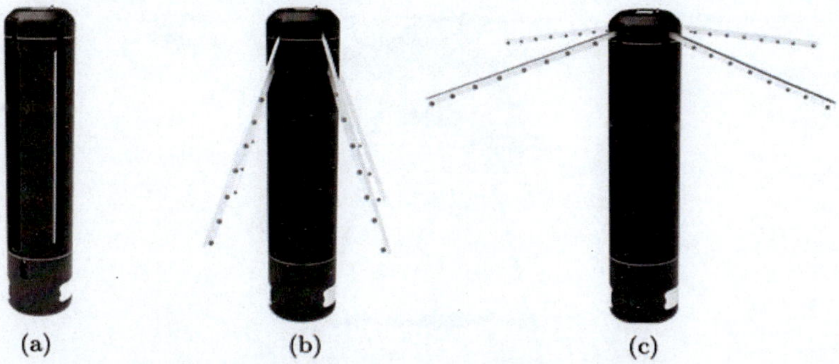

Fig. 1.1 Illustration of the airborne dipping and reconfigurable microphone array: (**a**) the four arms are in the original position and the array is a perfect cylindrical array; (**b**) the four arms are configured to 15° so the array becomes one of the conical frustum shapes; and (**c**) the four arms are configured to 90° so the array becomes one with a concentric circular topology

Fig. 1.2 Illustration of the airborne dipping and reconfigurable microphone array: the dipping array is in its home position

are controlled by the operator through a remote terminal, while in the latter, the geometry is optimized according to the tasks to accomplish. For detection and signal enhancement, the geometry is optimized to have the largest array gain for SNR improvement; but for localization, the geometry is optimized to have the highest possible resolution.

The release apparatus consists of a 60-m cable, a reeling machine, and a releasing control device. When the carrier is on the ground or in a flying mode, the apparatus is fully closed and the array is placed in its home position as illustrated in Fig. 1.2. This position can help protect the array from damage as well as safeguard the carrier from the impact of winds. Once the carrier is in the disaster area for operation, the array is released to a certain height as illustrated in Fig. 1.3. This, on the one hand, helps reduce the impact of propeller noise, thereby improving the SNR of the array observation signals and, on the other hand, makes the array close to the ground for optimal detection and localization performance.

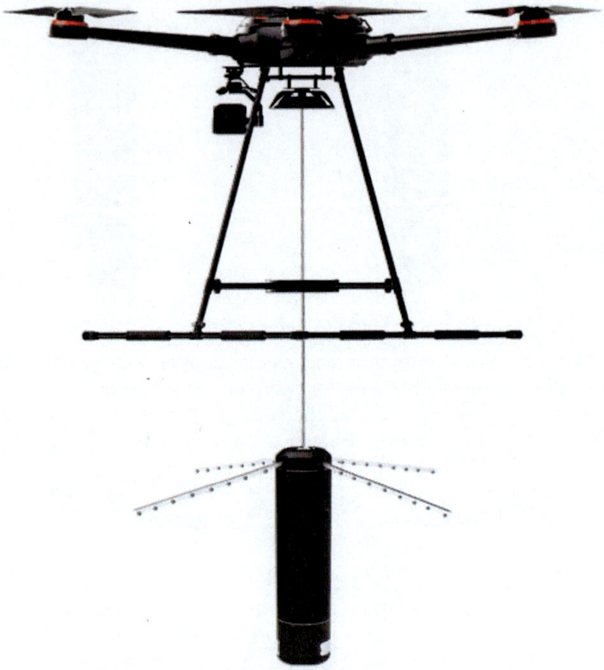

Fig. 1.3 Illustration of the airborne dipping and reconfigurable microphone array: the dipping array is on a released mode and reconfigured to a concentric circular topology

1.2.2 Linear Microphone Arrays for Smart Panels

Tele-education, which is an alternative mode of teaching and learning, has become a new normal for many teachers and students since the outbreak of the COVID-19 pandemic. One of most challenging issues in tele-education is how to deliver high-quality audio and video information to remote sites so that the remote participants can enjoy the same learning experience as the in-person students. To deal with the issue, our team developed a smart panel system, which consists of a linear microphone array, a high-definition and wide-angle camera, and a large touch screen as illustrated in Fig. 1.4.

The microphone array shown in Fig. 1.4 consists of 12 electret microphones arranged in a nonuniform way so that the camera can be placed in the middle of the upper frame for a more balanced video recording. Different processing algorithms are integrated into the system to process the microphone signals including source localization to determine the positions of the teacher (or any active speaker) in the podium, near-field adaptive beamforming to enhance the teacher's speech, multichannel echo cancellation to enable full duplex communications between different classrooms, sound field separation to separate sound signals in the podium region from those in the rest part of the classroom, and feedback/howling control

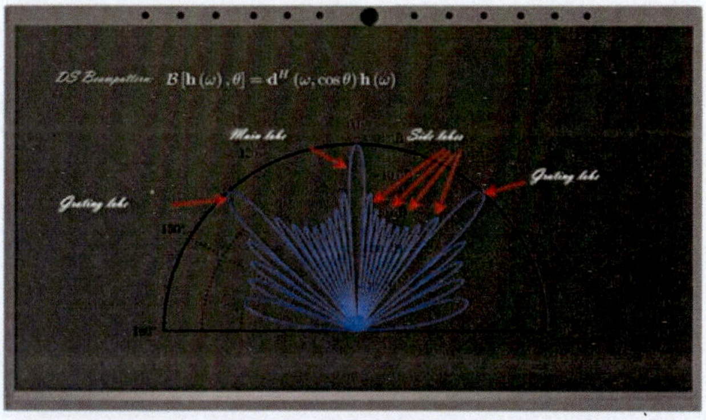

Fig. 1.4 Illustration of the smart panel system equipped with a linear microphone array for classroom sound acquisition

so that the microphone array outputs can be played back through local loudspeakers without howling. The array can achieve decimeter level of localization accuracy and improve the SNR by 10 dB or more. It also produces multichannel outputs for stereophonic sound reproduction in remote classrooms.

Note that the previous two examples show only some possibilities of using microphone arrays to solve important acoustic problems. The number of applications is enormous and still growing quickly, which cannot be enumerated completely.

1.3 Major Addressed Problems

Although a number of edited/monograph books and many technical papers have been published to address the problems encountered in microphone arrays, and much progress has been made in this field of research, there are still many tasks to be understood/investigated and solutions to be developed. This motivated us to write this book in which we cover the following major problems:

- In an array system, all the microphones measure the same sound field but each from its own viewpoint. As a result, signals from different sensors contain both complementary and redundant information. While the complementary information is needed, the redundancy may greatly affect the efficiency in storing and transmitting the signals as well as the performance of the subsequent processors. A legitimate question one may ask then is how to remove the redundancy while preserving the useful information, thereby reducing the dimensionality of the array signals, which will be investigated in this book.
- It is well known that speech signals as well as room acoustic impulse responses (i.e., the acoustic system) are of low rank in nature, and this is even more so in the multichannel scenario associated with arrays. An important question then

arises as how to take the low-rank property into consideration in microphone array processing, which will be discussed in this book.

- One of the major tasks in microphone array processing is to reduce the impact of noise, thereby enhancing the signals of interest. It is known that in the single-channel case, noise reduction is achieved by introducing speech distortion, and the amount of this distortion is generally proportional to the level of noise reduction [24]. A critical issue then arises as whether and how signal enhancement can be achieved without adding distortion to the speech signals of interest. This will also be covered in this book.
- Typically, beamforming takes the microphone array observations as its inputs and generates one output. This multiple-input/single-output (MISO) structure may be optimal if the output serves as the input for subsequent processing. But it is suboptimal if the output is for a human listener as in speech communication as this processing structure does not take into account the properties of the human binaural hearing system. How to optimize beamformers by jointly considering speech enhancement and binaural hearing properties is an important problem to be investigated.
- Microphone arrays are designed to measure the sound field, either the pressure field or the differential fields of different orders. While many works have been published to discuss the relationship between beamforming and sound field measurement, how this relationship can be used to optimize beamformers for high-fidelity sound acquisition still needs further investigation.

Note that the investigation of the aforementioned problems is by no means comprehensive and complete. We hope that this study will serve as a stepping stone for readers to take as they embark on their own journey of study.

1.4 Organization of the Work

This book attempts to cover the most basic concepts, fundamental principles, as well as some new advancements in microphone array processing. The material discussed in the text is arranged into ten chapters, including this one.

Single microphone techniques have been dominantly used in all applications that require sound acquisition and processing. Though this trend will likely continue for the foreseeable future, more and more speech communication, sound recording, and human-machine speech interface systems start to use microphone arrays since single microphone methods suffer from a number of great limitations. Chapter 2 presents an analysis of these limitations, thereby clarifying the necessity and advantages of using microphone arrays from a signal enhancement and spatial filtering perspective. What is also covered in this part is the importance of spatial information through which microphone arrays offer more flexibility in processing speech signals and achieving much better speech enhancement performance.

After a brief but insightful discussion on the limitations of single microphone processing and the importance of using spatial information in Chap. 2, Chap. 3 presents a concise overview of the most fundamental concepts in microphone array processing. It starts with the signal model of a general three-dimensional array topology. With this model, it is explained how to conduct linear beamforming, based on which a number of useful performance measures are derived. A large family of fixed and adaptive beamformers are then deduced and some signal statistics estimators are also presented. Different array geometries are subsequently compared and some examples are finally presented to illustrate the performance of different beamformers.

In a microphone array system, the observation signals at different sensors consist of complementary as well as redundant information. As a result, processing is generally needed to remove some of the redundancy, which is expected to not only reduce the complexity but also improve the robustness and performance of the subsequent multiple-channel processing, e.g., beamforming. Principal component analysis (PCA) is one of the most popular and useful dimensionality reduction techniques; it transforms a random signal vector of high dimension to one of much smaller dimension with little loss of the useful information. Chapter 4 discusses the principle of PCA and shows how it can be applied to noise reduction and beamforming.

Speech signals and room acoustic impulse responses (i.e., the acoustic system) are of low rank in nature, and so are the microphone observations. This low-rank feature leads to a type of redundant information different from that of multiple channels. Therefore, it is important to take into account the low-rank property in microphone array processing, which is addressed in Chap. 5. We explain the principle of low-rank processing in beamforming and derive a class of low-rank fixed and adaptive beamformers.

One of the most important objectives of beamforming is to recover the signals of interest without introducing any signal distortion, which leads to the idea of distortionless beamforming. Chapter 6 studies this principle and presents a family of distortionless beamformers. We also discuss how to make these beamformers robust so that they can be implemented in real-world applications.

While most works in the literature address the problem of beamforming from either the signal processing or optimization perspectives, there is another very different but fundamental approach to the problem, i.e., through examining and measuring the physical sound fields including the sound pressure field and the differential sound pressure fields of different orders. This unique way of studying the problem leads to a better understanding not only of the beamforming process but also to a new category of beamforming, which is known as the differential beamforming. Differential beamformers have many prominent properties such as high directional gains and frequency-invariant spatial responses, which make them very useful for high-fidelity speech signal acquisition and processing. Chapter 7 focuses on differential beamforming. It starts with the so-called first-order linear difference equations and then derives a class of differential beamformers.

Another major task of beamforming is to reduce noise, thereby enhancing the signals of interest, which can be achieved through different ways including beamforming as discussed in the previous chapters. Chapter 8 is concerned with another way to achieve noise reduction, i.e., adaptive noise cancellation, which first finds or creates a noise reference and then uses it to adaptively cancel the noise from the microphone observation signals. The analysis in this chapter shows that adaptive noise cancellation is not only just another approach to speech enhancement but also helps to gain many insights into distortionless adaptive beamforming.

Chapter 9 addresses the problem of binaural beamforming from a fresh perspective, where the properties of the human binaural hearing system are taken into account in the beamforming process. After binaural beamforming, the speech signal of interest and the unwanted noise are rendered in different positions in the perceptual space to maximize intelligibility. Note that both the objective and principle of binaural beamforming in this chapter are very different from those in conventional binaural beamforming methods published in the literature, which attempt to preserve the sound source realism while achieving noise reduction. The methods presented in this part are designed to maximize the spatial separation between the desired speech and unwanted noise in the perceptual space to help the human binaural hearing system better process the signals.

Finally, Chap. 10 studies beamforming in the context of very large arrays, i.e., arrays that contain a large number of microphones. Conventional beamforming in this context may not be practical or optimal in terms of performance, complexity, robustness, and statistics estimation. This chapter tackles the problem from a low-rank beamforming perspective, which leads to flexible beamforming methods that can achieve either simply better performance or better compromise among array gain, robustness, and complexity, which are important performance metrics that are difficult, if not impossible, to be optimized at the same time.

References

1. M.R. Schroeder, Apparatus for suppressing noise and distortion in communication signals. U.S. Patent No 3,180,936, filed Dec. 1, 1960, issued Apr. 27, 1965
2. M. Branstein, D. Ward (eds.), *Microphone Arrays: Signal Processing Techniques and Applications* (Springer, Berlin, 2001)
3. J. Benesty, T. Gänsler, D.R. Morgan, M.M. Sondhi, S.L. Gay, *Advances in Network and Acoustic Echo Cancellation* (Springer, Berlin, 2001)
4. J. Benesty, S. Makino, J. Chen (eds.), *Speech Enhancement* (Springer, Berlin, 2005)
5. J. Benesty, M.M. Sondhi, Y. Huang (eds.), *Springer Handbook of Speech Processing* (Springer, Berlin, 2007)
6. E. Hänsler, G. Schmidt (eds.), *Topics in Acoustic Echo and Noise Control: Selected Methods for the Cancellation of Acoustical Echoes, the Reduction of Background Noise, and Speech Processing* (Springer, Berlin, 2006)
7. S. Makino, T.-W. Lee, H. Sawada (eds.), *Blind Speech Separation* (Springer, Berlin, 2006)
8. Y. Huang, J. Benesty, J. Chen, *Acoustic MIMO Signal Processing* (Springer, Berlin, 2006)
9. J. Benesty, J. Chen, Y. Huang, *Microphone Array Signal Processing* (Springer, Berlin, 2008)

10. J. Benesty, J. Chen, Y. Huang, I. Cohen, *Noise Reduction in Speech Processing* (Springer, Berlin, 2009)
11. J. Benesty, J. Chen, E. Habets, *Speech Enhancement in the STFT Domain* (Springer, Berlin, 2011)
12. J. Benesty, C. Paleologu, T. Gänsler, S. Ciochină, *A Perspective on Stereophonic Acoustic Echo Cancellation* (Springer, Berlin, 2009)
13. J. Benesty, J. Chen, *Study and Design of Differential Microphone Arrays* (Springer, Berlin, 2012)
14. M.R. Bai, J.-G. Ih, J. Benesty, *Acoustic Array Systems: Theory, Implementation, and Application* (Wiley-IEEE Press, Singapore, 2013)
15. J. Benesty, J.R. Jensen, M.G. Christensen, J. Chen, *Speech Enhancement: A Signal Subspace Perspective* (Academic, Cambridge, 2014)
16. J. Benesty, J. Chen, I. Cohen, *Design of Circular Differential Microphone Arrays* (Springer, Switzerland, 2015)
17. J. Benesty, J. Chen, C. Pan, *Fundamentals of Differential Beamforming* (Springer, Switzerland, 2016)
18. D.P. Jarrett, E.A.P. Habets, P.A. Naylor, *Theory and Applications of Spherical Microphone Array Processing* (Springer, Switzerland, 2017)
19. J. Benesty, I. Cohen, J. Chen, *Fundamentals of Signal Enhancement and Array Signal Processing* (Wiley-IEEE Press, Singapore, 2018)
20. B. Rafaely, *Fundamentals of Spherical Array Processing* (Springer, Switzerland, 2019)
21. J. Benesty, I. Cohen, J. Chen, *Array Processing-Kronecker Product Beamforming* (Springer, Switzerland, 2019)
22. J. Benesty, I. Cohen, J. Chen, *Array Beamforming with Linear Difference Equations* (Springer, Switzerland, 2021)
23. J. Paterson, H. Lee (eds.), *3D Audio (Perspectives on Music Production)* (Routledge, New York, 2022)
24. J. Chen, J. Benesty, Y. Huang, S. Doclo, New insights into the noise reduction Wiener filter. IEEE Trans. Audio Speech Lang. Process. **14**(4), 1218–1234 (2006)

Chapter 2
Limitations of Single Microphone Processing

Single microphone processing is a very popular technique in speech enhancement. It has been studied for several decades and has been implemented in many systems. This method has, obviously, great limitations. In this chapter, we explain why and show the importance of spatial information (with just two microphones) to fully understand why it leads to greater flexibility and much better performance.

2.1 Signal Model

Let us consider the single-channel case in which a single microphone picks up the sound in its environment:

$$y(t) = x(t) + v(t), \tag{2.1}$$

where $y(t)$ is the observed signal at the discrete-time index t, $x(t)$ is the zero-mean desired signal, and $v(t)$ is the zero-mean additive noise. It is assumed that $x(t)$ and $v(t)$ are uncorrelated, which is almost always the case in practice. Then, the objective of single microphone processing, in this well-defined context, is to estimate the desired signal from the observations in the best possible way [1, 2]. This is often called noise reduction or speech/signal enhancement.

Usually, it is more convenient and efficient to work in the frequency domain. Therefore, (2.1) becomes

$$Y(f) = X(f) + V(f), \tag{2.2}$$

where $Y(f)$, $X(f)$, and $V(f)$ are the frequency-domain representations of $y(t)$, $x(t)$, and $v(t)$, respectively, at the frequency index f. Since the zero-mean signals $X(f)$ and $V(f)$ are assumed to be incoherent (i.e., uncorrelated in the time domain),

J. Benesty et al., *Microphone Arrays*, Springer Topics in Signal Processing 22,
https://doi.org/10.1007/978-3-031-36974-2_2

the variance of $Y(f)$ is

$$\begin{aligned}\phi_Y(f) &= E\left[|Y(f)|^2\right] \\ &= \phi_X(f) + \phi_V(f),\end{aligned} \tag{2.3}$$

where $\phi_X(f) = E\left[|X(f)|^2\right]$ and $\phi_V(f) = E\left[|V(f)|^2\right]$ are the variances of $X(f)$ and $V(f)$, respectively, with $E[\cdot]$ denoting mathematical expectation. From (2.3), we find that the input signal-to-noise ratio (SNR) is

$$\text{iSNR}(f) = \frac{\phi_X(f)}{\phi_V(f)}. \tag{2.4}$$

2.2 Optimal Noise Reduction

An estimate of $X(f)$ can be obtained by multiplying $Y(f)$ with a complex-valued gain, $H(f)$, i.e.,

$$\begin{aligned}Z(f) &= H(f)Y(f) \\ &= H(f)\left[X(f) + V(f)\right] \\ &= X_{\text{fd}}(f) + V_{\text{rn}}(f),\end{aligned} \tag{2.5}$$

where $Z(f)$ is the frequency-domain representation of the time-domain signal $z(t)$, which is an estimate of $x(t)$,

$$X_{\text{fd}}(f) = H(f)X(f) \tag{2.6}$$

is the filtered desired signal, and

$$V_{\text{rn}}(f) = H(f)V(f) \tag{2.7}$$

is the residual noise. The variance of $Z(f)$ can then be written as

$$\begin{aligned}\phi_Z(f) &= E\left[|Z(f)|^2\right] \\ &= |H(f)|^2 \phi_Y(f) \\ &= \phi_{X_{\text{fd}}}(f) + \phi_{V_{\text{rn}}}(f),\end{aligned} \tag{2.8}$$

where

$$\phi_{X_{\text{fd}}}(f) = |H(f)|^2 \phi_X(f) \tag{2.9}$$

is the variance of the filtered desired signal and

$$\phi_{V_{\text{rn}}}(f) = |H(f)|^2 \phi_V(f) \tag{2.10}$$

is the variance of the residual noise. We deduce that the output SNR is

$$\begin{aligned} \text{oSNR}\left[H(f)\right] &= \frac{\phi_{X_{\text{fd}}}(f)}{\phi_{V_{\text{rn}}}(f)} \\ &= \text{iSNR}(f). \end{aligned} \tag{2.11}$$

Clearly, with just a complex gain, it is impossible to improve the (narrowband) output SNR;[1] this is a major shortcoming of single microphone processing and a fundamental difference with the multichannel case, where the (narrowband) output SNR can be improved (see Sect. 2.3 and Chap. 3). Also, from the filtered desired signal, we see that the only possibility for the gain to not distort the desired signal is to take it equal to 1, i.e., $H(f) = 1$; however, in this particular case, there is no noise reduction. This is another important limitation of single microphone processing, where its actual number of degrees of freedom is only 1.

Other useful definitions are the noise reduction factor:

$$\begin{aligned} \xi\left[H(f)\right] &= \frac{\phi_V(f)}{\phi_{V_{\text{rn}}}(f)} \\ &= \frac{1}{|H(f)|^2} \end{aligned} \tag{2.12}$$

and the desired signal distortion index:

$$\begin{aligned} \upsilon\left[H(f)\right] &= \frac{E\left[|X_{\text{fd}}(f) - X(f)|^2\right]}{\phi_X(f)} \\ &= |H(f) - 1|^2 . \end{aligned} \tag{2.13}$$

Now, let us define the mean-squared error (MSE) criterion between the estimated and desired signals:

$$\begin{aligned} \mathcal{J}\left[H(f)\right] &= E\left[|Z(f) - X(f)|^2\right] \\ &= E\left[|H(f)Y(f) - X(f)|^2\right] \\ &= |1 - H(f)|^2 \phi_X(f) + |H(f)|^2 \phi_V(f) \end{aligned}$$

[1] However, the broadband output SNR can be improved [3].

$$= \upsilon\left[H(f)\right]\phi_X(f) + \frac{\phi_V(f)}{\xi\left[H(f)\right]}. \tag{2.14}$$

The minimization of $\mathcal{J}\left[H(f)\right]$ leads to the optimal Wiener gain:

$$\begin{aligned} H_{\mathrm{W}}(f) &= \frac{\phi_X(f)}{\phi_Y(f)} \\ &= 1 - \frac{\phi_V(f)}{\phi_Y(f)} \\ &= \frac{\mathrm{iSNR}(f)}{1 + \mathrm{iSNR}(f)}. \end{aligned} \tag{2.15}$$

We observe that this gain is always real, positive, and smaller than 1. We also observe how this gain depends only on the input SNR, which means that for low values of $\mathrm{iSNR}(f)$, the Wiener gain has values close to 0, meaning that it can have a disastrous effect on the desired signal. This is another major limitation of single microphone processing. One way to compromise between noise reduction and desired signal distortion is via the tradeoff gain:

$$H_{\mathrm{T},\mu}(f) = \frac{\mathrm{iSNR}(f)}{\mu(f) + \mathrm{iSNR}(f)}, \tag{2.16}$$

where $\mu(f) \geq 0$. However, this tradeoff gain is also very limited in practice, and there is a real need to increase the number of degrees of freedom in the noise reduction problem.

2.3 Importance of Spatial Information

One apparent way to increase the number of degrees of freedom is by exploring the spatial information. For that, we need to have more than one sensor in order to capture different perspectives of the surrounding environment.

Let us consider the situation in which two microphones are in two different positions to capture the sound. Similar to the single-channel case, the signal picked up by the first microphone is

$$Y_1(f) = X(f) + V_1(f). \tag{2.17}$$

Assuming that the desired signal, $X(f)$, is a point source, the signal picked up by the second microphone can be written as

$$Y_2(f) = e^{-\jmath\tau(f)}X(f) + V_2(f), \tag{2.18}$$

where $\jmath$ is the imaginary unit and $\tau(f)$ represents the phase between the two sensors.

Now, we can form the signal:

$$Y_{12}(f) = e^{\jmath\tau(f)} Y_2(f) - Y_1(f). \tag{2.19}$$

Assuming that $e^{\jmath\tau(f)}$ is known or can be estimated, (2.19) can be rewritten as

$$\begin{aligned} Y_{12}(f) &= e^{\jmath\tau(f)} V_2(f) - V_1(f) \\ &= V_{12}(f), \end{aligned} \tag{2.20}$$

which depends on the noise only. Therefore, a good estimate of the desired signal, $X(f)$, is

$$\begin{aligned} Z_{12}(f) &= Y_1(f) - G(f)Y_{12}(f) \\ &= X(f) + V_1(f) - G(f)\left[e^{\jmath\tau(f)} V_2(f) - V_1(f)\right] \\ &= X(f) + V_{12,\text{rn}}(f), \end{aligned} \tag{2.21}$$

where $G(f)$ is a complex-valued gain and

$$V_{12,\text{rn}}(f) = V_1(f) - G(f)V_{12}(f) \tag{2.22}$$

is the residual noise. We see that with this approach, we have at least three advantages as compared to the single-channel case. First, since the desired signal is not filtered, this technique is distortionless. Second, thanks to the gain $G(f)$ some noise reduction is possible. Third, by exploiting the nature of the noise field, it is easy to estimate the different statistics required in speech enhancement algorithms [4, 5].

The variance of $Z_{12}(f)$ is

$$\phi_{Z_{12}}(f) = \phi_X(f) + \phi_{V_{12,\text{rn}}}(f), \tag{2.23}$$

where

$$\begin{aligned} \phi_{V_{12,\text{rn}}}(f) = \phi_{V_1}(f) - G^*(f)\phi_{V_1V_{12}}(f) - G(f)\phi^*_{V_1V_{12}}(f) \\ + |G(f)|^2 \phi_{V_{12}}(f) \end{aligned} \tag{2.24}$$

is the variance of the residual noise, with the superscript * being the complex-conjugate operator, and

$$\phi_{V_1}(f) = E\left[|V_1(f)|^2\right],$$

$$\phi_{V_1 V_{12}}(f) = E\left[V_1(f) V_{12}^*(f)\right],$$
$$\phi_{V_{12}}(f) = E\left[|V_{12}(f)|^2\right].$$

As a result, the (narrowband) noise reduction factor is

$$\xi\left[G(f)\right] = \frac{\phi_{V_1}(f)}{\phi_{V_{12,\mathrm{rm}}}(f)} \tag{2.25}$$

and the (narrowband) output SNR is

$$\begin{aligned} \mathrm{oSNR}\left[G(f)\right] &= \frac{\phi_X(f)}{\phi_{V_{12,\mathrm{rm}}}(f)} \\ &\neq \mathrm{iSNR}(f). \end{aligned} \tag{2.26}$$

As we can see, this time, the (narrowband) output SNR can really be improved. In fact, it can be maximized by simply minimizing its denominator, i.e., $\phi_{V_{12,\mathrm{rm}}}(f)$. We obtain the maximum SNR gain (MSNRG):

$$G_{\mathrm{MSNRG}}(f) = \frac{\phi_{V_1 V_{12}}(f)}{\phi_{V_{12}}(f)}, \tag{2.27}$$

which is fundamentally different from the Wiener gain since, as it should be expected, it does not depend on the input SNR but rather on the levels of coherence between the two noises $V_1(f)$ and $V_{12}(f)$. Substituting (2.27) into (2.26), we get

$$\begin{aligned} \mathrm{oSNR}\left[G_{\mathrm{MSNRG}}(f)\right] &= \frac{\phi_X(f)}{\phi_{V_1}(f) - \dfrac{\left|\phi_{V_1 V_{12}}(f)\right|^2}{\phi_{V_{12}}(f)}} \\ &= \frac{\phi_X(f)}{\phi_{V_1}(f)} \times \frac{1}{1 - \dfrac{\left|\phi_{V_1 V_{12}}(f)\right|^2}{\phi_{V_1}(f)\phi_{V_{12}}(f)}} \\ &= \mathrm{iSNR}(f) \times \frac{1}{1 - \left|\gamma_{V_1 V_{12}}(f)\right|^2}, \end{aligned} \tag{2.28}$$

where

$$\left|\gamma_{V_1 V_{12}}(f)\right|^2 = \frac{\left|\phi_{V_1 V_{12}}(f)\right|^2}{\phi_V(f)\phi_{V_{12}}(f)} \tag{2.29}$$

is the magnitude-squared coherence function (MSCF) between $V_1(f)$ and $V_{12}(f)$. In (2.28) and (2.29), it is assumed that $\phi_V(f) = \phi_{V_1}(f) = \phi_{V_2}(f)$. As a result, the

SNR gain is

$$\mathcal{G}\left[G_{\text{MSNRG}}(f)\right] = \frac{\text{oSNR}\left[G_{\text{MSNRG}}(f)\right]}{\text{iSNR}(f)} = \frac{1}{1-\left|\gamma_{V_1V_{12}}(f)\right|^2} > 1. \tag{2.30}$$

We can see that $H_{\text{W}}(f)$ and $G_{\text{MSNRG}}(f)$ work very much differently. The former is just a smooth gain that attenuates the observed signal based on the input SNR,[2] while the latter, because of the two degrees of freedom of the approach, it does not distort the desired signal and is able to reduce the additive noise based on $\left|\gamma_{V_1V_{12}}(f)\right|^2$; the higher is the value of this MSCF, the more is noise reduction.

Another insightful way to express (2.30) is

$$\mathcal{G}\left[G_{\text{MSNRG}}(f)\right] = 2\frac{1-\Re\left[e^{-\jmath\tau(f)}\gamma_{V_1V_2}(f)\right]}{1-\left|\gamma_{V_1V_2}(f)\right|^2}, \tag{2.31}$$

where $\Re[\cdot]$ is the real part of a complex number,

$$\gamma_{V_1V_2}(f) = \frac{\phi_{V_1V_2}(f)}{\phi_V(f)} \tag{2.32}$$

is the coherence function between $V_1(f)$ and $V_2(f)$, with

$$\phi_{V_1V_2}(f) = E\left[V_1(f)V_2^*(f)\right],$$

and $\mathcal{G}\left[G_{\text{MSNRG}}(f)\right] = 2$ if and only if $\gamma_{V_1V_2}(f) = 0$. In some noise contexts, this gain can be as high as 4, without distorting the desired signal.

2.4 Illustrative Examples

Let us start by studying an example of the single-channel case with a complex-valued gain for noise reduction. We consider the tradeoff gain, $H_{\text{T},\mu}(f)$, given in (2.16), for different values of $\mu(f)$, where the Wiener gain is a special case by taking $\mu(f) = 1$, i.e., $H_{\text{T},1}(f) = H_{\text{W}}(f)$. For better comparison, we also define the normalized MSE (NMSE) as

[2] It is good to recall that $\mathcal{G}\left[H_{\text{W}}(f)\right] = 1$.

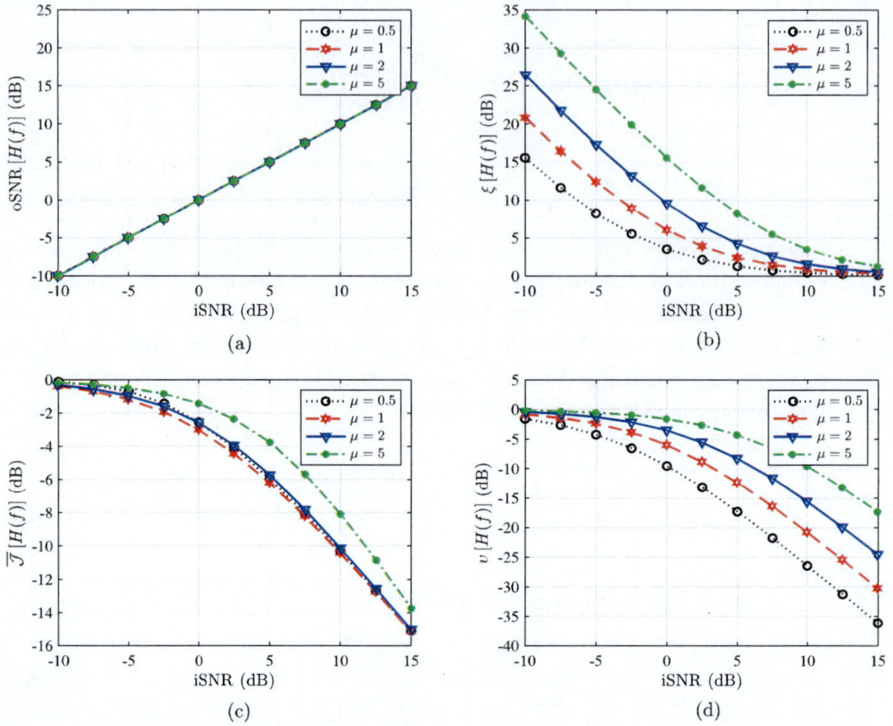

Fig. 2.1 Performance of the tradeoff gain, $H_{\mathrm{T},\mu}(f)$, as a function of the input SNR, for different values of $\mu(f)$: (**a**) output SNR, (**b**) noise reduction factor, (**c**) NMSE, and (**d**) desired signal distortion index. Note that for $\mu(f) = 1$, Wiener and tradeoff gains are identical

$$\begin{aligned}\overline{\mathcal{J}}\left[H(f)\right] &= \frac{\mathcal{J}\left[H(f)\right]}{\phi_X(f)} \\ &= \upsilon\left[H(f)\right] + \frac{1}{\mathrm{iSNR}(f) \times \xi\left[H(f)\right]}.\end{aligned} \tag{2.33}$$

Figure 2.1 shows plots of the output SNR, $\mathrm{oSNR}\left[H(f)\right]$, noise reduction factor, $\xi\left[H(f)\right]$, NMSE, $\overline{\mathcal{J}}\left[H(f)\right]$, and desired signal distortion index, $\upsilon\left[H(f)\right]$, as a function of the input SNR, for several values of $\mu(f)$, where the variance of the desired signal, $\phi_X(f)$, is set to 1 and the variance of the noise, $\phi_V(f)$, is computed according to the given input SNR. It can be seen that the single-channel tradeoff gain does not improve the narrowband output SNR and introduces significant desired signal distortion. It can also be observed that the higher is the value of $\mu(f)$, the higher is the noise reduction factor, but at the cost of higher NMSE and higher desired signal distortion index. This clearly demonstrates the limitations of single microphone processing with just a complex gain, which cannot improve the narrowband SNR and inevitably distorts the desired signal.

Now, let us give some examples of noise reduction with two microphones, whose interelement spacing is $\delta = 0.01$ m, by exploring the spatial information. We assume that the desired source signal propagates from the endfire direction, i.e., 0°, so that the phase between the two sensors is $\tau(f) = 2\pi f\delta/c$. We consider an environment with diffuse and white noises; then, the noise signals picked up by the two microphones are

$$V_1(f) = V_{1,\mathrm{d}}(f) + V_{1,\mathrm{w}}(f),$$
$$V_2(f) = V_{2,\mathrm{d}}(f) + V_{2,\mathrm{w}}(f),$$

where $V_{1,\mathrm{d}}(f)$ and $V_{1,\mathrm{w}}(f)$ denote the diffuse and white noises, respectively, picked up by the first microphone, and $V_{2,\mathrm{d}}(f)$ and $V_{2,\mathrm{w}}(f)$ are the diffuse and white noises, respectively, picked up by the second microphone. Without loss of generality, we assume that the variances of the same kind of noise are equal at the two sensors, i.e., $\phi_{V_{1,\mathrm{d}}}(f) = \phi_{V_{2,\mathrm{d}}}(f) = \phi_{V_\mathrm{d}}(f)$ and $\phi_{V_{1,\mathrm{w}}}(f) = \phi_{V_{2,\mathrm{w}}}(f) = \phi_{V_\mathrm{w}}(f)$, where $\phi_{V_{1,\mathrm{d}}}(f)$, $\phi_{V_{2,\mathrm{d}}}(f)$, $\phi_{V_{1,\mathrm{w}}}(f)$, and $\phi_{V_{2,\mathrm{w}}}(f)$ are the variances of $V_{1,\mathrm{d}}(f)$, $V_{2,\mathrm{d}}(f)$, $V_{1,\mathrm{w}}(f)$, and $V_{2,\mathrm{w}}(f)$, respectively. In this case, we have

$$\phi_{V_1V_{12}}(f) = e^{-\jmath\tau(f)}\mathrm{sinc}\left(\frac{2\pi f\delta}{c}\right)\phi_{V_\mathrm{d}}(f) - \phi_{V_1}(f),$$

$$\phi_{V_{12}}(f) = 2\phi_{V_1}(f) - 2\cos\left[\tau(f)\right]\mathrm{sinc}\left(\frac{2\pi f\delta}{c}\right)\phi_{V_\mathrm{d}}(f),$$

$$\phi_{V_1V_2}(f) = \mathrm{sinc}\left(\frac{2\pi f\delta}{c}\right)\phi_{V_\mathrm{d}}(f),$$

where $\mathrm{sinc}(x) = \sin x/x$ and $c = 340$ m/s is the speed of sound in air. To demonstrate the performance of the MSNRG, $G_{\mathrm{MSNRG}}(f)$, given in (2.27), we consider different noise conditions by setting $\phi_{V_\mathrm{w}}(f) = \alpha\phi_{V_\mathrm{d}}(f)$, where $\alpha > 0$ is related to the ratio between the powers of the white and diffuse noises. Figure 2.2 shows plots of the output SNR, $\mathrm{oSNR}\left[G(f)\right]$, noise reduction factor, $\xi\left[G(f)\right]$, coherence function between $V_1(f)$ and $V_2(f)$, $\gamma_{V_1V_2}(f)$, and SNR gain, $\mathcal{G}\left[G(f)\right]$, as a function of the input SNR, for different values of α, where the variance of the desired signal, $\phi_X(f)$, is set to 1 and the variances of the noises, $\phi_{V_\mathrm{w}}(f)$ and $\phi_{V_\mathrm{d}}(f)$, are computed according to the specified input SNR and values of α. It is clearly seen that MSNRG improves the output SNR. It should also be noted that, in this scenario, the desired signal distortion index is always equal to 0 (not shown in the figure). For a given input SNR, the coherence function between $V_1(f)$ and $V_2(f)$ decreases as α increases. This is easy to understand as a larger value of α means that noise signals at the two sensors are more white and less coherent. Consequently, the output SNR, noise reduction factor, and SNR gain decrease when α increases.

We also consider an environment with a point source noise and white noise; then, the noise signals picked up by the two microphones are

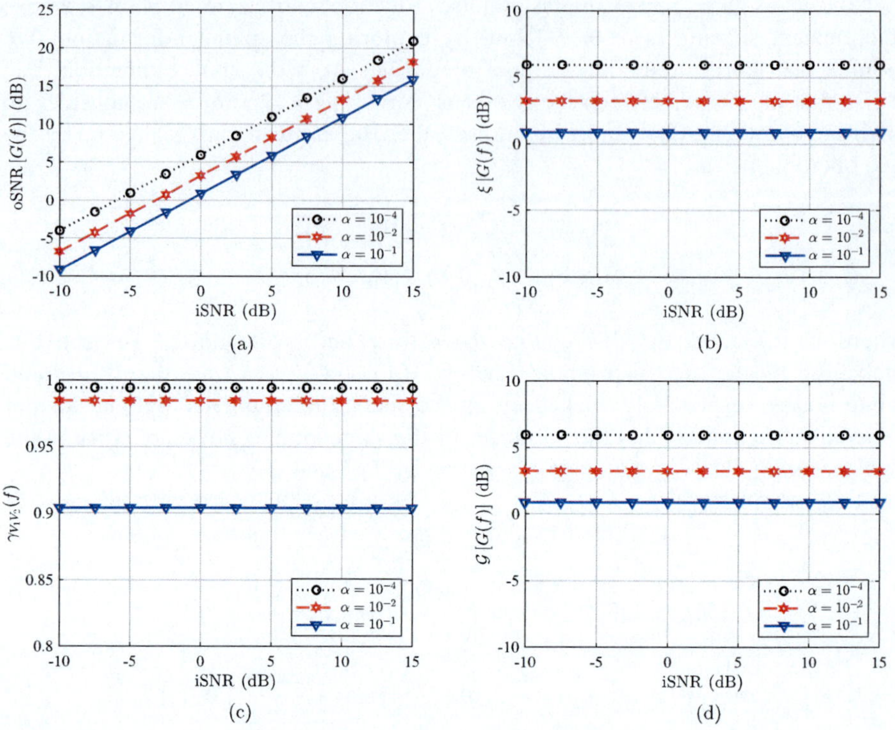

Fig. 2.2 Performance of the MSNRG, $G_{\mathrm{MSNRG}}(f)$, as a function of the input SNR, for different values of α, in an environment with diffuse and white noises: **(a)** output SNR, **(b)** noise reduction factor, **(c)** coherence function between $V_1(f)$ and $V_2(f)$, and **(d)** SNR gain

$$
\begin{aligned}
V_1(f) &= V_{\mathrm{p}}(f) + V_{1,\mathrm{w}}(f), \\
V_2(f) &= e^{-\jmath \tau_{\mathrm{p}}(f)} V_{\mathrm{p}}(f) + V_{2,\mathrm{w}}(f),
\end{aligned}
$$

where $V_{\mathrm{p}}(f)$ is a point source noise and $\tau_{\mathrm{p}}(f)$ represents the phase corresponding to this source between the two sensors. Without loss of generality, we still assume that the variances of the same kind of noise are equal at the two sensors. Hence, we have

$$
\begin{aligned}
\phi_{V_1 V_{12}}(f) &= e^{-\jmath \left[\tau(f) - \tau_{\mathrm{p}}(f)\right]} \phi_{V_{\mathrm{p}}}(f) - \phi_{V_1}(f), \\
\phi_{V_{12}}(f) &= 2\phi_{V_1}(f) - 2\cos\left[\tau(f) - \tau_{\mathrm{p}}(f)\right] \phi_{V_{\mathrm{p}}}(f), \\
\phi_{V_1 V_2}(f) &= e^{\jmath \tau_{\mathrm{p}}(f)} \phi_{V_{\mathrm{p}}}(f),
\end{aligned}
$$

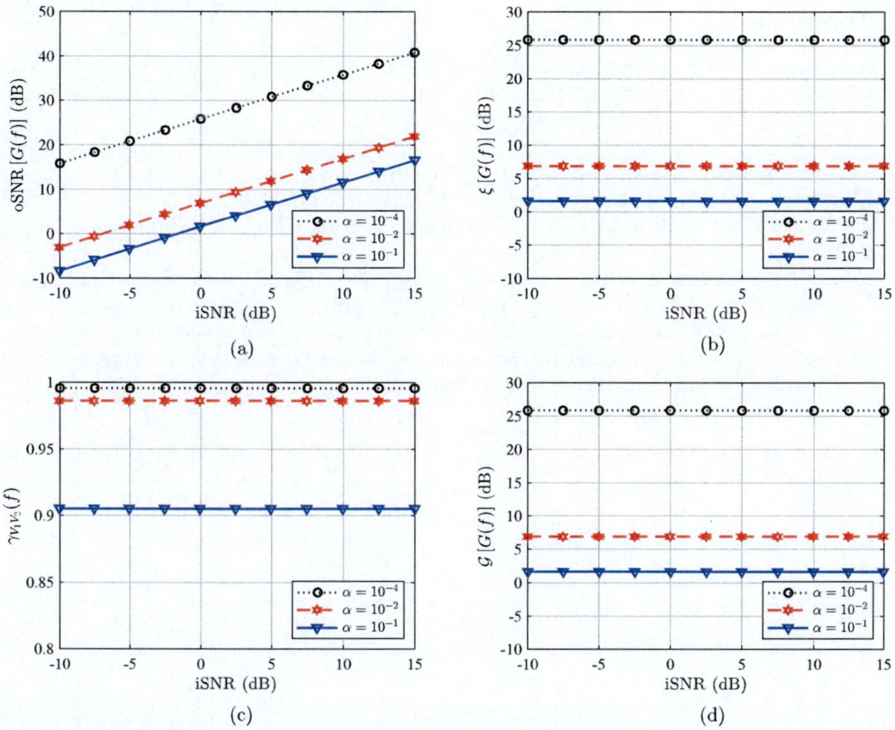

Fig. 2.3 Performance of the MSNRG, $G_{\text{MSNRG}}(f)$, as a function of the input SNR, for different values of α, in an environment with a point source noise and white noise: (**a**) output SNR, (**b**) noise reduction factor, (**c**) coherence function between $V_1(f)$ and $V_2(f)$, and (**d**) SNR gain

where $\phi_{V_\text{p}}(f)$ is the variance of $V_\text{p}(f)$. Assume that the desired source signal propagates from $0°$ and the point source noise propagates from $120°$, so that $\tau(f) = 2\pi f \delta/c$ and $\tau_\text{p}(f) \doteq -\pi f \delta/c$. We set $\phi_{V_\text{w}}(f) = \alpha\phi_{V_\text{p}}(f)$, where $\alpha > 0$ is related to the ratio between the powers of the white noise and the point source noise. By setting $\phi_X(f) = 1$, the variances $\phi_{V_\text{w}}(f)$ and $\phi_{V_\text{p}}(f)$ are computed according to the input SNR. Figure 2.3 shows plots of the output SNR, $\text{oSNR}\left[G(f)\right]$, noise reduction factor, $\xi\left[G(f)\right]$, coherence function between $V_1(f)$ and $V_2(f)$, $\gamma_{V_1V_2}(f)$, and SNR gain, $\mathcal{G}\left[G(f)\right]$, as a function of the input SNR, for different values of α. It can be seen that the MSNRG significantly improves the output SNR, and, for a given input SNR, the output SNR, noise reduction factor, MSCF, and SNR gain increase as the value of α decreases. The reason is that as the value of α decreases, the noise signals at the two microphones are more coherent; as a result, the spatial gain can suppress the noise more effectively.

References

1. J. Benesty, J. Chen, Y. Huang, S. Doclo, Study of the Wiener filter for noise reduction, in ed. by J. Benesty, S. Makino, J. Chen, *Speech Enhancement*, Chapter 2 (Springer, Berlin, 2005), pp. 9–41
2. J. Chen, J. Benesty, Y. Huang, S. Doclo, New insights into the noise reduction Wiener filter. IEEE Trans. Audio Speech Language Process. **14**, 1218–1234 (2006)
3. J. Benesty, J. Chen, Y. Huang, I. Cohen, *Noise Reduction in Speech Processing* (Springer, Berlin, 2009)
4. J. Benesty, I. Cohen, J. Chen, *Fundamentals of Signal Enhancement and Array Signal Processing* (Wiley-IEEE Press, Singapore, 2018)
5. G. Huang, J. Benesty, J. Chen, Study of the frequency-domain multichannel noise reduction problem with the Householder transformation, in *2017 IEEE International Conference on Acoustics, Speech and Signal Processing (ICASSP)* (2017), pp. 486–490

Chapter 3
Fundamentals of Microphone Array Processing

After a brief but insightful discussion on the limitations of single microphone processing and the unquestionable advantages of spatial information, in this chapter, we make a concise overview of the most fundamental concepts in microphone array processing. We start with the signal model by considering the general case of three-dimensional arrays. In this context, we explain how to conduct linear beamforming, from which very useful performance measures are derived. Then, a large family of fixed and adaptive beamformers are proposed as well as some signal statistics estimators. Finally, we explain how different array geometries can be compared.

3.1 Signal Model

We consider a three-dimensional (3-D) sensor array of $M \geq 3$ omnidirectional microphones, which are distributed in the 3-D space. It is assumed that the positions of all microphones are precisely known. Obviously, 3-D arrays encompass all possible geometries, including the ones in the two-dimensional (2-D) and one-dimensional (1-D) spaces. It is assumed that Microphone 1 of the array is the reference and coincides with the origin of the 3-D Cartesian coordinate system. An angle on the xy-plane, i.e., an angle from the x-axis to a point on this plane is referred to as the azimuth angle, and an angle from the z-axis to a point in the 3-D space is referred to as the elevation angle. Then, the coordinates of the sensors are given by

$$\mathbf{r}_m = r_m \begin{bmatrix} \sin\theta_m \cos\phi_m & \sin\theta_m \sin\phi_m & \cos\theta_m \end{bmatrix}^T, \tag{3.1}$$

for $m = 1, 2, \ldots, M$, where r_m is the distance from the mth microphone to the origin point (i.e., Microphone 1); θ_m and ϕ_m are the elevation and azimuth angles, respectively, of the mth array element; and the superscript ^{T} is the transpose

J. Benesty et al., *Microphone Arrays*, Springer Topics in Signal Processing 22,
https://doi.org/10.1007/978-3-031-36974-2_3

operator. Therefore, the distance between Microphones i and j is given by

$$\rho_{ij} = \left\| \mathbf{r}_i - \mathbf{r}_j \right\|, \tag{3.2}$$

for $i, j = 1, 2, \ldots, M$, where $\| \cdot \|$ is the Euclidean norm.

Let us consider a potential far-field source signal (plane wave) in the 3-D space, whose angular position is (θ, ϕ), which propagates in an anechoic acoustic environment at the speed of sound, i.e., $c = 340\,\text{m/s}$, and impinges on the above described 3-D array. Then, the steering vector of length M can be expressed as [1, 2]

$$\mathbf{d}_{\theta,\phi}(f) = \left[e^{\jmath 2\pi f \mathbf{a}_{\theta,\phi}^T \mathbf{r}_1/c} \; e^{\jmath 2\pi f \mathbf{a}_{\theta,\phi}^T \mathbf{r}_2/c} \; \cdots \; e^{\jmath 2\pi f \mathbf{a}_{\theta,\phi}^T \mathbf{r}_M/c} \right]^T, \tag{3.3}$$

where

$$\mathbf{a}_{\theta,\phi} = \left[\sin\theta\cos\phi \;\; \sin\theta\sin\phi \;\; \cos\theta \right]^T. \tag{3.4}$$

Now, let the desired source signal propagates from the known position $(\theta_\text{s}, \phi_\text{s})$; then, we can arrange an observed signal vector of length M in the frequency domain as [3, 4]

$$\begin{aligned} \mathbf{y}(f) &= \left[Y_1(f) \;\; Y_2(f) \;\; \cdots \;\; Y_M(f) \right]^T \\ &= \mathbf{x}(f) + \mathbf{v}(f) \\ &= \mathbf{d}_{\theta_\text{s},\phi_\text{s}}(f) X(f) + \mathbf{v}(f), \end{aligned} \tag{3.5}$$

where $Y_m(f)$ is the mth ($m = 1, 2, \ldots, M$) microphone signal of the 3-D array, $\mathbf{d}_{\theta_\text{s},\phi_\text{s}}(f)$ is the steering vector at $(\theta, \phi) = (\theta_\text{s}, \phi_\text{s})$ (direction of the desired source), $X(f)$ is the (zero-mean) desired signal, $\mathbf{v}(f)$ is the (zero-mean) additive noise signal vector defined similarly to $\mathbf{y}(f)$, and $X(f)$ and $\mathbf{v}(f)$ are incoherent. In the rest of this work, in order to simplify the notation, we drop the dependence on the frequency, f. We deduce that the covariance matrix of $\mathbf{y}$ is

$$\begin{aligned} \mathbf{\Phi}_\mathbf{y} &= E\left(\mathbf{y}\mathbf{y}^H\right) \\ &= \phi_X \mathbf{d}_{\theta_\text{s},\phi_\text{s}} \mathbf{d}_{\theta_\text{s},\phi_\text{s}}^H + \mathbf{\Phi}_\mathbf{v}, \end{aligned} \tag{3.6}$$

where the superscript ^{H} is the conjugate-transpose operator, $\phi_X = E\left(|X|^2\right)$ is the variance of X, and $\mathbf{\Phi}_\mathbf{v} = E\left(\mathbf{v}\mathbf{v}^H\right)$ is the covariance matrix of $\mathbf{v}$. For a small and compact 3-D array, we may assume that the variance of the noise is the same at all sensors, i.e., $\phi_V = \phi_{V_1} = \phi_{V_2} = \cdots = \phi_{V_M}$, with $\phi_{V_m} = E\left(|V_m|^2\right)$, $m = 1, 2, \ldots, M$; in this case, we can express (3.6) as

$$\mathbf{\Phi}_\mathbf{y} = \phi_X \mathbf{d}_{\theta_\text{s},\phi_\text{s}} \mathbf{d}_{\theta_\text{s},\phi_\text{s}}^H + \phi_V \mathbf{\Gamma}_\mathbf{v}, \tag{3.7}$$

where

$$\mathbf{\Gamma}_{\mathbf{v}} = \frac{\mathbf{\Phi}_{\mathbf{v}}}{\phi_V} \tag{3.8}$$

is the coherence matrix of the noise and $\phi_Y = \phi_{Y_1} = \phi_{Y_2} = \cdots = \phi_{Y_M}$, with $\phi_{Y_m} = E\left(|Y_m|^2\right)$, $m = 1, 2, \ldots, M$. We deduce that the input SNR is

$$\begin{aligned} \text{iSNR} &= \frac{\text{tr}\left(\phi_X \mathbf{d}_{\theta_s,\phi_s} \mathbf{d}_{\theta_s,\phi_s}^H\right)}{\text{tr}\left(\phi_V \mathbf{\Gamma}_{\mathbf{v}}\right)} \\ &= \frac{\phi_X}{\phi_V}, \end{aligned} \tag{3.9}$$

where $\text{tr}(\cdot)$ denotes the trace of a square matrix.

3.2 Beamforming

The easiest and most practical way to perform (linear) beamforming in the frequency domain is by applying a complex-valued filter, $\mathbf{h}$ of length M, to the observed signal vector, $\mathbf{y}$, i.e.,

$$\begin{aligned} Z &= \sum_{m=1}^{M} H_m^* Y_m \\ &= \mathbf{h}^H \mathbf{y} \\ &= X_{\text{fd}} + V_{\text{rn}}, \end{aligned} \tag{3.10}$$

where Z is the so-called beamformer output signal , which, in fact, is an estimate of the desired signal, X,

$$\mathbf{h} = \begin{bmatrix} H_1 & H_2 & \cdots & H_M \end{bmatrix}^T \tag{3.11}$$

is the beamforming weight vector or beamformer of length M,

$$X_{\text{fd}} = X \mathbf{h}^H \mathbf{d}_{\theta_s,\phi_s} \tag{3.12}$$

is the filtered desired signal, and

$$V_{\text{rn}} = \mathbf{h}^H \mathbf{v} \tag{3.13}$$

is the residual noise.

Since the two terms on the right-hand side of (3.10) are incoherent, the variance of Z is the sum of two variances:

$$\begin{aligned}\phi_Z &= \mathbf{h}^H \boldsymbol{\Phi}_\mathbf{y} \mathbf{h} \\ &= \phi_{X_{\mathrm{fd}}} + \phi_{V_{\mathrm{rn}}},\end{aligned} \tag{3.14}$$

where

$$\phi_{X_{\mathrm{fd}}} = \phi_X \left| \mathbf{h}^H \mathbf{d}_{\theta_\mathrm{s},\phi_\mathrm{s}} \right|^2, \tag{3.15}$$

$$\phi_{V_{\mathrm{rn}}} = \mathbf{h}^H \boldsymbol{\Phi}_\mathbf{v} \mathbf{h}. \tag{3.16}$$

In many real-world applications, it is desired to contain speech distortion, which is due to the filtering action on the observations. Therefore, it is possible to solicit the distortionless constraint on the beamformer, i.e.,

$$\mathbf{h}^H \mathbf{d}_{\theta_\mathrm{s},\phi_\mathrm{s}} = 1, \tag{3.17}$$

meaning that any signal arriving from $(\theta_\mathrm{s}, \phi_\mathrm{s})$ will pass through the beamformer undistorted from a theoretical point of view.

3.3 Performance Measures

In this section, the most fundamental measures for microphone array processing are explained. They are extremely useful not only in the evaluation of all kinds of beamformers but also in the derivation of the most interesting ones.

Each beamformer has a pattern of directional sensitivity, i.e., it has different sensitivities from sounds arriving from different directions. The beampattern or directivity pattern describes the sensitivity of the beamformer to a plane wave (source signal) impinging on the array from the direction (θ, ϕ). Mathematically, it is defined as

$$\begin{aligned}\mathcal{B}_{\theta,\phi}(\mathbf{h}) &= \mathbf{d}_{\theta,\phi}^H \mathbf{h} \\ &= \sum_{m=1}^{M} H_m e^{-\jmath 2\pi f \mathbf{a}_{\theta,\phi}^T \mathbf{r}_m / c}.\end{aligned} \tag{3.18}$$

Usually, $\left|\mathcal{B}_{\theta,\phi}(\mathbf{h})\right|^2$, which is called the power pattern [2], is illustrated with a polar plot.

According to (3.14), the output SNR can be defined as

$$\begin{aligned} \text{oSNR}(\mathbf{h}) &= \phi_X \frac{\left|\mathbf{h}^H \mathbf{d}_{\theta_s,\phi_s}\right|^2}{\mathbf{h}^H \mathbf{\Phi}_{\mathbf{v}} \mathbf{h}} \\ &= \frac{\phi_X}{\phi_V} \times \frac{\left|\mathbf{h}^H \mathbf{d}_{\theta_s,\phi_s}\right|^2}{\mathbf{h}^H \mathbf{\Gamma}_{\mathbf{v}} \mathbf{h}}. \end{aligned} \tag{3.19}$$

From the definitions of the input and output SNRs, we deduce that the SNR gain is

$$\begin{aligned} \mathcal{G}(\mathbf{h}) &= \frac{\text{oSNR}(\mathbf{h})}{\text{iSNR}} \\ &= \frac{\left|\mathbf{h}^H \mathbf{d}_{\theta_s,\phi_s}\right|^2}{\mathbf{h}^H \mathbf{\Gamma}_{\mathbf{v}} \mathbf{h}} \end{aligned} \tag{3.20}$$

and from the Cauchy-Schwarz inequality, we easily get

$$\mathcal{G}(\mathbf{h}) \leq \mathbf{d}_{\theta_s}^H \mathbf{\Gamma}_{\mathbf{v}}^{-1} \mathbf{d}_{\theta_s,\phi_s}, \ \forall \mathbf{h}. \tag{3.21}$$

As a result, the maximum SNR gain is

$$\mathcal{G}_{\max,\theta_s,\phi_s,\mathbf{v}} = \mathbf{d}_{\theta_s,\phi_s}^H \mathbf{\Gamma}_{\mathbf{v}}^{-1} \mathbf{d}_{\theta_s,\phi_s}, \tag{3.22}$$

which is frequency, (desired source signal) direction, and noise statistics dependent.

The most convenient way to evaluate the sensitivity of the array to some of its imperfections, such as sensor noise, with a specific beamformer, is via the so-called white noise gain (WNG), which is defined by taking $\mathbf{\Gamma}_{\mathbf{v}} = \mathbf{I}_M$ in (3.20), where $\mathbf{I}_M$ is the $M \times M$ identity matrix, i.e.,

$$\mathcal{W}(\mathbf{h}) = \frac{\left|\mathbf{h}^H \mathbf{d}_{\theta_s,\phi_s}\right|^2}{\mathbf{h}^H \mathbf{h}}. \tag{3.23}$$

From the Cauchy-Schwarz inequality, we get

$$\mathcal{W}(\mathbf{h}) \leq M, \ \forall \mathbf{h}. \tag{3.24}$$

As a result, the maximum WNG is

$$\mathcal{W}_{\max} = M, \tag{3.25}$$

which is frequency and (desired source signal) direction independent.

Another important measure, which quantifies the ability of the beamformer in suppressing spatial noise from directions other than the look direction, is the directivity factor (DF). It can be expressed as

$$\mathcal{D}(\mathbf{h}) = \frac{\left|\mathbf{h}^H \mathbf{d}_{\theta_s,\phi_s}\right|^2}{\mathbf{h}^H \mathbf{\Gamma}_\mathrm{d} \mathbf{h}}, \tag{3.26}$$

where the elements of the $M \times M$ matrix $\mathbf{\Gamma}_\mathrm{d}$ are

$$(\mathbf{\Gamma}_\mathrm{d})_{ij} = \mathrm{sinc}\left(\frac{2\pi f \rho_{ij}}{c}\right), \tag{3.27}$$

for $i, j = 1, 2, \ldots, M$. Again, by invoking the Cauchy-Schwarz inequality, we find that

$$\mathcal{D}(\mathbf{h}) \leq \mathbf{d}_{\theta_s,\phi_s}^H \mathbf{\Gamma}_\mathrm{d}^{-1} \mathbf{d}_{\theta_s,\phi_s}, \ \forall \mathbf{h}. \tag{3.28}$$

As a result, the maximum DF is

$$\mathcal{D}_{\max,\theta_s,\phi_s} = \mathbf{d}_{\theta_s,\phi_s}^H \mathbf{\Gamma}_\mathrm{d}^{-1} \mathbf{d}_{\theta_s,\phi_s}, \tag{3.29}$$

which is frequency and (desired source signal) direction dependent. The DF can also be expressed as [2]

$$\mathcal{D}(\mathbf{h}) = \frac{\left|\mathcal{B}_{\theta_s,\phi_s}(\mathbf{h})\right|^2}{\frac{1}{4\pi}\int_0^{\pi}\int_0^{2\pi} \left|\mathcal{B}_{\theta,\phi}(\mathbf{h})\right|^2 \sin\theta d\theta d\phi}. \tag{3.30}$$

One can verify that (3.26) are (3.30) are identical.

Now, let us define the error signal between the estimated and desired source signals:

$$\begin{aligned} \mathcal{E} &= Z - X \\ &= X_\mathrm{fd} + V_\mathrm{rn} - X \\ &= \mathcal{E}_\mathrm{ds} + \mathcal{E}_\mathrm{rn}, \end{aligned} \tag{3.31}$$

where

$$\mathcal{E}_\mathrm{ds} = \left(\mathbf{h}^H \mathbf{d}_{\theta_s,\phi_s} - 1\right) X \tag{3.32}$$

is the desired signal distortion due to the beamformer and

$$\mathcal{E}_\mathrm{rn} = \mathbf{h}^H \mathbf{v} = V_\mathrm{rn} \tag{3.33}$$

represents the residual noise. Since X and $\mathbf{v}$ are incoherent, so are $\mathcal{E}_\mathrm{ds}$ and $\mathcal{E}_\mathrm{rn}$. As a result, the MSE criterion can be expressed as

$$\begin{aligned}\mathcal{J}(\mathbf{h}) &= E\left(|\mathcal{E}|^2\right) \\ &= E\left(|\mathcal{E}_{\mathrm{ds}}|^2\right) + E\left(|\mathcal{E}_{\mathrm{rn}}|^2\right) \\ &= \mathcal{J}_{\mathrm{ds}}(\mathbf{h}) + \mathcal{J}_{\mathrm{rn}}(\mathbf{h}) \\ &= \phi_X + \mathbf{h}^H \mathbf{\Phi}_{\mathbf{y}} \mathbf{h} - \phi_X \mathbf{h}^H \mathbf{d}_{\theta_s,\phi_s} - \phi_X \mathbf{d}_{\theta_s,\phi_s}^H \mathbf{h},\end{aligned} \tag{3.34}$$

where

$$\begin{aligned}\mathcal{J}_{\mathrm{ds}}(\mathbf{h}) &= \phi_X \left|\mathbf{h}^H \mathbf{d}_{\theta_s,\phi_s} - 1\right|^2 \\ &= \phi_X \upsilon(\mathbf{h})\end{aligned} \tag{3.35}$$

and

$$\begin{aligned}\mathcal{J}_{\mathrm{rn}}(\mathbf{h}) &= \mathbf{h}^H \mathbf{\Phi}_{\mathbf{v}} \mathbf{h} \\ &= \frac{\phi_V}{\xi(\mathbf{h})},\end{aligned} \tag{3.36}$$

with

$$\upsilon(\mathbf{h}) = \left|\mathbf{h}^H \mathbf{d}_{\theta_s,\phi_s} - 1\right|^2 \tag{3.37}$$

being the desired signal distortion index and

$$\xi(\mathbf{h}) = \frac{\phi_V}{\mathbf{h}^H \mathbf{\Phi}_{\mathbf{v}} \mathbf{h}} \tag{3.38}$$

being the noise reduction factor. Another useful definition to quantify distortion of the desired signal is the desired signal reduction factor:

$$\xi_{\mathrm{d}}(\mathbf{h}) = \frac{1}{\left|\mathbf{h}^H \mathbf{d}_{\theta_s,\phi_s}\right|^2}. \tag{3.39}$$

3.4 Examples of Conventional Beamformers

In this section, we briefly present some examples of optimal beamformers. We discuss both fixed and adaptive beamformers. While many more can be found in the rich literature of beamforming, we only discuss the most conventional ones.

The delay-and-sum (DS) beamformer is derived by maximizing the WNG. We easily get the optimal filter:

$$\mathbf{h}_{\mathrm{DS}} = \frac{\mathbf{d}_{\theta_s,\phi_s}}{\mathbf{d}_{\theta_s,\phi_s}^H \mathbf{d}_{\theta_s,\phi_s}} = \frac{\mathbf{d}_{\theta_s,\phi_s}}{M}, \tag{3.40}$$

which is also called the maximum WNG (MWNG) beamformer. Therefore, with this beamformer, the WNG and DF are, respectively,

$$\mathcal{W}(\mathbf{h}_{\mathrm{DS}}) = M = \mathcal{W}_{\max} \tag{3.41}$$

and

$$\mathcal{D}(\mathbf{h}_{\mathrm{DS}}) = \frac{M^2}{\mathbf{d}_{\theta_s,\phi_s}^H \mathbf{\Gamma}_{\mathrm{d}} \mathbf{d}_{\theta_s,\phi_s}}. \tag{3.42}$$

Since $\mathbf{d}_{\theta_s,\phi_s}^H \mathbf{\Gamma}_{\mathrm{d}} \mathbf{d}_{\theta_s,\phi_s} \leq M^2$, we have $\mathcal{D}(\mathbf{h}_{\mathrm{DS}}) \geq 1$. While the DS beamformer maximizes the WNG, it never amplifies the diffuse noise since $\mathcal{D}(\mathbf{h}_{\mathrm{DS}}) \geq 1$.

The maximum DF (MDF) beamformer, as its name implies, maximizes the DF. Then, the MDF beamformer is [5]

$$\mathbf{h}_{\mathrm{MDF}} = \frac{\mathbf{\Gamma}_{\mathrm{d}}^{-1} \mathbf{d}_{\theta_s,\phi_s}}{\mathbf{d}_{\theta_s,\phi_s}^H \mathbf{\Gamma}_{\mathrm{d}}^{-1} \mathbf{d}_{\theta_s,\phi_s}}. \tag{3.43}$$

We deduce that the WNG and DF are, respectively,

$$\mathcal{W}(\mathbf{h}_{\mathrm{MDF}}) = \frac{\left(\mathbf{d}_{\theta_s,\phi_s}^H \mathbf{\Gamma}_{\mathrm{d}}^{-1} \mathbf{d}_{\theta_s,\phi_s}\right)^2}{\mathbf{d}_{\theta_s,\phi_s}^H \mathbf{\Gamma}_{\mathrm{d}}^{-2} \mathbf{d}_{\theta_s,\phi_s}} \tag{3.44}$$

and

$$\mathcal{D}(\mathbf{h}_{\mathrm{MDF}}) = \mathbf{d}_{\theta_s,\phi_s}^H \mathbf{\Gamma}_{\mathrm{d}}^{-1} \mathbf{d}_{\theta_s,\phi_s} = \mathcal{D}_{\max,\theta_s,\phi_s}. \tag{3.45}$$

While the MDF beamformer maximizes the DF, it may amplify the white noise, especially at low frequencies. In some particular contexts, the MDF beamformer is called the superdirective beamformer as it leads to the highest possible gain [4, 5].

There are different techniques to compromise between WNG and DF, but the most obvious one is through the following robust beamformer [6]:

$$\mathbf{h}_{\mathrm{R},\epsilon} = \frac{(\mathbf{\Gamma}_{\mathrm{d}} + \epsilon \mathbf{I}_M)^{-1} \mathbf{d}_{\theta_{\mathrm{s}},\phi_{\mathrm{s}}}}{\mathbf{d}_{\theta_{\mathrm{s}},\phi_{\mathrm{s}}}^H (\mathbf{\Gamma}_{\mathrm{d}} + \epsilon \mathbf{I}_M)^{-1} \mathbf{d}_{\theta_{\mathrm{s}},\phi_{\mathrm{s}}}}, \tag{3.46}$$

where $\epsilon \geq 0$ is the regularization parameter. This parameter tries to find some kind of compromise between WNG and DF. We observe that a small ϵ leads to a large DF and a low WNG, while a large ϵ gives a low DF and a large WNG. We have

$$\mathbf{h}_{\mathrm{R},0} = \mathbf{h}_{\mathrm{MDF}},$$

$$\mathbf{h}_{\mathrm{R},\infty} = \mathbf{h}_{\mathrm{DS}}.$$

One of the most useful adaptive beamformers in sensor arrays is the so-called minimum variance distortionless response (MVDR) [4, 7, 8]. It is obtained from the maximization of the SNR gain given in (3.20), where the distortionless constraint is applied to the filter. We easily get

$$\begin{aligned} \mathbf{h}_{\mathrm{MVDR}} &= \frac{\mathbf{\Gamma}_{\mathbf{v}}^{-1} \mathbf{d}_{\theta_{\mathrm{s}},\phi_{\mathrm{s}}}}{\mathbf{d}_{\theta_{\mathrm{s}},\phi_{\mathrm{s}}}^H \mathbf{\Gamma}_{\mathbf{v}}^{-1} \mathbf{d}_{\theta_{\mathrm{s}},\phi_{\mathrm{s}}}} \\ &= \frac{\mathbf{\Phi}_{\mathbf{y}}^{-1} \mathbf{d}_{\theta_{\mathrm{s}},\phi_{\mathrm{s}}}}{\mathbf{d}_{\theta_{\mathrm{s}},\phi_{\mathrm{s}}}^H \mathbf{\Phi}_{\mathbf{y}}^{-1} \mathbf{d}_{\theta_{\mathrm{s}},\phi_{\mathrm{s}}}}, \end{aligned} \tag{3.47}$$

which leads to the maximum SNR gain:

$$\begin{aligned} \mathcal{G}(\mathbf{h}_{\mathrm{MVDR}}) &= \mathbf{d}_{\theta_{\mathrm{s}},\phi_{\mathrm{s}}}^H \mathbf{\Gamma}_{\mathbf{v}}^{-1} \mathbf{d}_{\theta_{\mathrm{s}},\phi_{\mathrm{s}}} \\ &= \mathcal{G}_{\max,\theta_{\mathrm{s}},\phi_{\mathrm{s}},\mathbf{v}}. \end{aligned} \tag{3.48}$$

This distortionless beamformer is adaptive because it depends on the statistics of the noise or the statistics of the observed signals. In practice, the two (equivalent) forms of $\mathbf{h}_{\mathrm{MVDR}}$ shown in (3.47) will behave differently and their performances will greatly depend on the accuracy of the signal statistics estimation. Another interesting and practical form of the MVDR beamformer is

$$\begin{aligned} \mathbf{h}_{\mathrm{MVDR}} &= D_{\theta_{\mathrm{s}},\phi_{\mathrm{s}},1} \frac{\mathbf{\Phi}_{\mathbf{v}}^{-1} \mathbf{d}_{\theta_{\mathrm{s}},\phi_{\mathrm{s}}} \mathbf{d}_{\theta_{\mathrm{s}},\phi_{\mathrm{s}}}^H \mathbf{i}_1}{\mathrm{tr}\left(\mathbf{\Phi}_{\mathbf{v}}^{-1} \mathbf{d}_{\theta_{\mathrm{s}},\phi_{\mathrm{s}}} \mathbf{d}_{\theta_{\mathrm{s}},\phi_{\mathrm{s}}}^H\right)} \\ &= D_{\theta_{\mathrm{s}},\phi_{\mathrm{s}},1} \frac{\mathbf{\Phi}_{\mathbf{v}}^{-1} \mathbf{\Phi}_{\mathbf{x}} \mathbf{i}_1}{\mathrm{tr}\left(\mathbf{\Phi}_{\mathbf{v}}^{-1} \mathbf{\Phi}_{\mathbf{x}}\right)} \\ &= D_{\theta_{\mathrm{s}},\phi_{\mathrm{s}},1} \frac{\left(\mathbf{\Phi}_{\mathbf{v}}^{-1} \mathbf{\Phi}_{\mathbf{y}} - \mathbf{I}_M\right) \mathbf{i}_1}{\mathrm{tr}\left(\mathbf{\Phi}_{\mathbf{v}}^{-1} \mathbf{\Phi}_{\mathbf{y}}\right) - M}, \end{aligned} \tag{3.49}$$

where $D_{\theta_s,\phi_s,1}$ is the first element of $\mathbf{d}_{\theta_s,\phi_s}$ and $\mathbf{i}_1$ is the first column of $\mathbf{I}_M$. From a practical point of view, the term $D_{\theta_s,\phi_s,1}$ in (3.49) is not important and can be neglected if desired.

While the MVDR filter maximizes the narrowband output SNR as defined in (3.19), it does not maximize its fullband counterpart [4]. As a result, noise reduction with this approach may not be enough in practice unless a large number of microphones are deployed. One manner to deal with this is to relax the distortionless constraint and, instead, minimize the MSE criterion in (3.34), from which we get the classical Wiener beamformer [4]:

$$\begin{aligned}\mathbf{h}_W &= \phi_X \mathbf{\Phi}_\mathbf{y}^{-1}\mathbf{d}_{\theta_s,\phi_s} \\ &= \frac{\phi_X \mathbf{\Phi}_\mathbf{v}^{-1}\mathbf{d}_{\theta_s,\phi_s}}{1+\phi_X \mathbf{d}_{\theta_s,\phi_s}^H \mathbf{\Phi}_\mathbf{v}^{-1}\mathbf{d}_{\theta_s,\phi_s}} \\ &= D_{\theta_s,\phi_s,1}\frac{\left(\mathbf{\Phi}_\mathbf{v}^{-1}\mathbf{\Phi}_\mathbf{y}-\mathbf{I}_M\right)\mathbf{i}_1}{\text{tr}\left(\mathbf{\Phi}_\mathbf{v}^{-1}\mathbf{\Phi}_\mathbf{y}\right)-M+1}.\end{aligned} \tag{3.50}$$

While the Wiener beamformer maximizes the narrowband output SNR, it certainly does not maximize its broadband counterpart; however, it definitely makes this latter greater than the one with the MVDR beamformer [4]. Distortion is obviously expected and is increased when the input SNR is decreased. However, if we increase the number of sensors, we decrease distortion. Substituting (3.50) into (3.34), we obtain the minimum MSE (MMSE):

$$\mathcal{J}\left(\mathbf{h}_W\right) = \phi_X\left(1-\mathbf{d}_{\theta_s,\phi_s}^H \mathbf{\Phi}_\mathbf{y}^{-1}\mathbf{d}_{\theta_s,\phi_s}\right) \tag{3.51}$$

and the expected distortion is

$$\mathbf{h}_W^H\mathbf{d}_{\theta_s,\phi_s} = \phi_X - \mathcal{J}\left(\mathbf{h}_W\right) < 1. \tag{3.52}$$

One straightforward way to compromise between the performance of the MVDR and Wiener beamformers is via the so-called tradeoff beamformer [4]:

$$\mathbf{h}_{T,\mu} = D_{\theta_s,\phi_s,1}\frac{\left(\mathbf{\Phi}_\mathbf{v}^{-1}\mathbf{\Phi}_\mathbf{y}-\mathbf{I}_M\right)\mathbf{i}_1}{\text{tr}\left(\mathbf{\Phi}_\mathbf{v}^{-1}\mathbf{\Phi}_\mathbf{y}\right)-M+\mu}, \tag{3.53}$$

where $\mu \geq 0$ is a tuning parameter, which can be frequency dependent. We can see that for

- $\mu = 1$, $\mathbf{h}_{T,1} = \mathbf{h}_W$, which is the Wiener beamformer;
- $\mu = 0$, $\mathbf{h}_{T,0} = \mathbf{h}_{MVDR}$, which is the MVDR beamformer;
- $\mu > 1$, results in a beamformer with low residual noise (from a broadband perspective) at the expense of high desired signal distortion (as compared to Wiener); and

- $\mu < 1$, results in a beamformer with high residual noise and low desired signal distortion (as compared to Wiener).

Now, assume that we have one interference impinging on the array from the direction $(\theta_\mathrm{i}, \phi_\mathrm{i}) \neq (\theta_\mathrm{s}, \phi_\mathrm{s})$ and would like to place null in that direction with a beamformer $\mathbf{h}$, and, meanwhile, recover the desired source coming from the direction $(\theta_\mathrm{s}, \phi_\mathrm{s})$. The combination of these two constraints leads to the constraint equation:

$$\mathbf{C}_{\mathrm{s,i}}^H \mathbf{h} = \begin{bmatrix} 1 \\ 0 \end{bmatrix}, \tag{3.54}$$

where

$$\mathbf{C}_{\mathrm{s,i}} = \begin{bmatrix} \mathbf{d}_{\theta_\mathrm{s},\phi_\mathrm{s}} & \mathbf{d}_{\theta_\mathrm{i},\phi_\mathrm{i}} \end{bmatrix} \tag{3.55}$$

is the constraint matrix of size $M \times 2$, whose *two* columns are linearly independent, and $\mathbf{d}_{\theta_\mathrm{i},\phi_\mathrm{i}}$ is the steering vector in the direction $(\theta_\mathrm{i}, \phi_\mathrm{i})$. From the minimization of the residual noise subject to (3.54), we find the linearly constrained minimum variance (LCMV) beamformer [9, 10]:

$$\begin{aligned} \mathbf{h}_{\mathrm{LCMV}} &= \mathbf{\Phi}_\mathbf{v}^{-1} \mathbf{C}_{\mathrm{s,i}} \left(\mathbf{C}_{\mathrm{s,i}}^H \mathbf{\Phi}_\mathbf{v}^{-1} \mathbf{C}_{\mathrm{s,i}} \right)^{-1} \begin{bmatrix} 1 \\ 0 \end{bmatrix} \\ &= \mathbf{\Phi}_\mathbf{y}^{-1} \mathbf{C}_{\mathrm{s,i}} \left(\mathbf{C}_{\mathrm{s,i}}^H \mathbf{\Phi}_\mathbf{y}^{-1} \mathbf{C}_{\mathrm{s,i}} \right)^{-1} \begin{bmatrix} 1 \\ 0 \end{bmatrix}. \end{aligned} \tag{3.56}$$

In general, we have

$$\mathcal{G}(\mathbf{h}_{\mathrm{LCMV}}) \leq \mathcal{G}(\mathbf{h}_{\mathrm{MVDR}}).$$

3.5 Input SNR Estimation

In most adaptive beamformers, the input SNR or some sort of a function depending on it needs to be estimated in order to implement the filter properly [11, 12]. In this section, we show how to do this with the Wiener beamformer.

One can check that the Wiener filter in (3.50) can be rewritten as

$$\mathbf{h}_\mathrm{W} = H_\mathrm{W} \mathbf{\Gamma}_\mathbf{y}^{-1} \mathbf{d}_{\theta_\mathrm{s},\phi_\mathrm{s}}, \tag{3.57}$$

where

$$H_\mathrm{W} = \frac{\mathrm{iSNR}}{1 + \mathrm{iSNR}} \tag{3.58}$$

is the Wiener gain (see Chap. 2), with $0 \leq H_{\text{W}} \leq 1$, and

$$\mathbf{\Gamma}_{\mathbf{y}} = \frac{\mathbf{\Phi}_{\mathbf{y}}}{\phi_Y} \tag{3.59}$$

is the coherence matrix of $\mathbf{y}$. We see from (3.57) that we need to estimate iSNR or H_{W}, but it is much wiser to estimate the latter as it is bounded.

We can also express (3.59) as

$$\begin{aligned}\mathbf{\Gamma}_{\mathbf{y}} &= \frac{\phi_X}{\phi_Y}\mathbf{d}_{\theta_s,\phi_s}\mathbf{d}_{\theta_s,\phi_s}^H + \frac{\phi_V}{\phi_Y}\mathbf{\Gamma}_{\mathbf{v}} \\ &= \frac{\text{iSNR}}{1+\text{iSNR}}\mathbf{d}_{\theta_s,\phi_s}\mathbf{d}_{\theta_s,\phi_s}^H + \frac{1}{1+\text{iSNR}}\mathbf{\Gamma}_{\mathbf{v}} \\ &= H_{\text{W}}\mathbf{d}_{\theta_s,\phi_s}\mathbf{d}_{\theta_s,\phi_s}^H + (1-H_{\text{W}})\,\mathbf{\Gamma}_{\mathbf{v}}.\end{aligned} \tag{3.60}$$

By pre- and post-multiplying both sides of (3.60) by $\mathbf{d}_{\theta_s,\phi_s}^H$ and $\mathbf{d}_{\theta_s,\phi_s}$, respectively, we obtain

$$H_{\text{W}} = \frac{\mathbf{d}_{\theta_s,\phi_s}^H\left(\mathbf{\Gamma}_{\mathbf{y}} - \mathbf{\Gamma}_{\mathbf{v}}\right)\mathbf{d}_{\theta_s,\phi_s}}{M^2 - \mathbf{d}_{\theta_s,\phi_s}^H\mathbf{\Gamma}_{\mathbf{v}}\mathbf{d}_{\theta_s,\phi_s}}. \tag{3.61}$$

In practice, it is easy to estimate $\mathbf{\Gamma}_{\mathbf{y}}$ directly from the observations, which we denote by $\widehat{\mathbf{\Gamma}}_{\mathbf{y}}$, and assuming that we are in the presence of the spherically isotropic noise, we can replace $\mathbf{\Gamma}_{\mathbf{v}}$ by $\mathbf{\Gamma}_{\text{d}}$. As a result, a possible estimator of the Wiener gain is

$$\widehat{H}_{\text{W}} = \frac{\mathbf{d}_{\theta_s,\phi_s}^H\left(\widehat{\mathbf{\Gamma}}_{\mathbf{y}} - \mathbf{\Gamma}_{\text{d}}\right)\mathbf{d}_{\theta_s,\phi_s}}{M^2 - \mathbf{d}_{\theta_s,\phi_s}^H\mathbf{\Gamma}_{\text{d}}\mathbf{d}_{\theta_s,\phi_s}}. \tag{3.62}$$

Hence, the Wiener beamformer in (3.57) can be properly implemented.

3.6 Comparison of Different Array Geometries

In practice, it is often desirable to compare in a fair way the performances (including steering) of different array geometries with the same number of microphones. However, it is not clear at all how to do this. In this section, we explain a reasonable approach.

Due to its structure, the steering vector in (3.3) can be decomposed as

$$\mathbf{d}_{\theta,\phi} = \mathbf{d}_{\theta,\phi}^{\text{x}} \circ \mathbf{d}_{\theta,\phi}^{\text{y}} \circ \mathbf{d}_{\theta}^{\text{z}}, \tag{3.63}$$

where $\circ$ is the Hadamard product,

$$\mathbf{d}_{\theta,\phi}^{\mathrm{x}} = \left[e^{\jmath \delta_1^{\mathrm{x}} \varpi_{\theta,\phi}^{\mathrm{x}}} \;\; e^{\jmath \delta_2^{\mathrm{x}} \varpi_{\theta,\phi}^{\mathrm{x}}} \;\; \cdots \;\; e^{\jmath \delta_M^{\mathrm{x}} \varpi_{\theta,\phi}^{\mathrm{x}}} \right]^T \tag{3.64}$$

is the steering vector along the x-axis,

$$\mathbf{d}_{\theta,\phi}^{\mathrm{y}} = \left[e^{\jmath \delta_1^{\mathrm{y}} \varpi_{\theta,\phi}^{\mathrm{y}}} \;\; e^{\jmath \delta_2^{\mathrm{y}} \varpi_{\theta,\phi}^{\mathrm{y}}} \;\; \cdots \;\; e^{\jmath \delta_M^{\mathrm{y}} \varpi_{\theta,\phi}^{\mathrm{y}}} \right]^T \tag{3.65}$$

is the steering vector along the y-axis, and

$$\mathbf{d}_{\theta}^{\mathrm{z}} = \left[e^{\jmath \delta_1^{\mathrm{z}} \varpi_{\theta}^{\mathrm{z}}} \;\; e^{\jmath \delta_2^{\mathrm{z}} \varpi_{\theta}^{\mathrm{z}}} \;\; \cdots \;\; e^{\jmath \delta_M^{\mathrm{z}} \varpi_{\theta}^{\mathrm{z}}} \right]^T \tag{3.66}$$

is the steering vector along the z-axis, with

$$\begin{aligned}
\delta_m^{\mathrm{x}} &= r_m \sin\theta_m \cos\phi_m, \\
\delta_m^{\mathrm{y}} &= r_m \sin\theta_m \sin\phi_m, \\
\delta_m^{\mathrm{z}} &= r_m \cos\theta_m,
\end{aligned}$$

for $m = 1, 2, \ldots, M$, and

$$\begin{aligned}
\varpi_{\theta,\phi}^{\mathrm{x}} &= \frac{2\pi f \sin\theta \cos\phi}{c}, \\
\varpi_{\theta,\phi}^{\mathrm{y}} &= \frac{2\pi f \sin\theta \sin\phi}{c}, \\
\varpi_{\theta}^{\mathrm{z}} &= \frac{2\pi f \cos\theta}{c}.
\end{aligned}$$

Since Microphone 1 is the reference and coincides with the origin of the three-dimensional Cartesian coordinate system, it is clear that we have $r_1 = 0$ and $\delta_1^{\mathrm{x}} = \delta_1^{\mathrm{y}} = \delta_1^{\mathrm{z}} = 0$. It is important to notice that $\mathbf{d}_{\theta,\phi}^{\mathrm{x}}$, $\mathbf{d}_{\theta,\phi}^{\mathrm{y}}$, and $\mathbf{d}_{\theta}^{\mathrm{z}}$ are steering vectors corresponding to nonuniform linear arrays (NULAs), which are 1-D arrays, while $\mathbf{d}_{\theta,\phi}^{\mathrm{x}} \circ \mathbf{d}_{\theta,\phi}^{\mathrm{y}}$, $\mathbf{d}_{\theta,\phi}^{\mathrm{y}} \circ \mathbf{d}_{\theta}^{\mathrm{z}}$, and $\mathbf{d}_{\theta,\phi}^{\mathrm{x}} \circ \mathbf{d}_{\theta}^{\mathrm{z}}$ are steering vectors corresponding to 2-D arrays. Therefore, we have shown that the steering vector associated with any 3-D array can be decomposed as the Hadamard product of three steering vectors corresponding to linear arrays along the x-axis, y-axis, and z-axis. Other three useful ways to express (3.63) are

$$\begin{aligned}
\mathbf{d}_{\theta,\phi} &= \mathbf{D}_{\theta,\phi}^{\mathrm{yz}} \mathbf{d}_{\theta,\phi}^{\mathrm{x}} \\
&= \mathbf{D}_{\theta,\phi}^{\mathrm{xz}} \mathbf{d}_{\theta,\phi}^{\mathrm{y}} \\
&= \mathbf{D}_{\theta,\phi}^{\mathrm{xy}} \mathbf{d}_{\theta}^{\mathrm{z}},
\end{aligned} \tag{3.67}$$

where

$$\mathbf{D}_{\theta,\phi}^{\text{yz}} = \text{diag}\left(\mathbf{d}_{\theta,\phi}^{\text{y}} \circ \mathbf{d}_{\theta}^{\text{z}}\right),$$

$$\mathbf{D}_{\theta,\phi}^{\text{xz}} = \text{diag}\left(\mathbf{d}_{\theta,\phi}^{\text{x}} \circ \mathbf{d}_{\theta}^{\text{z}}\right),$$

$$\mathbf{D}_{\theta,\phi}^{\text{xy}} = \text{diag}\left(\mathbf{d}_{\theta,\phi}^{\text{x}} \circ \mathbf{d}_{\theta,\phi}^{\text{y}}\right)$$

are diagonal matrices whose main diagonal elements are the components of the vectors $\mathbf{d}_{\theta,\phi}^{\text{y}} \circ \mathbf{d}_{\theta}^{\text{z}}$, $\mathbf{d}_{\theta,\phi}^{\text{x}} \circ \mathbf{d}_{\theta}^{\text{z}}$, and $\mathbf{d}_{\theta,\phi}^{\text{x}} \circ \mathbf{d}_{\theta,\phi}^{\text{y}}$, respectively. It is obvious that $\left(\mathbf{D}_{\theta,\phi}^{\text{yz}}\right)^{-1} = \left(\mathbf{D}_{\theta,\phi}^{\text{yz}}\right)^{*}$, $\left(\mathbf{D}_{\theta,\phi}^{\text{xz}}\right)^{-1} = \left(\mathbf{D}_{\theta,\phi}^{\text{xz}}\right)^{*}$, and $\left(\mathbf{D}_{\theta,\phi}^{\text{xy}}\right)^{-1} = \left(\mathbf{D}_{\theta,\phi}^{\text{xy}}\right)^{*}$. Also, we have $\left(\mathbf{D}_{\theta,\phi}^{\text{yz}}\right)^{*} \mathbf{D}_{\theta,\phi}^{\text{yz}} = \left(\mathbf{D}_{\theta,\phi}^{\text{xz}}\right)^{*} \mathbf{D}_{\theta,\phi}^{\text{xz}} = \left(\mathbf{D}_{\theta,\phi}^{\text{xy}}\right)^{*} \mathbf{D}_{\theta,\phi}^{\text{xy}} = \mathbf{I}_M$.

Consequently, from the previous developments, we can express the signal model in (3.5) as

$$\begin{aligned}\mathbf{y} &= \mathbf{D}_{\theta_\text{s},\phi_\text{s}}^{\text{yz}} \mathbf{d}_{\theta_\text{s},\phi_\text{s}}^{\text{x}} X + \mathbf{v} \\ &= \mathbf{D}_{\theta_\text{s},\phi_\text{s}}^{\text{xz}} \mathbf{d}_{\theta_\text{s},\phi_\text{s}}^{\text{y}} X + \mathbf{v} \\ &= \mathbf{D}_{\theta_\text{s},\phi_\text{s}}^{\text{xy}} \mathbf{d}_{\theta_\text{s}}^{\text{z}} X + \mathbf{v}.\end{aligned} \tag{3.68}$$

Now, left-multiplying both sides of (3.68) by $\left(\mathbf{D}_{\theta_\text{s},\phi_\text{s}}^{\text{yz}}\right)^{*}$, we convert the 3-D array signal model to a NULA (or 1-D array) signal model along the x-axis:

$$\begin{aligned}\mathbf{y}^{\text{x}} &= \left[\, Y_1^{\text{x}} \;\; Y_2^{\text{x}} \;\; \cdots \;\; Y_M^{\text{x}} \,\right]^T \\ &= \left(\mathbf{D}_{\theta_\text{s},\phi_\text{s}}^{\text{yz}}\right)^{*} \mathbf{y} \\ &= \mathbf{d}_{\theta_\text{s},\phi_\text{s}}^{\text{x}} X + \left(\mathbf{D}_{\theta_\text{s},\phi_\text{s}}^{\text{yz}}\right)^{*} \mathbf{v} \\ &= \mathbf{d}_{\theta_\text{s},\phi_\text{s}}^{\text{x}} X + \mathbf{v}^{\text{x}}.\end{aligned} \tag{3.69}$$

In the same way, left-multiplying both sides of (3.68) by $\left(\mathbf{D}_{\theta_\text{s},\phi_\text{s}}^{\text{xz}}\right)^{*}$, we convert the 3-D array signal model to a NULA signal model along the y-axis:

$$\begin{aligned}\mathbf{y}^{\text{y}} &= \left[\, Y_1^{\text{y}} \;\; Y_2^{\text{y}} \;\; \cdots \;\; Y_M^{\text{y}} \,\right]^T \\ &= \left(\mathbf{D}_{\theta_\text{s},\phi_\text{s}}^{\text{xz}}\right)^{*} \mathbf{y} \\ &= \mathbf{d}_{\theta_\text{s},\phi_\text{s}}^{\text{y}} X + \left(\mathbf{D}_{\theta_\text{s},\phi_\text{s}}^{\text{xz}}\right)^{*} \mathbf{v} \\ &= \mathbf{d}_{\theta_\text{s},\phi_\text{s}}^{\text{y}} X + \mathbf{v}^{\text{y}}.\end{aligned} \tag{3.70}$$

Finally, left-multiplying both sides of (3.68) by $\left(\mathbf{D}^{\text{xy}}_{\theta_\text{s},\phi_\text{s}}\right)^*$, we convert the 3-D array signal model to a NULA signal model along the z-axis:

$$\begin{aligned}\mathbf{y}^\text{z} &= \left[\, Y_1^\text{z} \;\; Y_2^\text{z} \;\; \cdots \;\; Y_M^\text{z} \,\right]^T \\ &= \left(\mathbf{D}^{\text{xy}}_{\theta_\text{s},\phi_\text{s}}\right)^* \mathbf{y} \\ &= \mathbf{d}^\text{z}_{\theta_\text{s}} X + \left(\mathbf{D}^{\text{xy}}_{\theta_\text{s},\phi_\text{s}}\right)^* \mathbf{v} \\ &= \mathbf{d}^\text{z}_{\theta_\text{s}} X + \mathbf{v}^\text{z}. \end{aligned} \tag{3.71}$$

Now, if we want to compare the performance of a 3-D array to the one of a 1-D array, we should construct the three above signal models from the 3-D array signal model. We see that all signal models are realized with exactly M microphones.

In the same way, we can form three different signal models for 2-D arrays:

$$\mathbf{y}^\text{xy} = \mathbf{d}^\text{x}_{\theta_\text{s},\phi_\text{s}} \circ \mathbf{d}^\text{y}_{\theta_\text{s},\phi_\text{s}} X + \mathbf{v}^\text{xy}, \tag{3.72}$$

$$\mathbf{y}^\text{yz} = \mathbf{d}^\text{y}_{\theta_\text{s},\phi_\text{s}} \circ \mathbf{d}^\text{z}_{\theta_\text{s},\phi_\text{s}} X + \mathbf{v}^\text{yz}, \tag{3.73}$$

$$\mathbf{y}^\text{xz} = \mathbf{d}^\text{x}_{\theta_\text{s},\phi_\text{s}} \circ \mathbf{d}^\text{z}_{\theta_\text{s},\phi_\text{s}} X + \mathbf{v}^\text{xz}. \tag{3.74}$$

Again, if we want to compare the performance of a 3-D array to the one of a 2-D array, we should construct the three above signal models from the 3-D array signal model.

3.7 Illustrative Examples

Having discussed the most fundamental concepts in microphone array processing and exposed a family of fixed and adaptive beamformers, we now study some examples of the presented beamformers. We consider a 3-D cube array, which consists of $M = M_0^3$ (with $M_0 \geq 3$) omnidirectional microphones, as shown in Fig. 3.1. This cube array is composed of M_0 parallel square planar uniform arrays with M_0^2 elements each and the interelement spacing along any axis of the Cartesian coordinate system is equal to δ.

Let us start by studying an example of the DS beamformer, $\mathbf{h}_\text{DS}$, given in (3.40), with $\delta = 1.0\,\text{cm}$ and $(\theta_\text{s}, \phi_\text{s}) = (80°, 0°)$. Figures 3.2 and 3.3 show plots of the power patterns, $\left|\mathcal{B}_{\theta,\phi}(\mathbf{h})\right|^2$, of this beamformer versus the azimuth angle for $\theta = \theta_\text{s}$, and versus the elevation angle for $\phi = \phi_\text{s}$, respectively, for $M_0 = 3$ and different frequencies. As seen and expected, the main beam of the power patterns points in the desired direction. It can also be observed that the power patterns are frequency dependent and close to the omnidirectionality response at

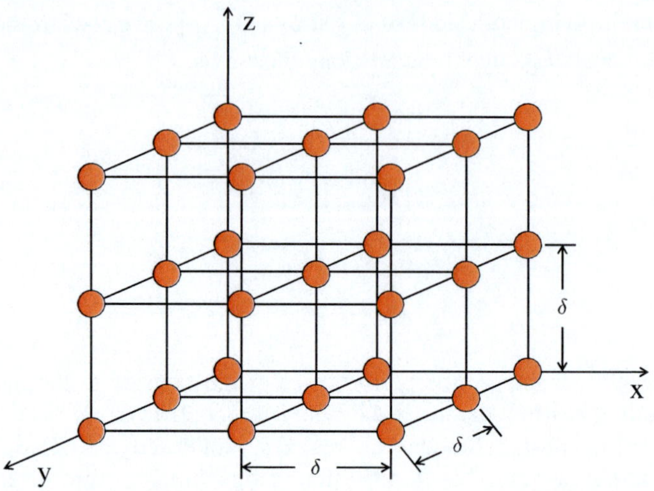

Fig. 3.1 Illustration of a 3-D cube array consisting of $M = M_0^3$ (with $M_0 \geq 3$) omnidirectional microphones in the Cartesian coordinate system. This cube array is composed of M_0 parallel square planar uniform arrays with M_0^2 elements each, and the interelement spacing along any axis of the Cartesian coordinate system is equal to δ

low frequencies. Figure 3.4 shows plots of the DF, $\mathcal{D}(\mathbf{h})$, and WNG, $\mathcal{W}(\mathbf{h})$, of the DS beamformer, as a function of frequency, for different numbers of sensors, $M = M_0^3$. As seen from this figure, both the WNG and DF increase when M_0 increases. However, the DF is low, particularly at low frequencies.

To demonstrate the performance of the MDF beamformer, $\mathbf{h}_{\text{MDF}}$, given in (3.43), we choose $\delta = 1.0\,\text{cm}$ and $(\theta_s, \phi_s) = (80°, 0°)$. Figures 3.5 and 3.6 show plots of the power patterns, $\left|\mathcal{B}_{\theta,\phi}(\mathbf{h})\right|^2$, of this beamformer versus the azimuth angle for $\theta = \theta_s$, and versus the elevation angle for $\phi = \phi_s$, respectively, for $M_0 = 3$ and different frequencies. As seen, the power patterns are almost frequency invariant and have much narrower main beams as compared to the ones with the DS beamformer. Figure 3.7 shows plots of the DF, $\mathcal{D}(\mathbf{h})$, and WNG, $\mathcal{W}(\mathbf{h})$, of the MDF beamformer, as a function of frequency, for different numbers of sensors, $M = M_0^3$. As seen, the DF is almost frequency invariant over the studied frequency band and is much higher than the DF of the DS beamformer. It can also be observed from Fig. 3.7 that the DF increases while the WNG decreases as the value of M_0 increases. The MDF beamformer has a very low WNG, particularly at low frequencies, which means that it may suffer from significant white noise amplification. Consequently, this beamformer can be very sensitive to sensor self-noise, mismatch among sensors, and other array imperfections; this lack of robustness is a big hurdle that prevents the MDF beamformer from being widely deployed in practical systems. How to improve the robustness of this beamformer is a major issue, which has been intensively studied in the literature [6, 13].

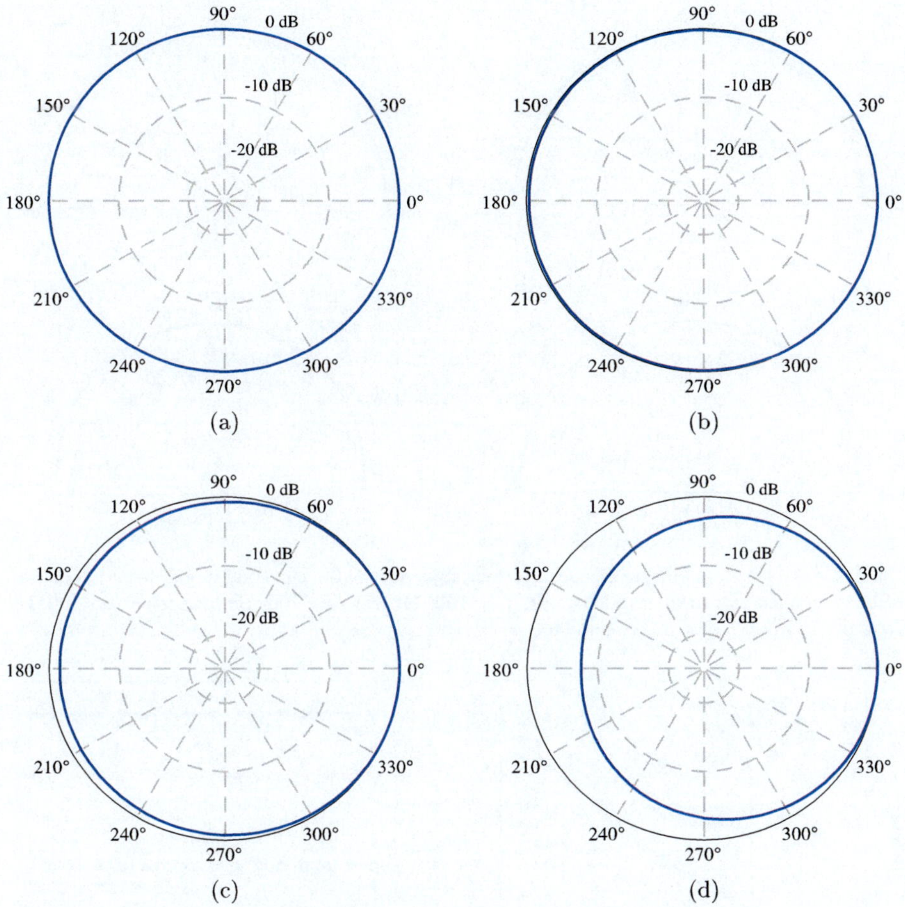

Fig. 3.2 Power patterns of the DS beamformer, $\mathbf{h}_{\rm DS}$, versus the azimuth angle, for $\theta = \theta_{\rm s}$ and different frequencies: (**a**) $f = 500\,$Hz, (**b**) $f = 1000\,$Hz, (**c**) $f = 2000\,$Hz, and (**d**) $f = 4000\,$Hz. Conditions: $M_0 = 3$, $\delta = 1.0\,$cm, and $(\theta_{\rm s}, \phi_{\rm s}) = (80^\circ, 0^\circ)$

We then demonstrate the performance of the robust beamformer, $\mathbf{h}_{\mathrm{R},\epsilon}$, given in (3.46), under the same conditions. Figures 3.8 and 3.9 show plots of the power patterns, $\left|\mathcal{B}_{\theta,\phi}\left(\mathbf{h}\right)\right|^2$, of this beamformer with $\epsilon = 10^{-4}$ versus the azimuth angle for $\theta = \theta_{\rm s}$, and versus the elevation angle for $\phi = \phi_{\rm s}$, respectively, for $M_0 = 3$ and different frequencies. It can be observed that the power patterns are frequency dependent, with its main beam wider than that of the MDF beamformer, but narrower than that of the DS beamformer. Figure 3.10 shows plots of the DF, $\mathcal{D}\left(\mathbf{h}\right)$, and WNG, $\mathcal{W}\left(\mathbf{h}\right)$, of the robust beamformer, as a function of frequency, for $M_0 = 3$ and different values of ϵ. As seen, a different value of ϵ gives a different compromise between WNG and DF, where a small ϵ leads to a large DF and a low WNG, while a large ϵ gives a low DF and a large WNG.

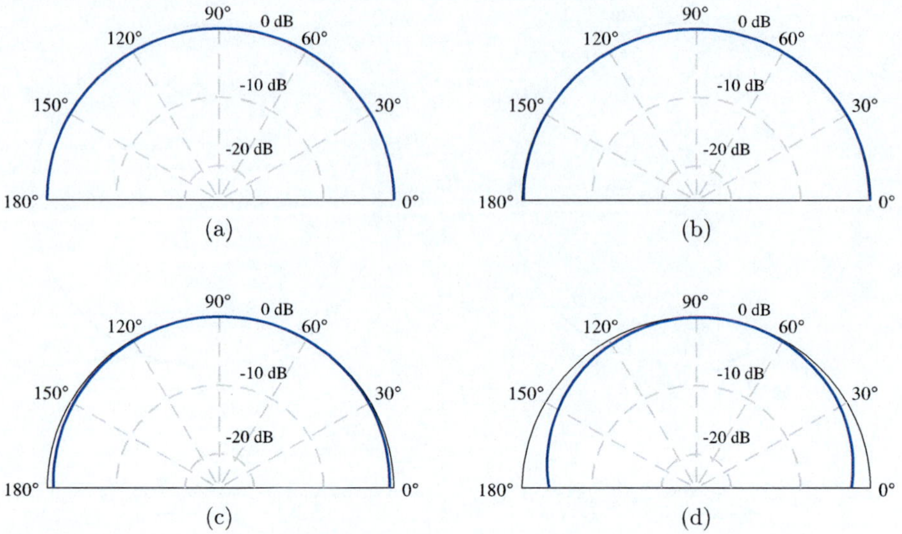

Fig. 3.3 Power patterns of the DS beamformer, $\mathbf{h}_{\mathrm{DS}}$, versus the elevation angle, for $\phi = \phi_{\mathrm{s}}$ and different frequencies: (**a**) $f = 500\,\mathrm{Hz}$, (**b**) $f = 1000\,\mathrm{Hz}$, (**c**) $f = 2000\,\mathrm{Hz}$, and (**d**) $f = 4000\,\mathrm{Hz}$. Conditions: $M_0 = 3$, $\delta = 1.0\,\mathrm{cm}$, and $(\theta_{\mathrm{s}}, \phi_{\mathrm{s}}) = (80^\circ, 0^\circ)$

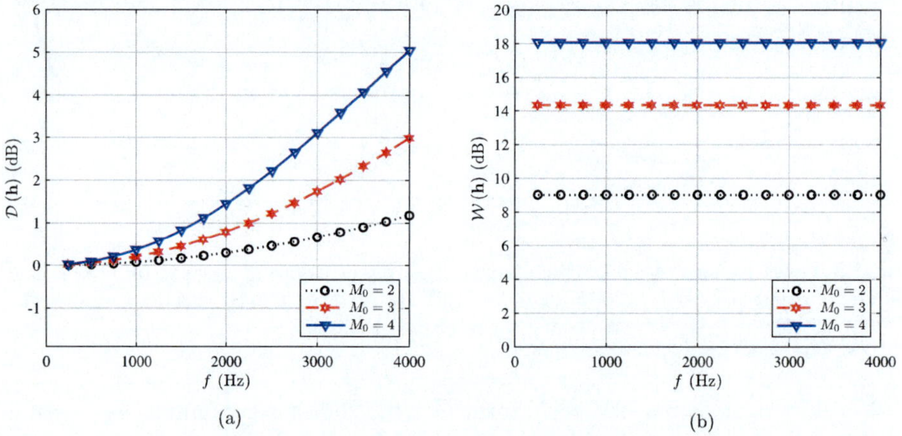

Fig. 3.4 DF and WNG of the DS beamformer, $\mathbf{h}_{\mathrm{DS}}$, as a function of frequency, for different numbers of sensors, $M = M_0^3$: (**a**) DF and (**b**) WNG. Conditions: $\delta = 1.0\,\mathrm{cm}$ and $(\theta_{\mathrm{s}}, \phi_{\mathrm{s}}) = (80^\circ, 0^\circ)$

Now, we study the performance of the adaptive beamformers. We consider an environment with point source noises and white noise, where the two statistically independent interferences impinge on the cube array from two different directions $\left(\theta_{\mathrm{i},1}, \phi_{\mathrm{i},1}\right)$ and $\left(\theta_{\mathrm{i},2}, \phi_{\mathrm{i},2}\right)$. The two interferences are incoherent with the desired signal, and their variances are the same and equal to ϕ_V. Then, the covariance matrix

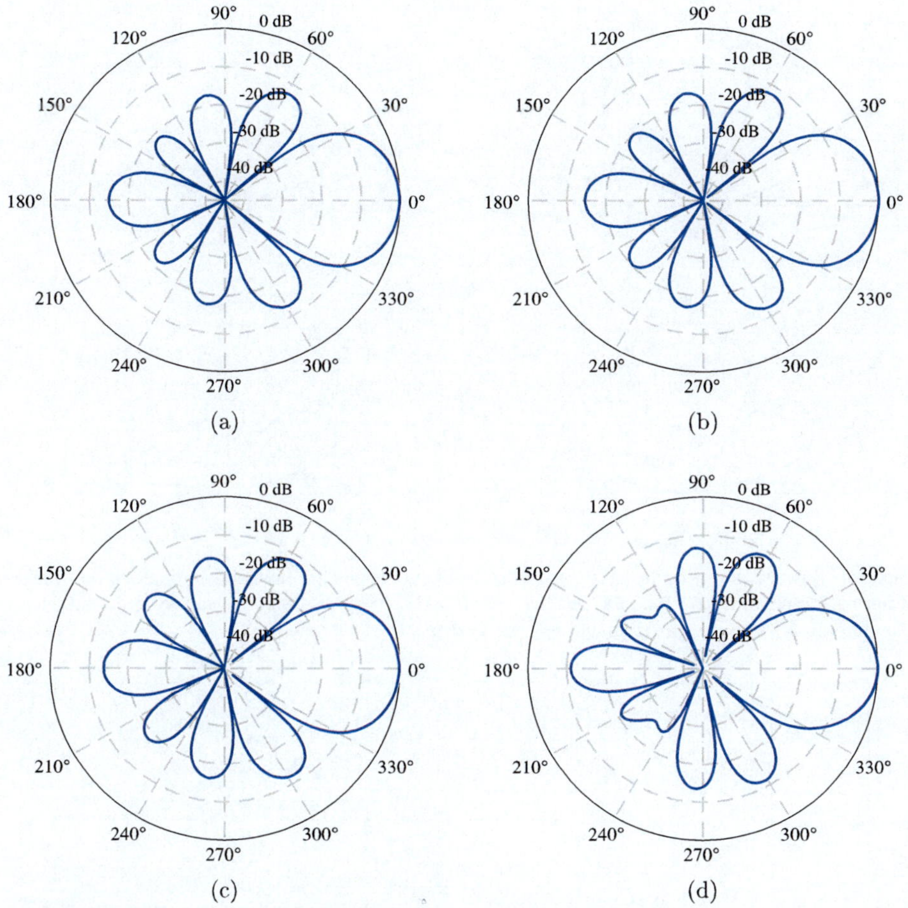

Fig. 3.5 Power patterns of the MDF beamformer, $\mathbf{h}_{\text{MDF}}$, versus the azimuth angle, for $\theta = \theta_s$ and different frequencies: (**a**) $f = 500$ Hz, (**b**) $f = 1000$ Hz, (**c**) $f = 2000$ Hz, and (**d**) $f = 4000$ Hz. Conditions: $M_0 = 3$, $\delta = 1.0$ cm, and $(\theta_s, \phi_s) = (80^\circ, 0^\circ)$

of the noise signal is

$$\mathbf{\Phi_v} = \phi_V \mathbf{d}_{\theta_{i,1},\phi_{i,1}} \mathbf{d}^H_{\theta_{i,1},\phi_{i,1}} + \phi_V \mathbf{d}_{\theta_{i,2},\phi_{i,2}} \mathbf{d}^H_{\theta_{i,2},\phi_{i,2}} + \phi_{V_w} \mathbf{I}_M, \tag{3.75}$$

where $\phi_{V_w} = \alpha \phi_V$ is the variance of the white noise and $\alpha > 0$ is related to the ratio between the powers of the white noise and point source noise. Unless otherwise specified, we choose $\delta = 1.0$ cm, $(\theta_s, \phi_s) = (80^\circ, 0^\circ)$, $\left(\theta_{i,1}, \phi_{i,1}\right) = (80^\circ, 150^\circ)$, $\left(\theta_{i,2}, \phi_{i,2}\right) = (80^\circ, 210^\circ)$, and $\alpha = 0.1$ in the following experiments. The variance of the desired signal, ϕ_X, is set to 1 and the variances of the noises, ϕ_V and ϕ_{V_w}, are computed according to the given input SNR and values of α.

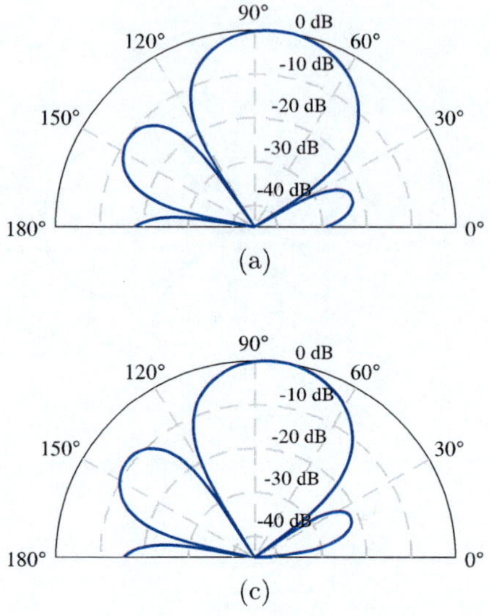

Fig. 3.6 Power patterns of the MDF beamformer, $\mathbf{h}_{\text{MDF}}$, versus the elevation angle, for $\phi = \phi_s$ and different frequencies: (**a**) $f = 500$ Hz, (**b**) $f = 1000$ Hz, (**c**) $f = 2000$ Hz, and (**d**) $f = 4000$ Hz. Conditions: $M_0 = 3$, $\delta = 1.0$ cm, and $(\theta_s, \phi_s) = (80°, 0°)$

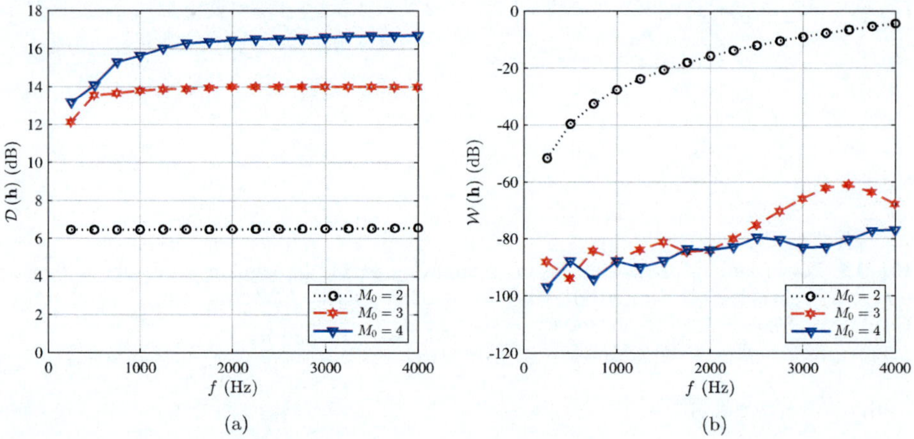

Fig. 3.7 DF and WNG of the MDF beamformer, $\mathbf{h}_{\text{MDF}}$, as a function of frequency, for different numbers of sensors, $M = M_0^3$: (**a**) DF and (**b**) WNG. Conditions: $\delta = 1.0$ cm and $(\theta_s, \phi_s) = (80°, 0°)$

We first study the performance of the Wiener beamformer, $\mathbf{h}_{\text{W}}$, given in (3.50). Figure 3.11 shows plots of the SNR gain, $\mathcal{G}(\mathbf{h})$; noise reduction factor, $\xi(\mathbf{h})$; desired signal reduction factor, $\xi_d(\mathbf{h})$; and desired signal distortion index, $\upsilon(\mathbf{h})$, of the Wiener beamformer, as a function of frequency, for iSNR $= 10$ dB and different numbers of sensors, $M = M_0^3$. Figure 3.12 shows plots of the SNR gain, $\mathcal{G}(\mathbf{h})$;

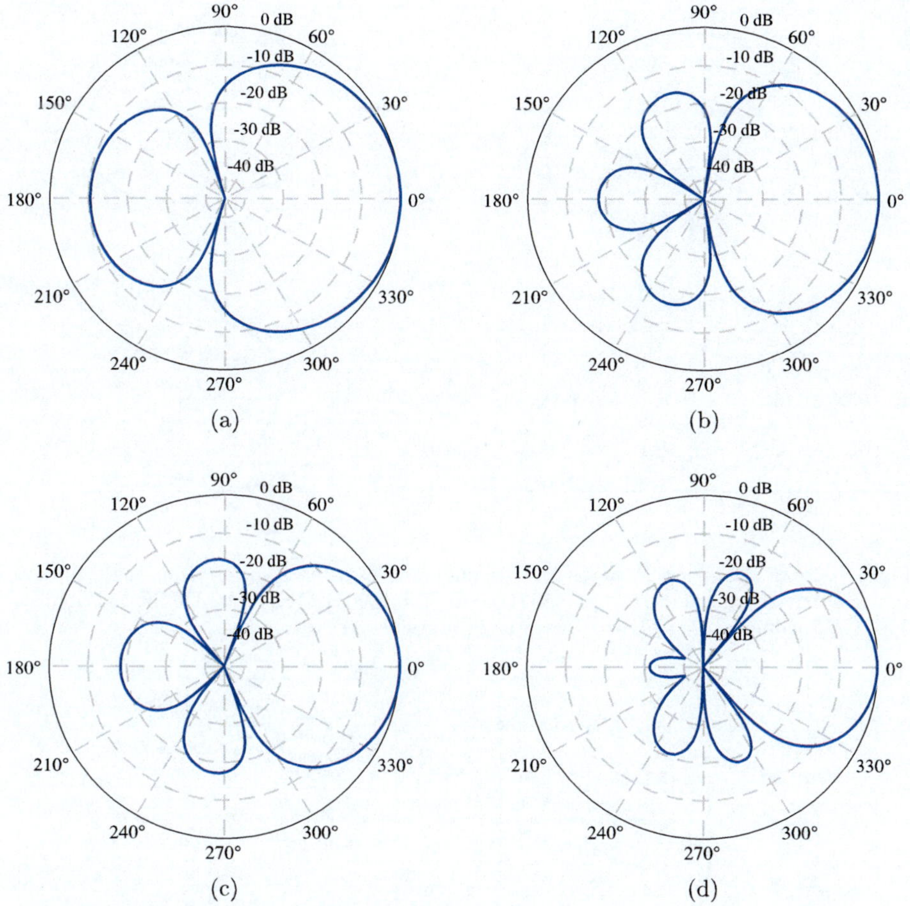

Fig. 3.8 Power patterns of the robust MDF beamformer, $\mathbf{h}_{\mathrm{R},\epsilon}$, versus the azimuth angle, for $\theta = \theta_s$ and different frequencies: (**a**) $f = 500$ Hz, (**b**) $f = 1000$ Hz, (**c**) $f = 2000$ Hz, and (**d**) $f = 4000$ Hz. Conditions: $M_0 = 3$, $\delta = 1.0$ cm, and $(\theta_s, \phi_s) = (80^\circ, 0^\circ)$

noise reduction factor, $\xi(\mathbf{h})$; desired signal reduction factor, $\xi_d(\mathbf{h})$; and desired signal distortion index, $\upsilon(\mathbf{h})$, of the Wiener beamformer, as a function of the input SNR, for $f = 1000$ Hz and different numbers of sensors, $M = M_0^3$. It is seen that the performance of the Wiener beamformer is improved when the number of sensors increases, i.e., the SNR gain and noise reduction factor increase with the value of M, while the desired signal reduction factor and desired signal distortion index decrease as the value of M increases.

We then study the performance of the tradeoff beamformer, $\mathbf{h}_{\mathrm{T},\mu}$, given in (3.53). Figure 3.13 shows plots of the SNR gain, $\mathcal{G}(\mathbf{h})$; noise reduction factor, $\xi(\mathbf{h})$; desired signal reduction factor, $\xi_d(\mathbf{h})$; and desired signal distortion index, $\upsilon(\mathbf{h})$, of the tradeoff beamformer, as a function of frequency, for $M_0 = 3$, iSNR $= 10$ dB, and

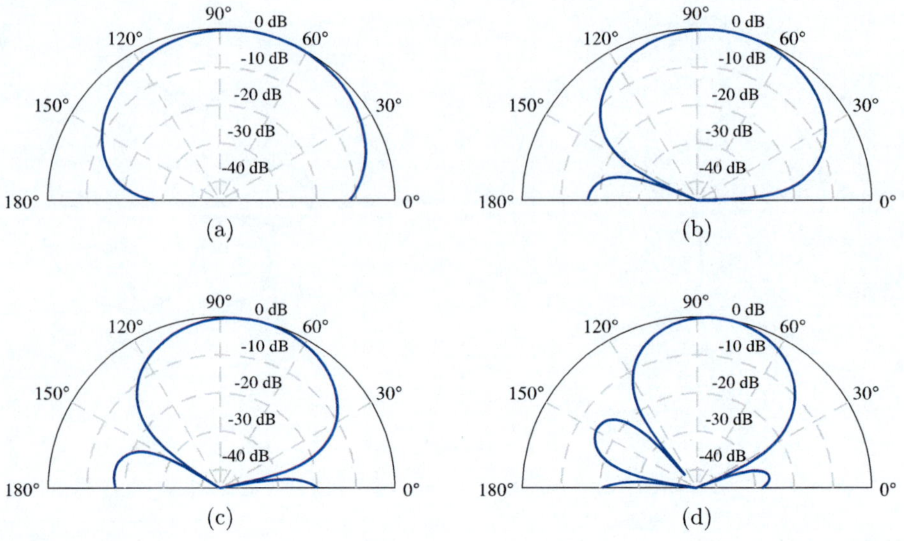

Fig. 3.9 Power patterns of the robust MDF beamformer, $\mathbf{h}_{\mathrm{R},\epsilon}$, versus the elevation angle, for $\phi = \phi_{\mathrm{s}}$ and different frequencies: **(a)** $f = 500$ Hz, **(b)** $f = 1000$ Hz, **(c)** $f = 2000$ Hz, and **(d)** $f = 4000$ Hz. Conditions: $M_0 = 3$, $\delta = 1.0$ cm, and $(\theta_{\mathrm{s}}, \phi_{\mathrm{s}}) = (80°, 0°)$

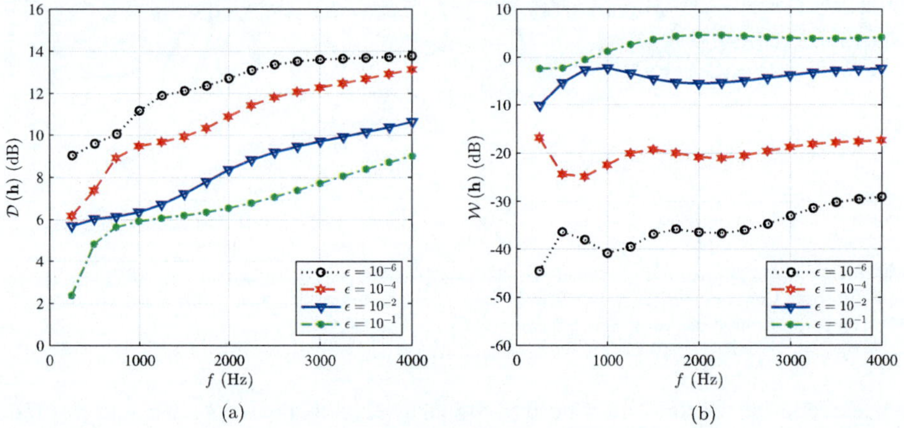

Fig. 3.10 DF and WNG of the robust MDF beamformer, $\mathbf{h}_{\mathrm{R},\epsilon}$, as a function of frequency, for different values of ϵ: **(a)** DF and **(b)** WNG. Conditions: $M_0 = 3$, $\delta = 1.0$ cm, and $(\theta_{\mathrm{s}}, \phi_{\mathrm{s}}) = (80°, 0°)$

different values of μ. As seen, the desired signal reduction factor and desired signal distortion index increase when μ increases, but the SNR gain is almost invariant with μ. The noise reduction factor increases when μ increases, but the increments are too small to be observed from Fig. 3.13. In fact, the SNR gain is independent of the parameter μ from a narrowband perspective. This is easily seen from (3.53),

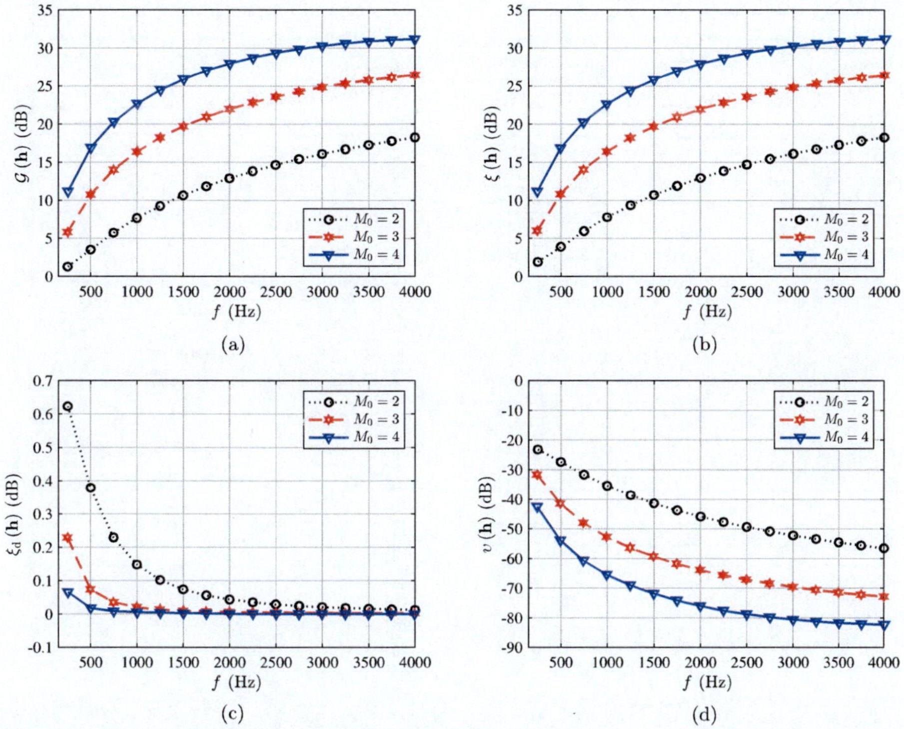

Fig. 3.11 Performance of the Wiener beamformer, $\mathbf{h}_{\mathrm{W}}$, as a function of frequency, for different numbers of sensors, $M = M_0^3$: (**a**) SNR gain, (**b**) noise reduction factor, (**c**) desired signal reduction factor, and (**d**) desired signal distortion index. Conditions: iSNR = 10 dB, $\delta = 1.0\,\mathrm{cm}$, $\alpha = 0.1$, and $(\theta_{\mathrm{s}}, \phi_{\mathrm{s}}) = (80^\circ, 0^\circ)$

where the denominator of the tradeoff beamformer, $\mathbf{h}_{\mathrm{T},\mu}$, consisting of μ, is a scalar and would not affect the (narrowband) SNR gain.

To better evaluate the performance of the tradeoff beamformer, we define some broadband performance metrics. The broadband input SNR and broadband output SNR are, respectively, defined as

$$\overline{\mathrm{iSNR}\,(\mathbf{h})} = \frac{\int_f \phi_X(f)df}{\int_f \phi_V(f)df} \tag{3.76}$$

and

$$\overline{\mathrm{oSNR}\,(\mathbf{h})} = \frac{\int_f \phi_{X_{\mathrm{fd}}}(f)df}{\int_f \phi_{V_{\mathrm{rn}}}(f)df}, \tag{3.77}$$

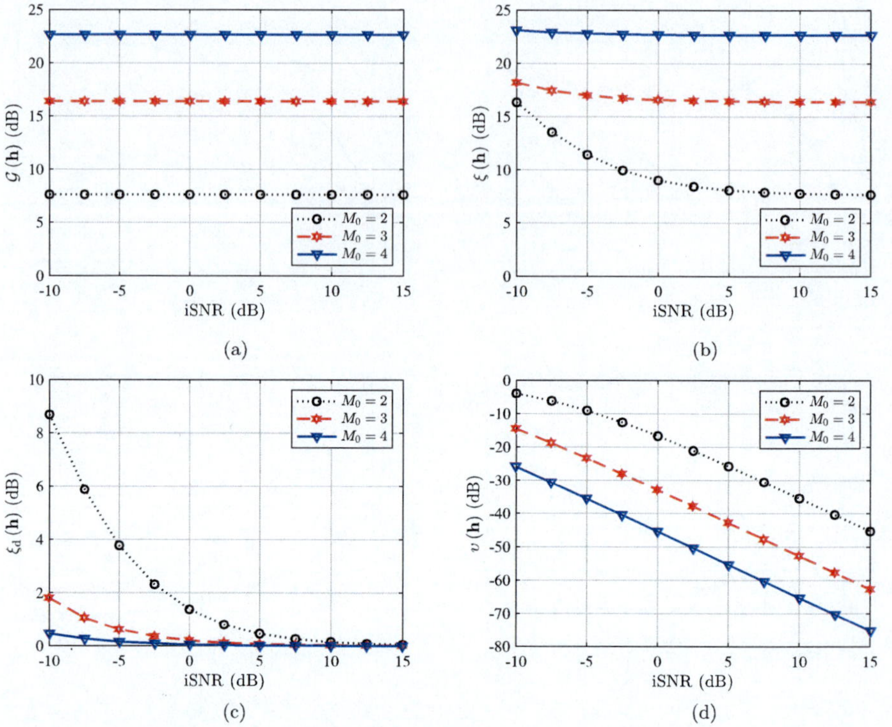

Fig. 3.12 Performance of the Wiener beamformer, $\mathbf{h}_W$, as a function of the input SNR, for different numbers of sensors, $M = M_0^3$: (**a**) SNR gain, (**b**) noise reduction factor, (**c**) desired signal reduction factor, and (**d**) desired signal distortion index. Conditions: $\delta = 1.0\,\text{cm}$, $f = 1000\,\text{Hz}$, $\alpha = 0.1$, and $(\theta_s, \phi_s) = (80°, 0°)$

where $\phi_X(f)$, $\phi_V(f)$, $\phi_{X_{fd}}(f)$, and $\phi_{V_{rn}}(f)$ are the variances of the desired signal, noise, filtered desired signal, and residual noise, respectively, at the frequency f. From the definitions of the broadband input and output SNRs, we deduce that the broadband SNR gain is

$$\overline{\mathcal{G}(\mathbf{h})} = \frac{\overline{\text{oSNR}(\mathbf{h})}}{\overline{\text{iSNR}(\mathbf{h})}}. \tag{3.78}$$

The broadband noise reduction factor is defined as

$$\overline{\xi(\mathbf{h})} = \frac{\int_f \phi_V(f)df}{\int_f \phi_{V_{rn}}(f)df}, \tag{3.79}$$

the broadband desired signal distortion index is defined as

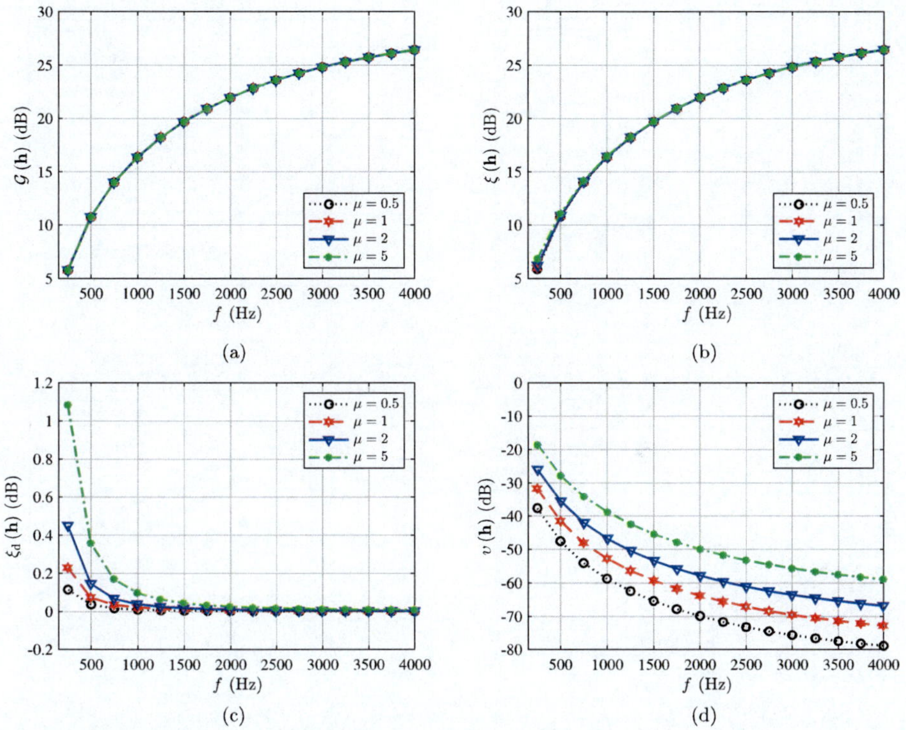

Fig. 3.13 Performance of the tradeoff beamformer, $\mathbf{h}_{\mathrm{T},\mu}$, as a function of frequency, for different values of μ: (**a**) SNR gain, (**b**) noise reduction factor, (**c**) desired signal reduction factor, and (**d**) desired signal distortion index. Conditions: $M_0 = 3$, $\delta = 1.0\,\mathrm{cm}$, iSNR = 10 dB, $\alpha = 0.1$, and $(\theta_{\mathrm{s}}, \phi_{\mathrm{s}}) = (80^\circ, 0^\circ)$

$$\overline{\upsilon\left(\mathbf{h}\right)} = \frac{\int_f \phi_X(f) \left|\mathbf{h}^H(f)\mathbf{d}_{\theta_{\mathrm{s}},\phi_{\mathrm{s}}}(f) - 1\right|^2 df}{\int_f \phi_X(f) df}, \tag{3.80}$$

and the broadband desired signal reduction factor is defined as

$$\overline{\xi_{\mathrm{d}}\left(\mathbf{h}\right)} = \frac{\int_f \phi_X(f) df}{\int_f \phi_{X_{\mathrm{fd}}}(f) df}. \tag{3.81}$$

Figure 3.14 shows plots of the broadband SNR gain, $\overline{\mathcal{G}\left(\mathbf{h}\right)}$; broadband noise reduction factor, $\overline{\xi\left(\mathbf{h}\right)}$; broadband desired signal reduction factor, $\overline{\xi\left(\mathbf{h}\right)}$; and broadband desired signal distortion index, $\overline{\upsilon\left(\mathbf{h}\right)}$, of the tradeoff beamformer, as a function of the input SNR, for $M_0 = 3$ and different values of μ. Note that for $\mu = 1$, the Wiener and tradeoff beamformers are identical. It can be seen that the SNR gain, noise reduction factor, desired signal reduction factor, and desired signal distortion index all increase as the value of μ increases, which indicates that the tradeoff

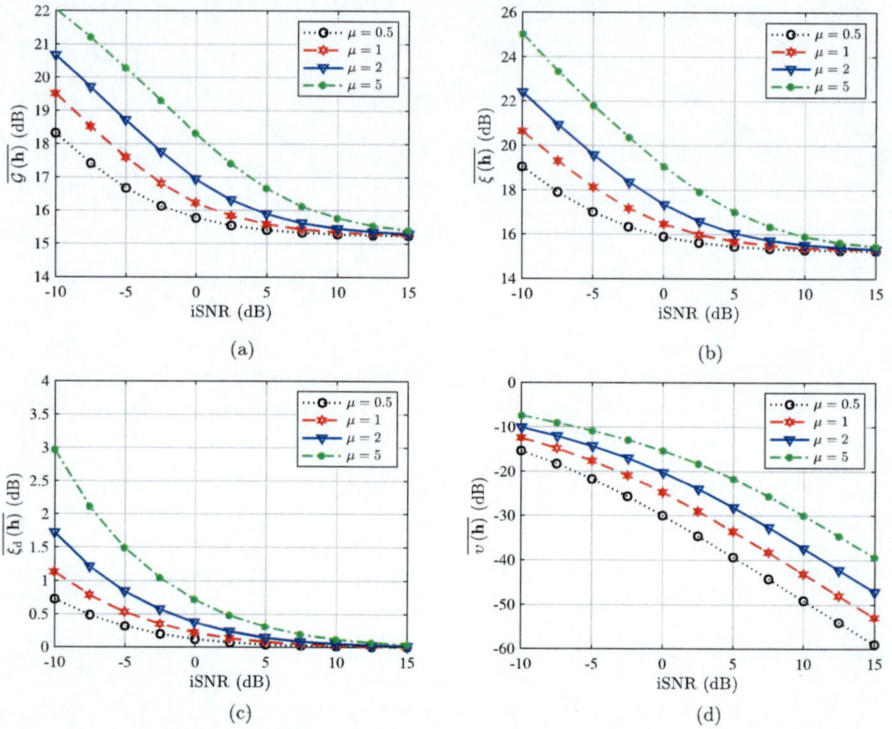

Fig. 3.14 Broadband performance of the tradeoff beamformer, $\mathbf{h}_{\mathrm{T},\mu}$, as a function of the input SNR, for different numbers of sensors, $M = M_0^3$: (**a**) broadband SNR gain, (**b**) broadband noise reduction factor, (**c**) broadband desired signal reduction factor, and (**d**) broadband desired signal distortion index. Conditions: $M_0 = 3$, $\delta = 1.0\,\mathrm{cm}$, $\alpha = 0.1$, and $(\theta_s, \phi_s) = (80°, 0°)$

beamformer achieves a good compromise between noise reduction and desired signal distortion from a broadband perspective.

We then demonstrate the performance of the LCMV beamformer, $\mathbf{h}_{\mathrm{LCMV}}$, given in (3.56), where two null constraints are set in the directions of the interferences, i.e., $(\theta_{i,1}, \phi_{i,1})$ and $(\theta_{i,2}, \phi_{i,2})$. Figures 3.15 and 3.16 show plots of the power patterns, $\left|\mathcal{B}_{\theta,\phi}(\mathbf{h})\right|^2$, of this beamformer versus the azimuth angle for $\theta = \theta_s$, and versus the elevation angle for $\phi = \phi_s$, respectively, for $M_0 = 3$, iSNR = 10 dB, and different frequencies. As seen, the main beam of the power patterns points in the direction of the desired signal, and there are nulls in the directions of the interferences. Figure 3.17 shows plots of the SNR gain, $\mathcal{G}(\mathbf{h})$; noise reduction factor, $\xi(\mathbf{h})$; desired signal reduction factor, $\xi_d(\mathbf{h})$; and desired signal distortion index, $\upsilon(\mathbf{h})$, of the LCMV beamformer, as a function of the input SNR, for $f = 1000$ Hz and different numbers of sensors, $M = M_0^3$. As seen, the SNR gain and noise reduction factor increase as the value of M_0 increases, while the desired signal reduction factor is always approximately equal to 1 (i.e., 0 dB) and the desired signal distortion

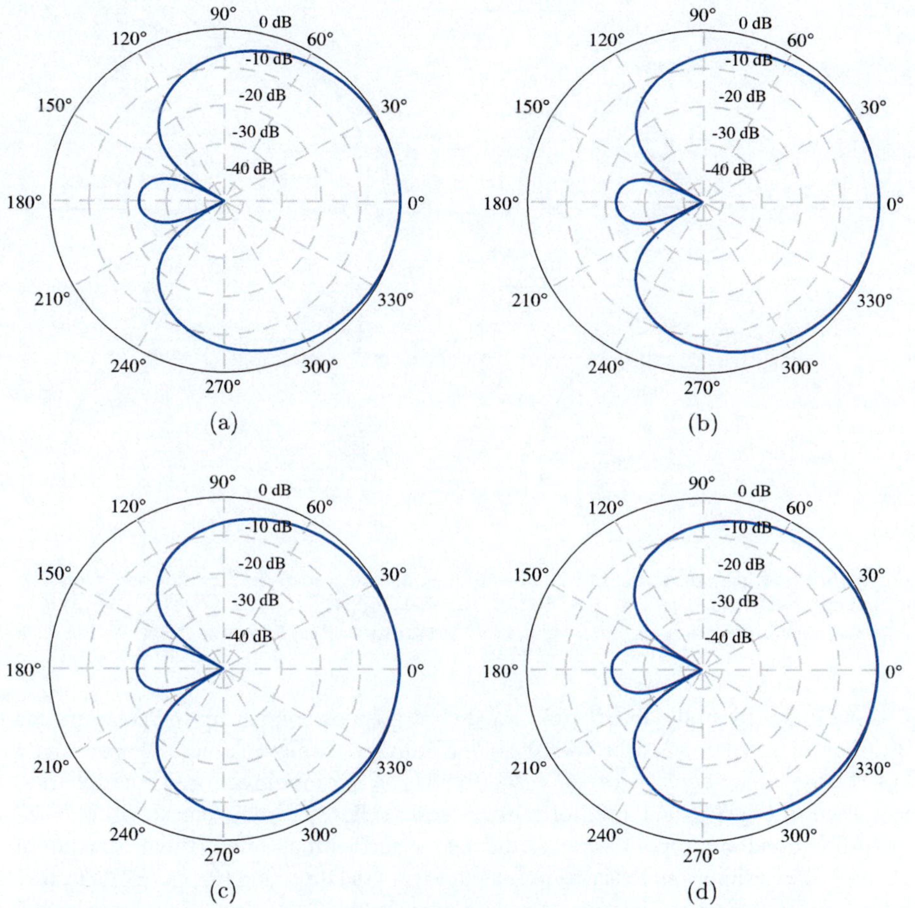

Fig. 3.15 Power patterns of the LCMV beamformer, $\mathbf{h}_{\mathrm{LCMV}}$, versus the azimuth angle, for $\theta = \theta_s$ and different frequencies: (**a**) $f = 500$ Hz, (**b**) $f = 1000$ Hz, (**c**) $f = 2000$ Hz, and (**d**) $f = 4000$ Hz. Conditions: $M_0 = 3$, $\delta = 1.0$ cm, iSNR = 10 dB, $\alpha = 0.1$, and $(\theta_s, \phi_s) = (80^\circ, 0^\circ)$

index is always approximately equal to 0 (i.e., $-\infty$ dB), thanks to the distortionless constraint.

To demonstrate the performance of the input SNR estimation with the estimated Wiener gain, $\widehat{H}_W$, given in (3.62), we consider an environment with two statistically independent point source noises and white noise, where the covariance matrix of the noise signal is given in (3.75). Figure 3.18 shows plots of the estimator, $\widehat{H}_W$, as a function of the input SNR, for $f = 1000$ Hz, different numbers of sensors, $M = M_0^3$, and different values of α, where the theoretical H_W is given in (3.58). As seen, the estimated values are closer to the theoretical ones as the number of sensors increases. It can also be observed that the estimated values are closer to the theoretical values when the value of α is small.

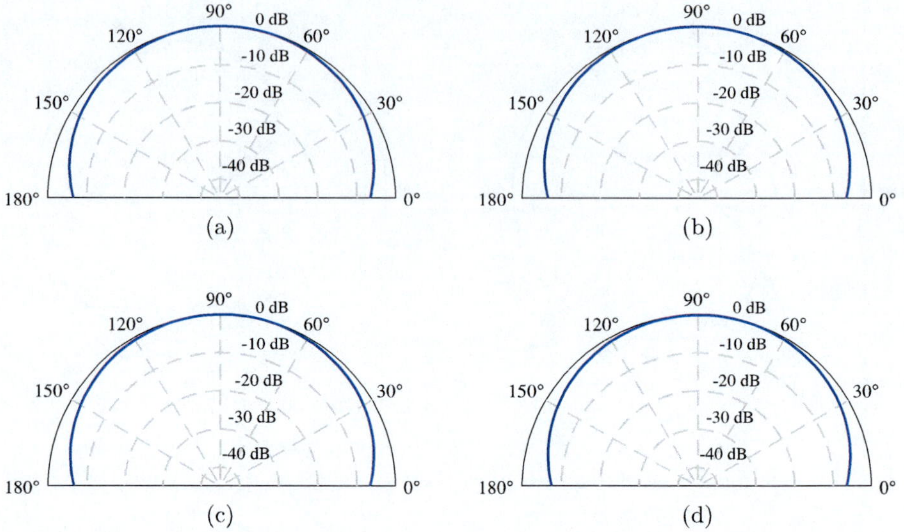

Fig. 3.16 Power patterns of the LCMV beamformer, $\mathbf{h}_{\text{LCMV}}$, versus the elevation angle, for $\phi = \phi_\text{s}$ and different frequencies: (**a**) $f = 500\,\text{Hz}$, (**b**) $f = 1000\,\text{Hz}$, (**c**) $f = 2000\,\text{Hz}$, and (**d**) $f = 4000\,\text{Hz}$. Conditions: $M_0 = 3$, $\delta = 1.0\,\text{cm}$, iSNR = 10 dB, $\alpha = 0.1$, and $(\theta_\text{s}, \phi_\text{s}) = (80^\circ, 0^\circ)$

Finally, let us study an example for the purpose of comparing the performance of different array geometries with the same number of microphones. We consider a cube array, which consists of $M = 27$, i.e., $M_0 = 3$, omnidirectional microphones, as shown in Fig. 3.2, and a uniform linear array (ULA), which consists of $M = 27$ omnidirectional microphones along the x-axis and with an interelement spacing of 1.0 cm. The desired source signal propagates from the direction $(\theta_\text{s}, \phi_\text{s}) = (90^\circ, 0^\circ)$. To compare their performance, we first convert the 3-D array signal models to 1-D array signal models along the x-axis according to (3.69) and then design the robust beamformer ($\epsilon = 10^{-4}$) with the converted 1-D array signal models. Figure 3.19 shows plots of the power patterns, $\left|\mathcal{B}_{\theta,\phi}(\mathbf{h})\right|^2$, of this beamformer versus the azimuth angle for $\theta = \theta_\text{s}$, at different frequencies, designed with the cube array and ULA. It is seen that with the ULA, we have much narrower main beams than with the cube array. Figure 3.20 shows plots of the DF, $\mathcal{D}(\mathbf{h})$, and WNG, $\mathcal{W}(\mathbf{h})$, of the robust MDF beamformer designed with the cube array and ULA, as a function of frequency. It is seen that the robust MDF beamformer designed with the ULA achieves higher DF and WNG. This shows that the robust MDF beamformer designed with a ULA achieves a better performance than a cube array with the same number of microphones if the steering direction is at the endfire of the ULA. However, it should be noted that the cube array has much more freedom to steer the beampattern and is expected to achieve a better performance in other steering directions [14, 15].

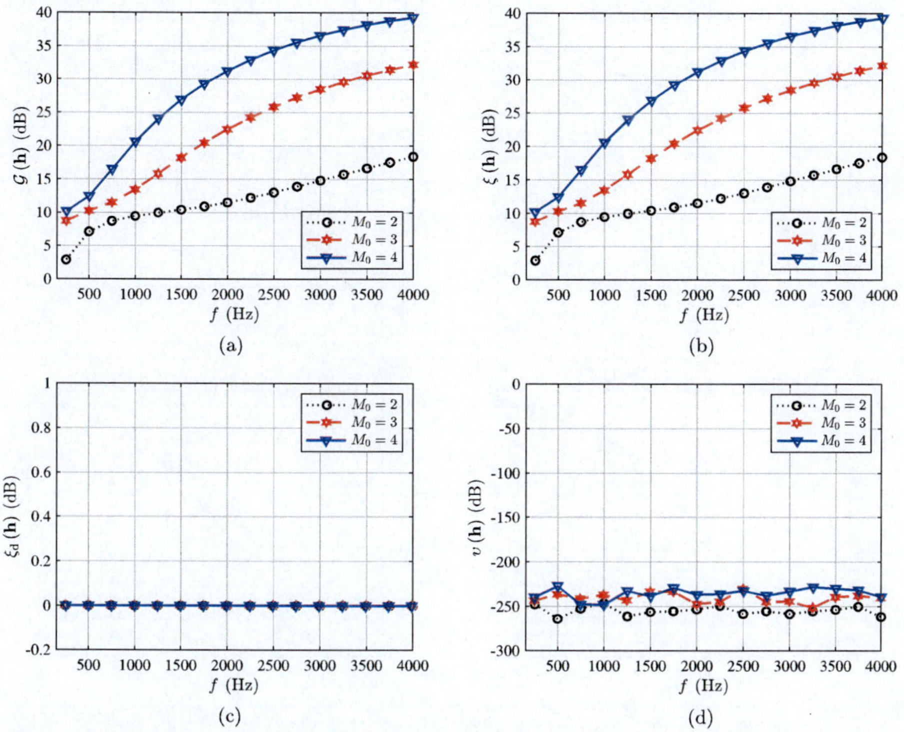

Fig. 3.17 Performance of the LCMV beamformer, $\mathbf{h}_{\text{LCMV}}$, as a function of frequency, for different numbers of sensors, $M = M_0^3$: (**a**) SNR gain, (**b**) noise reduction factor, (**c**) desired signal reduction factor, and (**d**) desired signal distortion index. Conditions: $\delta = 1.0$ cm, $f = 1000$ Hz, $\alpha = 0.1$, and $(\theta_s, \phi_s) = (80°, 0°)$

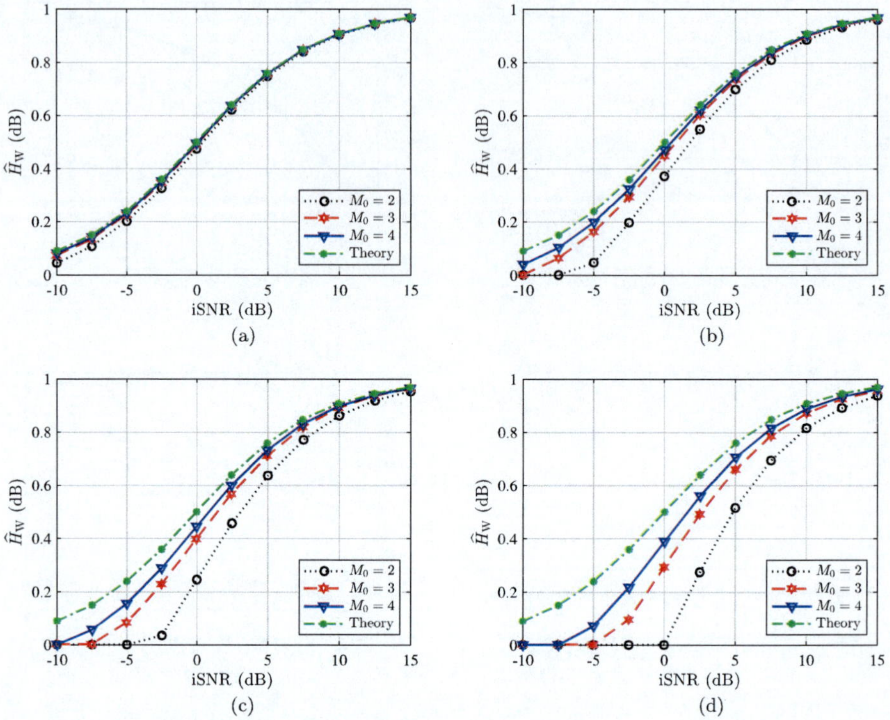

Fig. 3.18 The estimator, $\widehat{H}_W$, as a function of the input SNR, for different numbers of sensors, $M = M_0^3$, and different values of α: **(a)** $\alpha = 0.001$, **(b)** $\alpha = 0.005$, **(c)** $\alpha = 0.01$, and **(d)** $\alpha = 0.02$. Conditions: $\delta = 1.0$ cm, $f = 1000$ Hz, $\alpha = 0.1$, and $(\theta_s, \phi_s) = (80^\circ, 0^\circ)$

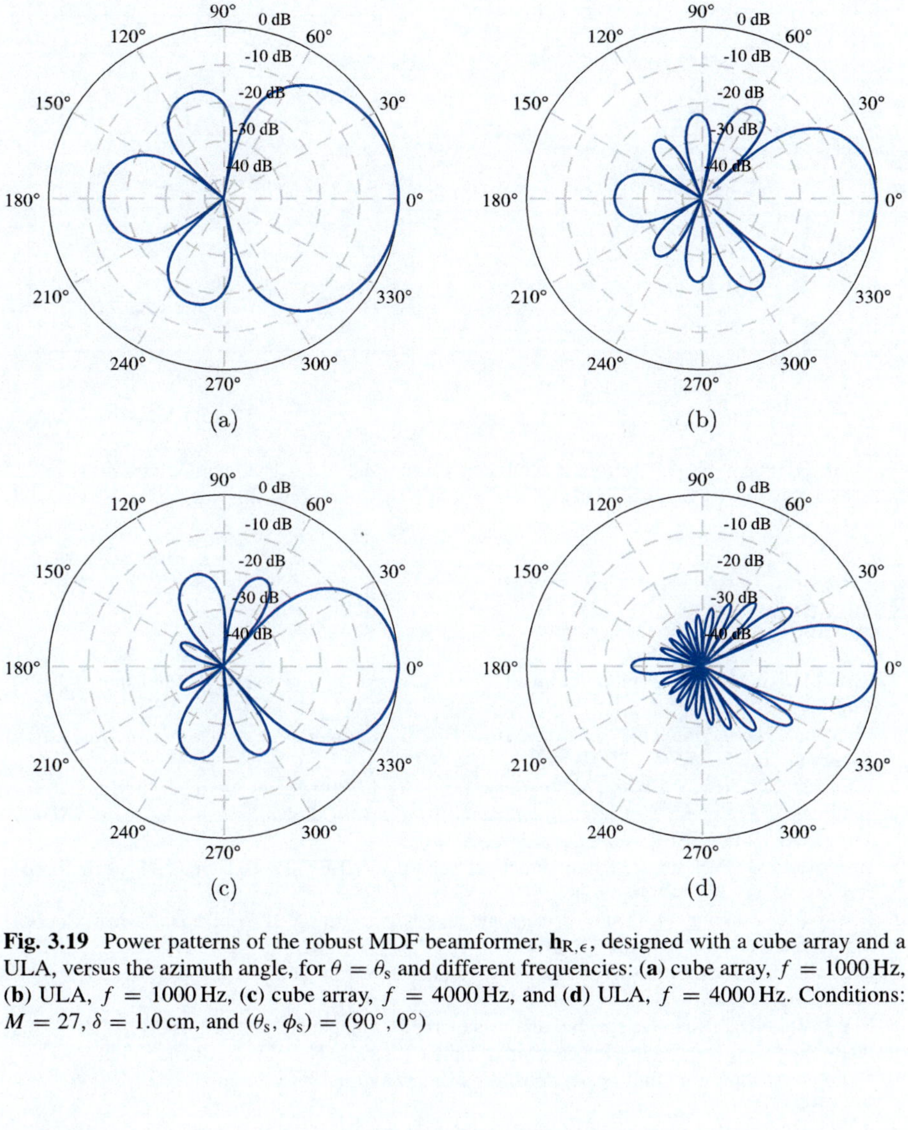

Fig. 3.19 Power patterns of the robust MDF beamformer, $\mathbf{h}_{R,\epsilon}$, designed with a cube array and a ULA, versus the azimuth angle, for $\theta = \theta_s$ and different frequencies: (**a**) cube array, $f = 1000$ Hz, (**b**) ULA, $f = 1000$ Hz, (**c**) cube array, $f = 4000$ Hz, and (**d**) ULA, $f = 4000$ Hz. Conditions: $M = 27$, $\delta = 1.0$ cm, and $(\theta_s, \phi_s) = (90°, 0°)$

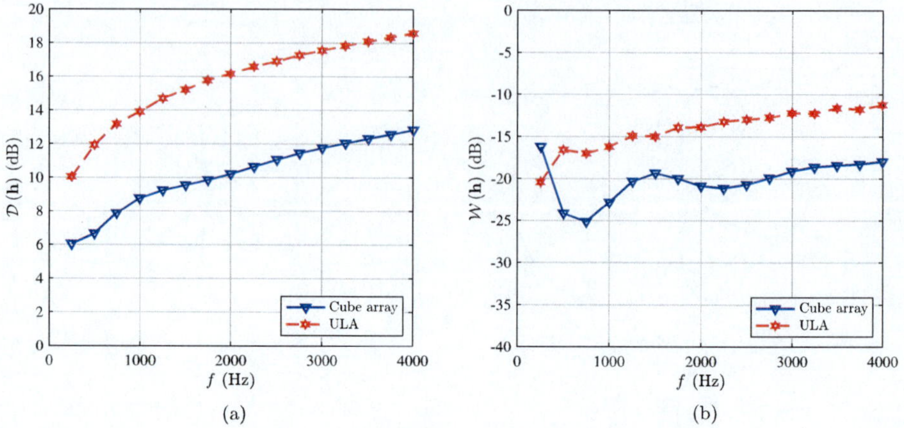

Fig. 3.20 DF and WNG of the robust MDF beamformer, $\mathbf{h}_{\mathrm{R},\epsilon}$, designed with a cube array and a ULA, as a function of frequency: (**a**) DF and (**b**) WNG. Conditions: $M = 27$, $\delta = 1.0$ cm, and $(\theta_s, \phi_s) = (90^\circ, 0^\circ)$

References

1. R.A. Monzingo, T.W. Miller, *Introduction to Adaptive Arrays* (SciTech Publishing, Delhi, 1980)
2. H. L. Van Trees, *Optimum Array Processing: Part IV of Detection, Estimation, and Modulation Theory* (Wiley, New York, 2002)
3. J. Benesty, J. Chen, Y. Huang, *Microphone Array Signal Processing* (Springer, Berlin, 2008)
4. J. Benesty, I. Cohen, J. Chen, *Fundamentals of Signal Enhancement and Array Signal Processing* (Wiley-IEEE Press, Singapore, 2018)
5. H. Cox, R.M. Zeskind, T. Kooij, Practical supergain. IEEE Trans. Acoust. Speech Signal Process. **ASSP-34**, 393–398 (1986)
6. H. Cox, R. Zeskind, M. Owen, Robust adaptive beamforming. IEEE Trans. Acoust. Speech Signal Process. **35**, 1365–1376 (1987)
7. J. Capon, High resolution frequency-wavenumber spectrum analysis. Proc. IEEE **57**, 1408–1418 (1969)
8. R.T. Lacoss, Data adaptive spectral analysis methods. Geophysics **36**, 661–675 (1971)
9. A. Booker, C.Y. Ong, Multiple constraint adaptive filtering. Geophysics **36**, 498–509 (1971)
10. O. Frost, An algorithm for linearly constrained adaptive array processing. Proc. IEEE **60**, 926–935 (1972)
11. R. Zelinski, A microphone array with adaptive post-filtering for noise reduction in reverberant rooms, in *International Conference on Acoustics, Speech, and Signal Processing*, vol. 5 (1988), pp. 2578–2581
12. S. Lefkimmiatis, P. Maragos, A generalized estimation approach for linear and nonlinear microphone array post-filters. Speech Commun. **49**, 657–666 (2007)
13. G. Huang, J. Benesty, J. Chen, Superdirective beamforming based on the Krylov matrix. IEEE/ACM Trans. Audio Speech Lang. Process. **24**, 2531–2543 (2016)
14. X. Wang, J. Benesty, J. Chen, G. Huang, I. Cohen, Beamforming with cube microphone arrays via Kronecker product decompositions. IEEE/ACM Trans. Audio Speech Lang. Process. **29**, 1774–1784 (2021)
15. J. Jin, G. Huang, X. Wang, J. Chen, J. Benesty, I. Cohen, Steering study of linear differential microphone arrays. IEEE/ACM Trans. Audio Speech Lang. Process. **29**, 158–170 (2020)

Chapter 4
Principal Component Analysis in Noise Reduction and Beamforming

Principal component analysis (PCA) is by far the most popular and useful dimensionality reduction technique that one can find in the literature. The objective of PCA is the reduction of the dimension of a random signal vector from M to P, where $P \ll M$, with little loss of the useful information. It does so by preserving the variability of the signal as much as possible, where the new variables are uncorrelated. In this chapter, we show how PCA can be applied to noise reduction and beamforming.

4.1 Brief Overview of PCA

In this section, we briefly explain how principal component analysis (PCA) works [1] and why dimensionality reduction is possible, thanks to the conventional eigenvalue decomposition.

Let

$$\mathbf{y} = \begin{bmatrix} Y_1 & Y_2 & \cdots & Y_M \end{bmatrix}^T \tag{4.1}$$

be a zero-mean complex-valued random signal vector of length M. Its covariance matrix is

$$\mathbf{\Phi}_{\mathbf{y}} = E\left(\mathbf{y}\mathbf{y}^H\right). \tag{4.2}$$

It is well known that this nonnegative-definite Hermitian matrix can be eigenvalue decomposed as

$$\mathbf{\Phi}_{\mathbf{y}} = \mathbf{U}\mathbf{\Lambda}\mathbf{U}^H, \tag{4.3}$$

J. Benesty et al., *Microphone Arrays*, Springer Topics in Signal Processing 22,
https://doi.org/10.1007/978-3-031-36974-2_4

where

$$\mathbf{U} = \begin{bmatrix} \mathbf{u}_1 & \mathbf{u}_2 & \cdots & \mathbf{u}_M \end{bmatrix} \tag{4.4}$$

is a unitary matrix, i.e., $\mathbf{U}\mathbf{U}^H = \mathbf{U}^H\mathbf{U} = \mathbf{I}_M$, and

$$\mathbf{\Lambda} = \text{diag}\left(\lambda_1, \lambda_2, \ldots, \lambda_M\right) \tag{4.5}$$

is a diagonal matrix, with $\lambda_1 \geq \lambda_2 \geq \cdots \geq \lambda_M \geq 0$. The nonnegative real-valued scalar λ_m (for $m = 1, 2, \ldots, M$) is the mth eigenvalue of $\mathbf{\Phi}_\mathbf{y}$ whose corresponding eigenvector is $\mathbf{u}_m$.

Let

$$\mathbf{h}_1 = \begin{bmatrix} H_{1,1} & H_{1,2} & \cdots & H_{1,M} \end{bmatrix}^T \tag{4.6}$$

be a complex-valued filter of length M that we apply to the random signal vector $\mathbf{y}$ in order to obtain the random signal:

$$Z_1 = \mathbf{h}_1^H \mathbf{y}. \tag{4.7}$$

In PCA, Z_1 is considered to be the first principal component if $\mathbf{h}_1$ is derived in such a way that the variance of Z_1, i.e.,

$$\begin{aligned} \phi_{Z_1} &= E\left(|Z_1|^2\right) \\ &= \mathbf{h}_1^H \mathbf{\Phi}_\mathbf{y} \mathbf{h}_1, \end{aligned} \tag{4.8}$$

is maximized under an appropriate constraint, so that it is finite. Then, our optimization problem is

$$\max_{\mathbf{h}_1} \mathbf{h}_1^H \mathbf{\Phi}_\mathbf{y} \mathbf{h}_1 \quad \text{subject to} \quad \mathbf{h}_1^H \mathbf{h}_1 = 1. \tag{4.9}$$

We easily find that the solution is

$$\mathbf{h}_1 = \mathbf{u}_1. \tag{4.10}$$

Therefore, we have $\phi_{Z_1} = \lambda_1$.

Let us form the second random signal:

$$Z_2 = \mathbf{h}_2^H \mathbf{y}, \tag{4.11}$$

where $\mathbf{h}_2$ is a complex-valued filter of length M defined similarly to $\mathbf{h}_1$. In PCA, Z_2 is the second principal component if $\mathbf{h}_2$ is found in such a way that the variance of Z_2, i.e., $\phi_{Z_2} = \mathbf{h}_2^H \mathbf{\Phi}_\mathbf{y} \mathbf{h}_2$, is the second largest one. Therefore, our optimization problem is

$$\max_{\mathbf{h}_2} \mathbf{h}_2^H \mathbf{\Phi}_\mathbf{y} \mathbf{h}_2 \quad \text{subject to} \quad \begin{cases} \mathbf{h}_2^H \mathbf{h}_2 = 1 \\ \mathbf{h}_2^H \mathbf{u}_1 = 0 \end{cases}, \tag{4.12}$$

for which the solution is

$$\mathbf{h}_2 = \mathbf{u}_2. \tag{4.13}$$

As a result, we have $\phi_{Z_2} = \lambda_2$ and $\phi_{Z_1 Z_2} = E\left(Z_1 Z_2^*\right) = 0$.

Clearly, any pth (for $p = 1, 2, \ldots, P$ and $1 \leq P \leq M$) principal component, $Z_p = \mathbf{h}_p^H \mathbf{y}$, is derived with the same strategy. We deduce the reduced random signal vector of length P:

$$\begin{aligned} \mathbf{z} &= \begin{bmatrix} Z_1 & Z_2 & \cdots & Z_P \end{bmatrix}^T \\ &= \begin{bmatrix} \mathbf{h}_1 & \mathbf{h}_2 & \cdots & \mathbf{h}_P \end{bmatrix}^H \mathbf{y} \\ &= \begin{bmatrix} \mathbf{u}_1 & \mathbf{u}_2 & \cdots & \mathbf{u}_P \end{bmatrix}^H \mathbf{y} \\ &= \mathbf{U}_{1:P}^H \mathbf{y}, \end{aligned} \tag{4.14}$$

where $\mathbf{U}_{1:P}$ is a matrix of size $M \times P$ containing the first P columns of $\mathbf{U}$. The covariance matrix of $\mathbf{z}$ is

$$\begin{aligned} \mathbf{\Phi}_\mathbf{z} &= \text{diag}\left(\phi_{Z_1}, \phi_{Z_2}, \ldots, \phi_{Z_P}\right) \\ &= \text{diag}\left(\lambda_1, \lambda_2, \ldots, \lambda_P\right). \end{aligned} \tag{4.15}$$

A good amount of dimensionality reduction happens when $P \ll M$ and the sum of the first P eigenvalues of $\mathbf{\Phi}_\mathbf{y}$ are much larger than the sum of its remaining $M - P$ eigenvalues, i.e.,

$$\sum_{p=1}^{P} \lambda_p \gg \sum_{i=P+1}^{M} \lambda_i, \tag{4.16}$$

so that little information is lost; in other words, we should have $\text{tr}\left(\mathbf{\Phi}_\mathbf{z}\right) \approx \text{tr}\left(\mathbf{\Phi}_\mathbf{y}\right)$. To better see this, let us compute the error signal between the low-rank random signal vector corresponding to $\mathbf{y}$, i.e., the reconstructed random signal vector, $\widehat{\mathbf{y}} = \mathbf{U}_{1:P}\mathbf{z}$, and $\mathbf{y}$:

$$\epsilon = E\left(\|\widehat{\mathbf{y}} - \mathbf{y}\|^2\right)$$

$$= E\left[\left\|\left(\mathbf{U}_{1:P}\mathbf{U}_{1:P}^H - \mathbf{I}_M\right)\mathbf{y}\right\|^2\right]$$
$$= \mathrm{tr}\left[\left(\mathbf{U}_{1:P}\mathbf{U}_{1:P}^H - \mathbf{I}_M\right)\mathbf{\Phi}_\mathbf{y}\left(\mathbf{U}_{1:P}\mathbf{U}_{1:P}^H - \mathbf{I}_M\right)\right]$$
$$= \mathrm{tr}\left(\mathbf{U}_{P+1:M}\mathbf{U}_{P+1:M}^H\mathbf{\Phi}_\mathbf{y}\mathbf{U}_{P+1:M}\mathbf{U}_{P+1:M}^H\right)$$
$$= \mathrm{tr}\left(\mathbf{U}_{P+1:M}^H\mathbf{\Phi}_\mathbf{y}\mathbf{U}_{P+1:M}\right)$$
$$= \sum_{i=P+1}^{M} \lambda_i, \tag{4.17}$$

where $\mathbf{U}_{P+1:M}$ is a matrix of size $M \times (M - P)$ containing the last $M - P$ columns of $\mathbf{U}$. Therefore, as long as $\mathrm{tr}\,(\mathbf{\Phi}_\mathbf{z}) \gg \epsilon$ [which is another way to write (4.16)], most of the information in the original data is kept in $\mathbf{z}$, and we can process this vector instead of the higher-dimensional one, $\mathbf{y}$. Obviously, the smaller is the value of P, the higher is the dimensionality reduction.

4.2 PCA for White Noise Reduction

In real-world applications, unfortunately, the signal vector $\mathbf{y}$ is never "clean"; it is always corrupted by some additive noise, so that, in general, we observe the signal vector:

$$\mathbf{y} = \mathbf{x} + \mathbf{v}, \tag{4.18}$$

where

$$\mathbf{x} = \begin{bmatrix} X_1 & X_2 & \cdots & X_M \end{bmatrix}^T$$

is actually the zero-mean clean or desired signal vector and

$$\mathbf{v} = \begin{bmatrix} V_1 & V_2 & \cdots & V_M \end{bmatrix}^T$$

is the zero-mean additive noise signal vector, which is uncorrelated with $\mathbf{x}$, i.e., $E\left(\mathbf{x}\mathbf{v}^H\right) = E\left(\mathbf{v}\mathbf{x}^H\right) = \mathbf{0}$. Then, the covariance matrix of $\mathbf{y}$ is

$$\mathbf{\Phi}_\mathbf{y} = \mathbf{\Phi}_\mathbf{x} + \mathbf{\Phi}_\mathbf{v}, \tag{4.19}$$

where $\mathbf{\Phi}_\mathbf{x}$ and $\mathbf{\Phi}_\mathbf{v}$ are the covariance matrices of $\mathbf{x}$ and $\mathbf{v}$, respectively. In this section, it is assumed that the noise is white, so that

$$\mathbf{\Phi_v} = \phi_V \mathbf{I}_M, \tag{4.20}$$

where ϕ_V is the variance of the white noise. Because of the structure of $\mathbf{\Phi_v}$, the covariance matrix $\mathbf{\Phi_y}$ can be decomposed as

$$\mathbf{\Phi_y} = \mathbf{U}\left(\mathbf{\Lambda} + \phi_V \mathbf{I}_M\right)\mathbf{U}^H, \tag{4.21}$$

where $\mathbf{U}$ and $\mathbf{\Lambda}$ are given in (4.4) and (4.5), respectively. From (4.19), we define the full-mode SNR of $\mathbf{y}$ in (4.18) as [2]

$$\begin{aligned} \mathrm{SNR}_{\mathbf{y}} &= \frac{\mathrm{tr}\left(\mathbf{\Phi_v}^{-1}\mathbf{\Phi_x}\right)}{M} \\ &= \frac{\sum_{m=1}^{M}\lambda_m}{M\phi_V} \\ &= \frac{\sum_{m=1}^{M}\mathrm{SNR}_m}{M}, \end{aligned} \tag{4.22}$$

where

$$\mathrm{SNR}_m = \frac{\lambda_m}{\phi_V},\ m = 1, 2, \ldots, M \tag{4.23}$$

is the mth spectral mode SNR, with

$$\mathrm{SNR}_1 \geq \mathrm{SNR}_2 \geq \cdots \geq \mathrm{SNR}_M.$$

The first fundamental question we ask in this section is how does white noise affect dimensionality reduction with PCA? It is obvious that the reduced dimension signal vector is

$$\begin{aligned} \mathbf{z} &= \mathbf{U}_{1:P}^H \mathbf{y} \\ &= \mathbf{U}_{1:P}^H \mathbf{x} + \mathbf{U}_{1:P}^H \mathbf{v}, \end{aligned} \tag{4.24}$$

whose covariance matrix is

$$\begin{aligned} \mathbf{\Phi_z} &= \mathrm{diag}\left(\phi_{Z_1}, \phi_{Z_2}, \ldots, \phi_{Z_P}\right) \\ &= \mathrm{diag}\left(\lambda_1 + \phi_V, \lambda_2 + \phi_V, \ldots, \lambda_P + \phi_V\right). \end{aligned} \tag{4.25}$$

As a result, the reconstructed desired signal vector is

$$\widehat{\mathbf{x}} = \mathbf{U}_{1:P}\mathbf{z} = \mathbf{U}_{1:P}\mathbf{U}_{1:P}^H \mathbf{y}. \tag{4.26}$$

Now, let us evaluate the error signal between $\widehat{\mathbf{x}}$ and $\mathbf{x}$. We have

$$
\begin{aligned}
\epsilon &= E\left(\|\widehat{\mathbf{x}} - \mathbf{x}\|^2\right) \\
&= E\left[\left\|\left(\mathbf{U}_{1:P}\mathbf{U}_{1:P}^H - \mathbf{I}_M\right)\mathbf{x} + \mathbf{U}_{1:P}\mathbf{U}_{1:P}^H\mathbf{v}\right\|^2\right] \\
&= \mathrm{tr}\left(\mathbf{U}_{P+1:M}^H \boldsymbol{\Phi}_{\mathbf{x}} \mathbf{U}_{P+1:M}\right) + \phi_V \mathrm{tr}\left(\mathbf{U}_{1:P}^H \mathbf{U}_{1:P}\right) \\
&= \sum_{i=P+1}^{M} \lambda_i + P\phi_V. \qquad (4.27)
\end{aligned}
$$

Therefore, if

$$
\mathrm{tr}\left(\boldsymbol{\Phi}_{\mathbf{z}}\right) = \sum_{p=1}^{P} \lambda_p + P\phi_V \gg \epsilon = \sum_{i=P+1}^{M} \lambda_i + P\phi_V, \qquad (4.28)
$$

which is identical to the condition in (4.16), $\mathbf{z}$ is almost as good as $\mathbf{y}$. We deduce that the additive white noise does not affect at all dimensionality reduction with PCA. Notice that the first term on the right-hand side of (4.27) is the model bias, while the second term is the model variance [3].

The second and last fundamental question we ask in this section is how does PCA contribute to white noise reduction? It is quite clear that the optimization problem to find the optimal filter in the pth principal component, $Z_p = \mathbf{h}_p^H \mathbf{y}$, is equivalent to maximizing the Rayleigh quotient (under the same constraints):

$$
\mathcal{R}\left(\mathbf{h}_p\right) = \frac{\mathbf{h}_p^H \boldsymbol{\Phi}_{\mathbf{y}} \mathbf{h}_p}{\mathbf{h}_p^H \mathbf{h}_p}, \qquad (4.29)
$$

which can be rewritten as

$$
\begin{aligned}
\mathcal{R}\left(\mathbf{h}_p\right) &= \frac{\mathbf{h}_p^H \boldsymbol{\Phi}_{\mathbf{x}} \mathbf{h}_p}{\mathbf{h}_p^H \mathbf{h}_p} + \phi_V \\
&= \phi_V \left[\mathrm{oSNR}\left(\mathbf{h}_p\right) + 1\right], \qquad (4.30)
\end{aligned}
$$

where

$$
\mathrm{oSNR}\left(\mathbf{h}_p\right) = \frac{\mathbf{h}_p^H \boldsymbol{\Phi}_{\mathbf{x}} \mathbf{h}_p}{\phi_V \mathbf{h}_p^H \mathbf{h}_p} \qquad (4.31)
$$

is the output SNR after the filtering operation with $\mathbf{h}_p$ or the SNR of the pth principal component, Z_p. Clearly, maximizing the Rayleigh quotient in (4.29) is equivalent to maximizing the output SNR in (4.31). After maximization, we obviously have $\mathbf{h}_p = \mathbf{u}_p$, $\phi_{Z_p} = \lambda_p + \phi_V = \mathcal{R}\left(\mathbf{h}_p\right)$, and

$$
\begin{aligned}
\text{oSNR}\left(\mathbf{h}_p\right) &= \frac{\lambda_p}{\phi_V} \\
&= \text{SNR}_p,
\end{aligned}
\tag{4.32}
$$

which is simply the pth spectral mode SNR. Furthermore, from (4.24) we find that the full-mode SNR of $\mathbf{z}$ is [2]

$$
\begin{aligned}
\text{SNR}_{\mathbf{z},P} &= \frac{\text{tr}\left[\left(\mathbf{U}_{1:P}^H \mathbf{\Phi}_{\mathbf{v}} \mathbf{U}_{1:P}\right)^{-1} \left(\mathbf{U}_{1:P}^H \mathbf{\Phi}_{\mathbf{x}} \mathbf{U}_{1:P}\right)\right]}{P} \\
&= \frac{\sum_{p=1}^{P} \lambda_p}{P\phi_V} \\
&= \frac{\sum_{p=1}^{P} \text{oSNR}\left(\mathbf{h}_p\right)}{P},
\end{aligned}
\tag{4.33}
$$

where

$$
\text{SNR}_{\mathbf{z},1} \geq \text{SNR}_{\mathbf{z},2} \geq \cdots \geq \text{SNR}_{\mathbf{z},M} = \text{SNR}_{\mathbf{y}}.
\tag{4.34}
$$

Obviously, the smaller is the value of P, the more is the white noise reduction in $\mathbf{z}$ as compared to $\mathbf{y}$. We can conclude that $\mathbf{z}$ is always less noisy than $\mathbf{y}$. In the extreme case of maximum dimensionality reduction, i.e., $P = 1$, the filter $\mathbf{h}_1 = \mathbf{u}_1$ of the only component Z_1, corresponds to the well-known maximum SNR filter, and Z_1 can be seen as an estimate of one of the elements of $\mathbf{x}$.

In general, noise reduction consists of two steps. The first one is dimensionality reduction with PCA as explained above, while the second step consists of finding the P scaling parameters since the P output SNRs in (4.31) are scale invariant. Let us assume that the aim in our noise reduction problem is to find a good estimate of the first element of $\mathbf{x}$, i.e., X_1. Then, by applying a complex-valued filter, $\overline{\mathbf{h}}$ containing the P scaling parameters, to $\mathbf{z}$, we get

$$
\begin{aligned}
\overline{Z} &= \overline{\mathbf{h}}^H \mathbf{z} \\
&= \overline{\mathbf{h}}^H \mathbf{U}_{1:P}^H \mathbf{x} + \overline{\mathbf{h}}^H \mathbf{U}_{1:P}^H \mathbf{v},
\end{aligned}
\tag{4.35}
$$

which is the estimate of X_1, and the output SNR is

$$
\text{oSNR}\left(\overline{\mathbf{h}}\right) = \frac{\overline{\mathbf{h}}^H \mathbf{\Lambda}_{1:P} \overline{\mathbf{h}}}{\phi_V \overline{\mathbf{h}}^H \overline{\mathbf{h}}},
\tag{4.36}
$$

where

$$
\mathbf{\Lambda}_{1:P} = \text{diag}\left(\lambda_1, \lambda_2, \ldots, \lambda_P\right).
\tag{4.37}
$$

We propose two options to find the optimal value of $\overline{\mathbf{h}}$. The first one is from the MSE criterion between X_1 and $\overline{Z}$, i.e.,

$$\mathcal{J}\left(\overline{\mathbf{h}}\right) = E\left(\left|X_1 - \overline{\mathbf{h}}^H \mathbf{z}\right|^2\right) \tag{4.38}$$

$$= \mathbf{i}_1^T \mathbf{\Phi}_\mathbf{x} \mathbf{i}_1 - \overline{\mathbf{h}}^H \mathbf{U}_{1:P}^H \mathbf{\Phi}_\mathbf{x} \mathbf{i}_1 - \mathbf{i}_1^T \mathbf{\Phi}_\mathbf{x} \mathbf{U}_{1:P} \overline{\mathbf{h}} + \overline{\mathbf{h}}^H \mathbf{\Phi}_\mathbf{z} \overline{\mathbf{h}},$$

where $\mathbf{i}_1$ is the first column of $\mathbf{I}_M$ and $\mathbf{\Phi}_\mathbf{z}$ is defined in (4.25). By minimizing the previous expression, we easily obtain the Wiener-type filter:

$$\overline{\mathbf{h}}_\mathrm{W} = \mathbf{\Phi}_\mathbf{z}^{-1} \mathbf{U}_{1:P}^H \mathbf{\Phi}_\mathbf{x} \mathbf{i}_1$$

$$= \left(\mathbf{\Lambda}_{1:P} + \phi_V \mathbf{I}_P\right)^{-1} \mathbf{U}_{1:P}^H \mathbf{\Phi}_\mathbf{x} \mathbf{i}_1, \tag{4.39}$$

where $\mathbf{I}_P$ is the $P \times P$ identity matrix. One can check that for $P = M$, $\overline{\mathbf{h}}_\mathrm{W}$ is equivalent to the conventional Wiener filter. The second option is to minimize the distortion-based MSE criterion:

$$\mathcal{J}_\mathrm{d}\left(\overline{\mathbf{h}}\right) = E\left(\left|X_1 - \overline{\mathbf{h}}^H \mathbf{U}_{1:P}^H \mathbf{x}\right|^2\right) \tag{4.40}$$

$$= \mathbf{i}_1^T \mathbf{\Phi}_\mathbf{x} \mathbf{i}_1 - \overline{\mathbf{h}}^H \mathbf{U}_{1:P}^H \mathbf{\Phi}_\mathbf{x} \mathbf{i}_1 - \mathbf{i}_1^T \mathbf{\Phi}_\mathbf{x} \mathbf{U}_{1:P} \overline{\mathbf{h}} + \overline{\mathbf{h}}^H \mathbf{\Lambda}_{1:P} \overline{\mathbf{h}},$$

from which we get the minimum-distortion filter:

$$\overline{\mathbf{h}}_\mathrm{MD} = \mathbf{\Lambda}_{1:P}^{-1} \mathbf{U}_{1:P}^H \mathbf{\Phi}_\mathbf{x} \mathbf{i}_1, \tag{4.41}$$

which is interesting to compare to $\overline{\mathbf{h}}_\mathrm{W}$. One can show that

$$\mathrm{oSNR}\left(\overline{\mathbf{h}}_\mathrm{W}\right) \geq \mathrm{SNR}_{\mathbf{z},P}. \tag{4.42}$$

Therefore, the second step with $\overline{Z}$ not only improves the SNR, but it also decreases distortion as compared to $\mathbf{z}$, with the Wiener-type filter. However, the role of $\overline{\mathbf{h}}_\mathrm{MD}$ is to essentially minimize distortion of the desired signal.

4.3 Noise-Dependent PCA for General Noise Reduction

What happens to PCA when noise is no longer white? Let us take the model in (4.18), i.e., $\mathbf{y} = \mathbf{x} + \mathbf{v}$, with covariance matrix $\mathbf{\Phi}_\mathbf{y} = \mathbf{\Phi}_\mathbf{x} + \mathbf{\Phi}_\mathbf{v} = \mathbf{U}\mathbf{\Lambda}\mathbf{U}^H$. Now, the covariance matrix $\mathbf{\Phi}_\mathbf{v}$ (assumed to have full rank) is not diagonal; as a result, the matrices $\mathbf{U}^H \mathbf{\Phi}_\mathbf{x} \mathbf{U}$ and $\mathbf{U}^H \mathbf{\Phi}_\mathbf{v} \mathbf{U}$ are no longer diagonal. In this case, the error signal between $\widehat{\mathbf{x}} = \mathbf{U}_{1:P}\mathbf{z} = \mathbf{U}_{1:P}\mathbf{U}_{1:P}^H \mathbf{y}$ and $\mathbf{x}$ is

$$
\begin{aligned}
\epsilon &= E\left(\|\widehat{\mathbf{x}} - \mathbf{x}\|^2\right) \\
&= E\left[\left\|\left(\mathbf{U}_{1:P}\mathbf{U}_{1:P}^H - \mathbf{I}_M\right)\mathbf{x} + \mathbf{U}_{1:P}\mathbf{U}_{1:P}^H\mathbf{v}\right\|^2\right] \\
&= \operatorname{tr}\left(\mathbf{U}_{P+1:M}^H\mathbf{\Phi}_{\mathbf{x}}\mathbf{U}_{P+1:M}\right) + \operatorname{tr}\left(\mathbf{U}_{1:P}^H\mathbf{\Phi}_{\mathbf{v}}\mathbf{U}_{1:P}\right) \\
&= \sum_{p=1}^{P}\lambda_p + \operatorname{tr}\left(\mathbf{U}_{P+1:M}^H\mathbf{\Phi}_{\mathbf{x}}\mathbf{U}_{P+1:M}\right) - \operatorname{tr}\left(\mathbf{U}_{1:P}^H\mathbf{\Phi}_{\mathbf{x}}\mathbf{U}_{1:P}\right).
\end{aligned}
\tag{4.43}
$$

This time, however, ϵ does not necessarily depend on the smallest eigenvalues of $\mathbf{\Phi}_{\mathbf{y}}$ and, as a consequence, the condition $\operatorname{tr}(\mathbf{\Phi}_{\mathbf{z}}) = \sum_{p=1}^{P}\lambda_p \gg \epsilon$ may no longer hold. Hence, (nonwhite) noise may have a disastrous effect on dimensionality reduction with PCA. Next, we explain how to deal with this problem.

The use of the joint diagonalization, of the matrices $\mathbf{\Phi}_{\mathbf{y}}$ and $\mathbf{\Phi}_{\mathbf{v}}$, is going to be very useful in the rest of this section, so we briefly discuss how it works. The two Hermitian matrices $\mathbf{\Phi}_{\mathbf{y}}$ and $\mathbf{\Phi}_{\mathbf{v}}$ can be jointly diagonalized as follows [4]:

$$\mathbf{T}^H\mathbf{\Phi}_{\mathbf{y}}\mathbf{T} = \mathbf{\Sigma}, \tag{4.44}$$

$$\mathbf{T}^H\mathbf{\Phi}_{\mathbf{v}}\mathbf{T} = \mathbf{I}_M, \tag{4.45}$$

where

$$\mathbf{T} = \begin{bmatrix} \mathbf{t}_1 & \mathbf{t}_2 & \cdots & \mathbf{t}_M \end{bmatrix} \tag{4.46}$$

is a full-rank square matrix (of size $M \times M$) and

$$\mathbf{\Sigma} = \operatorname{diag}(\sigma_1, \sigma_2, \ldots, \sigma_M) \tag{4.47}$$

is a diagonal matrix whose main elements are real and nonnegative. The eigenvalues of $\mathbf{\Phi}_{\mathbf{v}}^{-1}\mathbf{\Phi}_{\mathbf{y}}$ are ordered as $\sigma_1 \geq \sigma_2 \geq \cdots \geq \sigma_M > 0$. We also denote by $\mathbf{t}_1, \mathbf{t}_2, \ldots, \mathbf{t}_M$ the corresponding eigenvectors. From (4.44)–(4.45), it is easy to deduce that

$$\mathbf{T}^H\mathbf{\Phi}_{\mathbf{x}}\mathbf{T} = \mathbf{\Sigma} - \mathbf{I}_M \tag{4.48}$$

and

$$\mathbf{\Phi}_{\mathbf{v}}^{-1} = \mathbf{T}\mathbf{T}^H. \tag{4.49}$$

We will assume that $\mathbf{\Phi}_{\mathbf{v}}$ is known or, at least, we can have a good model for it. Then, we can whiten the noise with the simple procedure:

$$\begin{aligned}\mathbf{y}' &= \mathbf{\Phi}_{\mathbf{v}}^{-1/2}\mathbf{y} \\ &= \mathbf{\Phi}_{\mathbf{v}}^{-1/2}\mathbf{x} + \mathbf{\Phi}_{\mathbf{v}}^{-1/2}\mathbf{v} \\ &= \mathbf{x}' + \mathbf{v}'.\end{aligned} \tag{4.50}$$

Indeed, the covariance matrix of $\mathbf{y}'$ is $\mathbf{\Phi}_{\mathbf{y}'} = \mathbf{\Phi}_{\mathbf{x}'} + \mathbf{I}_M$, where $\mathbf{\Phi}_{\mathbf{x}'}$ is the covariance matrix of $\mathbf{x}'$. Now, instead of working with $\mathbf{y}$, we can equivalently work with $\mathbf{y}'$. In other words, we will perform dimensionality reduction with PCA on $\mathbf{y}'$ instead of $\mathbf{y}$. From (4.50), we find that the full-mode SNR of $\mathbf{y}'$ is

$$\begin{aligned}\text{SNR}_{\mathbf{y}'} &= \frac{\text{tr}\left[\left(\mathbf{\Phi}_{\mathbf{v}}^{-1/2}\mathbf{\Phi}_{\mathbf{v}}\mathbf{\Phi}_{\mathbf{v}}^{-1/2}\right)^{-1}\mathbf{\Phi}_{\mathbf{v}}^{-1/2}\mathbf{\Phi}_{\mathbf{x}}\mathbf{\Phi}_{\mathbf{v}}^{-1/2}\right]}{M} \\ &= \frac{\text{tr}\left(\mathbf{\Phi}_{\mathbf{v}}^{-1}\mathbf{\Phi}_{\mathbf{x}}\right)}{M} \\ &= \frac{\sum_{m=1}^{M}(\sigma_m - 1)}{M} \\ &= \frac{\sum_{m=1}^{M}\text{SNR}_m}{M},\end{aligned} \tag{4.51}$$

where

$$\text{SNR}_m = \sigma_m - 1,\ m = 1, 2, \ldots, M \tag{4.52}$$

is the mth spectral mode SNR, with

$$\text{SNR}_1 \geq \text{SNR}_2 \geq \cdots \geq \text{SNR}_M.$$

Let

$$Z'_p = \mathbf{h}'^H_p \mathbf{y} \tag{4.53}$$

be the pth principal component, where $\mathbf{h}'_p$ is a complex-valued filter of length M, and whose variance is $\phi_{Z'_p} = \mathbf{h}'^H_p \mathbf{\Phi}_{\mathbf{y}} \mathbf{h}'_p$. Following the standard PCA, we propose to derive $\mathbf{h}'_p$ from the optimization problem:

$$\max_{\mathbf{h}'_p} \mathbf{h}'^H_p \mathbf{\Phi}_{\mathbf{y}} \mathbf{h}'_p \quad \text{subject to} \quad \begin{cases} \mathbf{h}'^H_p \mathbf{\Phi}_{\mathbf{v}} \mathbf{h}'_p = 1 \\ \mathbf{h}'^H_p \mathbf{\Phi}_{\mathbf{v}} \mathbf{t}_1 = \cdots = \mathbf{h}'^H_p \mathbf{\Phi}_{\mathbf{v}} \mathbf{t}_{p-1} = 0 \end{cases}. \tag{4.54}$$

We find that the solution is

$$\mathbf{h}'_p = \mathbf{t}_p, \tag{4.55}$$

$\phi_{Z'_p} = \sigma_p$, and $\phi_{Z'_i Z'_j} = E\left(Z'_i Z'^*_j\right) = 0,\ i \neq j$. We deduce that the reduced random signal vector of length P is

$$\begin{aligned}
\mathbf{z}' &= \left[\, Z'_1 \ Z'_2 \ \cdots \ Z'_P \,\right]^T \\
&= \left[\, \mathbf{h}'_1 \ \mathbf{h}'_2 \ \cdots \ \mathbf{h}'_P \,\right]^H \mathbf{y} \\
&= \left[\, \mathbf{t}_1 \ \mathbf{t}_2 \ \cdots \ \mathbf{t}_P \,\right]^H \mathbf{y} \\
&= \mathbf{T}_{1:P}^H \mathbf{y},
\end{aligned} \tag{4.56}$$

where $\mathbf{T}_{1:P}$ is a matrix of size $M \times P$ containing the first P columns of $\mathbf{T}$. The covariance matrix of $\mathbf{z}'$ is

$$\begin{aligned}
\mathbf{\Phi}_{\mathbf{z}'} &= \text{diag}\left(\phi_{Z'_1}, \phi_{Z'_2}, \ldots, \phi_{Z'_P}\right) \\
&= \text{diag}\left(\sigma_1, \sigma_2, \ldots, \sigma_P\right),
\end{aligned} \tag{4.57}$$

with

$$\begin{aligned}
\text{tr}\left(\mathbf{\Phi}_{\mathbf{z}'}\right) &= \sum_{p=1}^{P} \sigma_p \\
&= \sum_{p=1}^{P} \left(\sigma_p - 1\right) + P.
\end{aligned} \tag{4.58}$$

Then, the reconstructed signal vector of $\mathbf{x}'$ with the proposed modified PCA is

$$\begin{aligned}
\widehat{\mathbf{x}}' &= \mathbf{\Phi}_{\mathbf{v}}^{1/2} \mathbf{T}_{1:P} \mathbf{z}' \\
&= \mathbf{\Phi}_{\mathbf{v}}^{1/2} \mathbf{T}_{1:P} \mathbf{T}_{1:P}^H \mathbf{y}.
\end{aligned} \tag{4.59}$$

Now, let us quantify the error signal between $\widehat{\mathbf{x}}'$ and $\mathbf{x}'$. We have

$$\begin{aligned}
\epsilon &= E\left(\left\| \widehat{\mathbf{x}}' - \mathbf{x}' \right\|^2\right) \\
&= E\left[\left\| \left(\mathbf{\Phi}_{\mathbf{v}}^{1/2} \mathbf{T}_{1:P} \mathbf{T}_{1:P}^H \mathbf{\Phi}_{\mathbf{v}}^{1/2} - \mathbf{I}_M\right) \mathbf{\Phi}_{\mathbf{v}}^{-1/2} \mathbf{x} + \mathbf{\Phi}_{\mathbf{v}}^{1/2} \mathbf{T}_{1:P} \mathbf{T}_{1:P}^H \mathbf{v} \right\|^2\right] \\
&= \text{tr}\left(\mathbf{\Phi}_{\mathbf{v}}^{1/2} \mathbf{T}_{P+1:M} \mathbf{T}_{P+1:M}^H \mathbf{\Phi}_{\mathbf{x}} \mathbf{T}_{P+1:M} \mathbf{T}_{P+1:M}^H \mathbf{\Phi}_{\mathbf{v}}^{1/2}\right) \\
&\quad + \text{tr}\left(\mathbf{\Phi}_{\mathbf{v}}^{1/2} \mathbf{T}_{1:P} \mathbf{T}_{1:P}^H \mathbf{\Phi}_{\mathbf{v}}^{1/2}\right)
\end{aligned}$$

$$
\begin{aligned}
&= \mathrm{tr}\left(\mathbf{T}_{P+1:M}^{H}\mathbf{\Phi}_{\mathbf{x}}\mathbf{T}_{P+1:M}\right) + \mathrm{tr}\left(\mathbf{T}_{1:P}^{H}\mathbf{\Phi}_{\mathbf{v}}\mathbf{T}_{1:P}\right) \\
&= \sum_{i=P+1}^{M} (\sigma_i - 1) + P,
\end{aligned} \tag{4.60}
$$

where $\mathbf{T}_{P+1:M}$ is a matrix of size $M \times (M - P)$ containing the last $M - P$ columns of $\mathbf{T}$. Therefore, if

$$
\mathrm{tr}\left(\mathbf{\Phi}_{\mathbf{z}'}\right) \gg \epsilon, \tag{4.61}
$$

i.e., $\sum_{p=1}^{P}\left(\sigma_p - 1\right) \gg \sum_{i=P+1}^{M} (\sigma_i - 1)$, $\mathbf{z}'$ is almost as good as $\mathbf{y}'$. It is worth noticing that this condition is equivalent to (4.28) with white noise. We conclude that dimensionality reduction of noisy data is possible with this proposed approach.

The optimization problem in (4.54) is equivalent to maximizing the generalized Rayleigh quotient (under the same constraints):

$$
\mathcal{R}\left(\mathbf{h}_p'\right) = \frac{\mathbf{h}_p'^{H}\mathbf{\Phi}_{\mathbf{y}}\mathbf{h}_p'}{\mathbf{h}_p'^{H}\mathbf{\Phi}_{\mathbf{v}}\mathbf{h}_p'}. \tag{4.62}
$$

The previous expression can be rewritten as

$$
\begin{aligned}
\mathcal{R}\left(\mathbf{h}_p'\right) &= \frac{\mathbf{h}_p'^{H}\mathbf{\Phi}_{\mathbf{x}}\mathbf{h}_p'}{\mathbf{h}_p'^{H}\mathbf{\Phi}_{\mathbf{v}}\mathbf{h}_p'} + 1 \\
&= \mathrm{oSNR}\left(\mathbf{h}_p'\right) + 1,
\end{aligned} \tag{4.63}
$$

where

$$
\mathrm{oSNR}\left(\mathbf{h}_p'\right) = \frac{\mathbf{h}_p'^{H}\mathbf{\Phi}_{\mathbf{x}}\mathbf{h}_p'}{\mathbf{h}_p'^{H}\mathbf{\Phi}_{\mathbf{v}}\mathbf{h}_p'} \tag{4.64}
$$

is the output SNR after the filtering operation with $\mathbf{h}_p'$ or the SNR of the pth principal component, Z_p'. Clearly, maximizing the generalized Rayleigh quotient in (4.62) is equivalent to maximizing the output SNR in (4.64) [5]. After maximization, we obviously have $\mathbf{h}_p' = \mathbf{t}_p$, $\phi_{Z_p}' = \sigma_p = \mathcal{R}\left(\mathbf{h}_p'\right)$, and

$$
\begin{aligned}
\mathrm{oSNR}\left(\mathbf{h}_p'\right) &= \sigma_p - 1 \\
&= \mathrm{SNR}_p,
\end{aligned} \tag{4.65}
$$

which is simply the pth spectral mode SNR. Furthermore, from (4.56) we find that the full-mode SNR of $\mathbf{z}'$ is

$$\begin{aligned}\mathrm{SNR}_{\mathbf{z}',P} &= \frac{\mathrm{tr}\left[\left(\mathbf{T}_{1:P}^H \mathbf{\Phi}_{\mathbf{v}} \mathbf{T}_{1:P}\right)^{-1} \left(\mathbf{T}_{1:P}^H \mathbf{\Phi}_{\mathbf{x}} \mathbf{T}_{1:P}\right)\right]}{P} \\ &= \frac{\sum_{p=1}^{P} (\sigma_p - 1)}{P} \\ &= \frac{\sum_{p=1}^{P} \mathrm{oSNR}\left(\mathbf{h}'_p\right)}{P},\end{aligned} \tag{4.66}$$

where

$$\mathrm{SNR}_{\mathbf{z}',1} \geq \mathrm{SNR}_{\mathbf{z}',2} \geq \cdots \geq \mathrm{SNR}_{\mathbf{z}',M} = \mathrm{SNR}_{\mathbf{y}'}. \tag{4.67}$$

Obviously, the smaller is the value of P, the more is the noise reduction in $\mathbf{z}'$ as compared to $\mathbf{y}'$. We can conclude that $\mathbf{z}'$ is always less noisy than $\mathbf{y}'$. In the extreme case of maximum dimensionality reduction, i.e., $P = 1$, the filter $\mathbf{h}'_1 = \mathbf{t}_1$ of the only component Z'_1, corresponds to the well-known maximum SNR filter, and Z'_1 can be seen as an estimate of one of the elements of $\mathbf{x}'$.

Dimensionality reduction with PCA can be seen as a necessary first step for noise reduction. The second step consists mostly of adjusting some scaling parameters in order to minimize distortion of the desired signal, which we consider here to be X_1, i.e., the first element of $\mathbf{x}$. Let $\overline{\mathbf{h}}'$ be a complex-valued filter of length P. Then, by applying $\overline{\mathbf{h}}'$ to $\mathbf{z}'$, we get an estimate of X_1, i.e.,

$$\begin{aligned}\overline{Z}' &= \overline{\mathbf{h}}'^H \mathbf{z}' \\ &= \overline{\mathbf{h}}'^H \mathbf{T}_{1:P}^H \mathbf{x} + \overline{\mathbf{h}}'^H \mathbf{T}_{1:P}^H \mathbf{v},\end{aligned} \tag{4.68}$$

from which we can define the output SNR as

$$\mathrm{oSNR}\left(\overline{\mathbf{h}}'\right) = \frac{\overline{\mathbf{h}}'^H \left(\mathbf{\Sigma}_{1:P} - \mathbf{I}_P\right) \overline{\mathbf{h}}'}{\overline{\mathbf{h}}'^H \overline{\mathbf{h}}'}, \tag{4.69}$$

where

$$\mathbf{\Sigma}_{1:P} = \mathrm{diag}\left(\sigma_1, \sigma_2, \ldots, \sigma_P\right). \tag{4.70}$$

Similar to the white noise case, we have two options to find the optimal value of $\overline{\mathbf{h}}'$. The first one is the from the MSE criterion between X_1 and $\overline{Z}'$, i.e.,

$$\begin{aligned}\mathcal{J}\left(\overline{\mathbf{h}}'\right) &= E\left(\left|X_1 - \overline{\mathbf{h}}'^H \mathbf{z}'\right|^2\right) \\ &= \mathbf{i}_1^T \mathbf{\Phi}_{\mathbf{x}} \mathbf{i}_1 - \overline{\mathbf{h}}'^H \mathbf{T}_{1:P}^H \mathbf{\Phi}_{\mathbf{x}} \mathbf{i}_1 - \mathbf{i}_1^T \mathbf{\Phi}_{\mathbf{x}} \mathbf{T}_{1:P} \overline{\mathbf{h}}' + \overline{\mathbf{h}}'^H \mathbf{\Phi}_{\mathbf{z}'} \overline{\mathbf{h}}',\end{aligned} \tag{4.71}$$

where $\mathbf{\Phi}_{\mathbf{z}'}$ is defined in (4.57). By minimizing the previous expression, we easily obtain the Wiener-type filter:

$$\begin{aligned}\overline{\mathbf{h}}'_{\mathrm{W}} &= \mathbf{\Phi}_{\mathbf{z}'}^{-1}\mathbf{T}_{1:P}^{H}\mathbf{\Phi}_{\mathbf{x}}\mathbf{i}_1 \\ &= \mathbf{\Sigma}_{1:P}^{-1}\mathbf{T}_{1:P}^{H}\mathbf{\Phi}_{\mathbf{x}}\mathbf{i}_1. \end{aligned} \tag{4.72}$$

One can verify that for $P = M$, $\overline{\mathbf{h}}'_{\mathrm{W}}$ is equivalent to the conventional Wiener filter. The second option is to minimize the distortion-based MSE criterion:

$$\begin{aligned}\mathcal{J}_{\mathrm{d}}\left(\overline{\mathbf{h}}'\right) &= E\left(\left|X_1 - \overline{\mathbf{h}}'^H\mathbf{T}_{1:P}^{H}\mathbf{x}\right|^2\right) \qquad (4.73)\\ &= \mathbf{i}_1^T\mathbf{\Phi}_{\mathbf{x}}\mathbf{i}_1 - \overline{\mathbf{h}}'^H\mathbf{T}_{1:P}^{H}\mathbf{\Phi}_{\mathbf{x}}\mathbf{i}_1 - \mathbf{i}_1^T\mathbf{\Phi}_{\mathbf{x}}\mathbf{T}_{1:P}\overline{\mathbf{h}}' + \overline{\mathbf{h}}'^H\left(\mathbf{\Sigma}_{1:P} - \mathbf{I}_P\right)\overline{\mathbf{h}}',\end{aligned}$$

from which we get the minimum-distortion filter:

$$\overline{\mathbf{h}}'_{\mathrm{MD}} = \left(\mathbf{\Sigma}_{1:P} - \mathbf{I}_P\right)^{-1}\mathbf{T}_{1:P}^{H}\mathbf{\Phi}_{\mathbf{x}}\mathbf{i}_1, \tag{4.74}$$

which is interesting to compare to $\overline{\mathbf{h}}'_{\mathrm{W}}$. One can show that

$$\mathrm{oSNR}\left(\overline{\mathbf{h}}'_{\mathrm{W}}\right) \geq \mathrm{SNR}_{\mathbf{z}',P}. \tag{4.75}$$

Therefore, the second step with $\overline{Z}'$ not only improves the SNR, but it also decreases distortion as compared to $\mathbf{z}'$, with the Wiener-type filter. However, the role of $\overline{\mathbf{h}}'_{\mathrm{MD}}$ is to essentially minimize distortion of the desired signal.

Finally, we end this section by saying that this work based on PCA is strongly related to the concept of optimal variable span linear filters developed in [6, 7].

4.4 Applications to Beamforming

There are different ways to apply PCA to beamforming. Here, we only explore one possible avenue.

We consider the signal model of Chap. 3, where a 3-D microphone array picks up the sound in its environment. Then, the observed signal vector of length M is

$$\begin{aligned}\mathbf{y} &= \begin{bmatrix} Y_1 & Y_2 & \cdots & Y_M \end{bmatrix}^T \\ &= \mathbf{x} + \mathbf{v} \\ &= \mathbf{d}_{\theta_s,\phi_s}X + \mathbf{v}, \end{aligned} \tag{4.76}$$

where

$$\mathbf{d}_{\theta_s,\phi_s} = \left[e^{\jmath 2\pi f \mathbf{a}_{\theta_s,\phi_s}^T \mathbf{r}_1/c} \ e^{\jmath 2\pi f \mathbf{a}_{\theta_s,\phi_s}^T \mathbf{r}_2/c} \ \cdots \ e^{\jmath 2\pi f \mathbf{a}_{\theta_s,\phi_s}^T \mathbf{r}_M/c} \right]^T \tag{4.77}$$

is the steering vector of length M. The covariance matrix of $\mathbf{y}$ is

$$\begin{aligned} \mathbf{\Phi}_{\mathbf{y}} &= E\left(\mathbf{y}\mathbf{y}^H\right) \\ &= \mathbf{\Phi}_{\mathbf{x}} + \mathbf{\Phi}_{\mathbf{v}} \\ &= \phi_X \mathbf{d}_{\theta_s,\phi_s}\mathbf{d}_{\theta_s,\phi_s}^H + \mathbf{\Phi}_{\mathbf{v}} \\ &= \phi_X \mathbf{d}_{\theta_s,\phi_s}\mathbf{d}_{\theta_s,\phi_s}^H + \phi_V \mathbf{\Gamma}_{\mathbf{v}}. \end{aligned} \tag{4.78}$$

In the case of the spherically isotropic (diffuse) noise field, (4.78) becomes

$$\mathbf{\Phi}_{\mathbf{y}} = \phi_X \mathbf{d}_{\theta_s,\phi_s}\mathbf{d}_{\theta_s,\phi_s}^H + \phi_{\mathrm{d}} \mathbf{\Gamma}_{\mathrm{d}}, \tag{4.79}$$

where ϕ_{d} is the variance of the diffuse noise and $\mathbf{\Gamma}_{\mathrm{d}}$ is defined in Chap. 3. As a consequence, the coherence matrix of $\mathbf{y}$ (from 4.79) is

$$\begin{aligned} \mathbf{\Gamma}_{\mathbf{y}} &= \frac{\mathbf{\Phi}_{\mathbf{y}}}{\phi_Y} \\ &= \frac{\mathrm{iSNR}_{\mathrm{d}}}{1+\mathrm{iSNR}_{\mathrm{d}}} \mathbf{d}_{\theta_s,\phi_s}\mathbf{d}_{\theta_s,\phi_s}^H + \frac{1}{1+\mathrm{iSNR}_{\mathrm{d}}} \mathbf{\Gamma}_{\mathrm{d}} \\ &= \frac{\mathrm{iSNR}_{\mathrm{d}}}{1+\mathrm{iSNR}_{\mathrm{d}}} \mathbf{\Gamma}_{\mathbf{x}} + \frac{1}{1+\mathrm{iSNR}_{\mathrm{d}}} \mathbf{\Gamma}_{\mathrm{d}}, \end{aligned} \tag{4.80}$$

where

$$\mathrm{iSNR}_{\mathrm{d}} = \frac{\phi_X}{\phi_{\mathrm{d}}} \tag{4.81}$$

is the input SNR in the presence of the diffuse noise and $\mathbf{\Gamma}_{\mathbf{x}} = \mathbf{\Phi}_{\mathbf{x}}/\phi_X = \mathbf{d}_{\theta_s,\phi_s}\mathbf{d}_{\theta_s,\phi_s}^H$ is the coherence matrix of $\mathbf{x}$, which does not depend on the signal.

Following some of the steps of the previous section, let us first jointly diagonalize the two matrices $\mathbf{\Gamma}_{\mathbf{y}}$ and $\mathbf{\Gamma}_{\mathrm{d}}$. We get

$$\mathbf{T}^H \mathbf{\Gamma}_{\mathbf{y}} \mathbf{T} = \mathbf{\Sigma}, \tag{4.82}$$

$$\mathbf{T}^H \mathbf{\Gamma}_{\mathrm{d}} \mathbf{T} = \mathbf{I}_M, \tag{4.83}$$

where

$$\mathbf{T} = \begin{bmatrix} \mathbf{t}_1 & \mathbf{t}_2 & \cdots & \mathbf{t}_M \end{bmatrix},$$
$$\boldsymbol{\Sigma} = \text{diag}\left(\sigma_1, \sigma_2, \ldots, \sigma_M\right),$$

with $\sigma_1 \geq \sigma_2 \geq \cdots \geq \sigma_M > 0$. Therefore, we also have

$$\mathbf{T}^H \boldsymbol{\Gamma}_{\mathbf{x}} \mathbf{T} = \left(1 + \frac{1}{\text{iSNR}_{\text{d}}}\right) \boldsymbol{\Sigma} - \frac{1}{\text{iSNR}_{\text{d}}} \mathbf{I}_M. \tag{4.84}$$

Beamforming based on PCA is made of two steps. The first one consists of finding the reduced random signal vector of length P, which is

$$\mathbf{z}' = \mathbf{T}_{1:P}^H \mathbf{y}, \tag{4.85}$$

where $\mathbf{T}_{1:P}$ is a matrix of size $M \times P$ containing the first P columns of $\mathbf{T}$. In the second step, we apply a complex-valued filter, $\overline{\mathbf{h}}'$ of length P, to $\mathbf{z}'$ to get an estimate of X, i.e.,

$$\begin{aligned} \overline{Z}' &= \overline{\mathbf{h}}'^H \mathbf{z}' \\ &= \overline{\mathbf{h}}'^H \mathbf{T}_{1:P}^H \mathbf{d}_{\theta_s,\phi_s} X + \overline{\mathbf{h}}'^H \mathbf{T}_{1:P}^H \mathbf{v}. \end{aligned} \tag{4.86}$$

Now, we need to redefine two important performance measures in this context; they are the WNG:

$$\mathcal{W}\left(\overline{\mathbf{h}}'\right) = \frac{\left|\overline{\mathbf{h}}'^H \mathbf{T}_{1:P}^H \mathbf{d}_{\theta_s,\phi_s}\right|^2}{\overline{\mathbf{h}}'^H \mathbf{T}_{1:P}^H \mathbf{T}_{1:P} \overline{\mathbf{h}}'} \tag{4.87}$$

and DF:

$$\begin{aligned} \mathcal{D}\left(\overline{\mathbf{h}}'\right) &= \frac{\left|\overline{\mathbf{h}}'^H \mathbf{T}_{1:P}^H \mathbf{d}_{\theta_s,\phi_s}\right|^2}{\overline{\mathbf{h}}'^H \mathbf{T}_{1:P}^H \boldsymbol{\Gamma}_{\text{d}} \mathbf{T}_{1:P} \overline{\mathbf{h}}'} \\ &= \frac{\left|\overline{\mathbf{h}}'^H \mathbf{T}_{1:P}^H \mathbf{d}_{\theta_s,\phi_s}\right|^2}{\overline{\mathbf{h}}'^H \overline{\mathbf{h}}'}. \end{aligned} \tag{4.88}$$

Then, from the previous two measures, it is easy to derive two interesting beamformers; they are the adaptive maximum WNG (AMWNG) beamformer:

$$\overline{\mathbf{h}}'_{\text{AMWNG}} = \frac{\left(\mathbf{T}_{1:P}^H \mathbf{T}_{1:P}\right)^{-1} \mathbf{T}_{1:P}^H \mathbf{d}_{\theta_s,\phi_s}}{\mathbf{d}_{\theta_s,\phi_s}^H \mathbf{T}_{1:P} \left(\mathbf{T}_{1:P}^H \mathbf{T}_{1:P}\right)^{-1} \mathbf{T}_{1:P}^H \mathbf{d}_{\theta_s,\phi_s}} \tag{4.89}$$

and the adaptive maximum DF (AMDF) beamformer:

$$\begin{aligned}\overline{\mathbf{h}}'_{\mathrm{AMDF}} &= \frac{\left(\mathbf{T}_{1:P}^H \boldsymbol{\Gamma}_{\mathrm{d}} \mathbf{T}_{1:P}\right)^{-1} \mathbf{T}_{1:P}^H \mathbf{d}_{\theta_{\mathrm{s}},\phi_{\mathrm{s}}}}{\mathbf{d}_{\theta_{\mathrm{s}},\phi_{\mathrm{s}}}^H \mathbf{T}_{1:P} \left(\mathbf{T}_{1:P}^H \boldsymbol{\Gamma}_{\mathrm{d}} \mathbf{T}_{1:P}\right)^{-1} \mathbf{T}_{1:P}^H \mathbf{d}_{\theta_{\mathrm{s}},\phi_{\mathrm{s}}}} \\ &= \frac{\mathbf{T}_{1:P}^H \mathbf{d}_{\theta_{\mathrm{s}},\phi_{\mathrm{s}}}}{\mathbf{d}_{\theta_{\mathrm{s}},\phi_{\mathrm{s}}}^H \mathbf{T}_{1:P} \mathbf{T}_{1:P}^H \mathbf{d}_{\theta_{\mathrm{s}},\phi_{\mathrm{s}}}}. \qquad (4.90)\end{aligned}$$

It is important to remember that both $\overline{\mathbf{h}}'_{\mathrm{AMWNG}}$ and $\overline{\mathbf{h}}'_{\mathrm{AMDF}}$ are adaptive beamformers, in general, since $\mathbf{T}_{1:P}$ depends on $\boldsymbol{\Gamma}_{\mathbf{y}}$.

In the same manner, instead of jointly digonalizing $\boldsymbol{\Gamma}_{\mathbf{y}}$ and $\boldsymbol{\Gamma}_{\mathrm{d}}$, we can jointly diagonalize $\boldsymbol{\Gamma}_{\mathbf{y}}$ and $\boldsymbol{\Gamma}_{\mathbf{v}}$ and redefine the gain in SNR as

$$\begin{aligned}\mathcal{G}\left(\overline{\mathbf{h}}'\right) &= \frac{\left|\overline{\mathbf{h}}'^H \mathbf{T}_{1:P}^H \mathbf{d}_{\theta_{\mathrm{s}},\phi_{\mathrm{s}}}\right|^2}{\overline{\mathbf{h}}'^H \mathbf{T}_{1:P}^H \boldsymbol{\Gamma}_{\mathbf{v}} \mathbf{T}_{1:P} \overline{\mathbf{h}}'} \\ &= \frac{\left|\overline{\mathbf{h}}'^H \mathbf{T}_{1:P}^H \mathbf{d}_{\theta_{\mathrm{s}},\phi_{\mathrm{s}}}\right|^2}{\overline{\mathbf{h}}'^H \overline{\mathbf{h}}'}, \qquad (4.91)\end{aligned}$$

where now $\mathbf{T}_{1:P}$ is deduced from the joint diagonalization of $\boldsymbol{\Gamma}_{\mathbf{y}}$ and $\boldsymbol{\Gamma}_{\mathbf{v}}$. Then, we find that the adaptive maximum SNR gain (AMSNRG) beamformer is

$$\overline{\mathbf{h}}'_{\mathrm{AMSNRG}} = \frac{\mathbf{T}_{1:P}^H \mathbf{d}_{\theta_{\mathrm{s}},\phi_{\mathrm{s}}}}{\mathbf{d}_{\theta_{\mathrm{s}},\phi_{\mathrm{s}}}^H \mathbf{T}_{1:P} \mathbf{T}_{1:P}^H \mathbf{d}_{\theta_{\mathrm{s}},\phi_{\mathrm{s}}}}. \qquad (4.92)$$

4.5 Illustrative Examples

Let us start by studying some examples of PCA applications to white noise reduction. We consider a 3-D cube array consisting of $M = M_0^3$ microphones, as shown in Fig. 3.1, in a room of size 6 m × 5 m × 3 m, where the reflection coefficients of all the six walls are assumed to be identical, are frequency independent, and are set to 0.9; the corresponding reverberation time, T_{60}, is approximately 500 ms. The reference point of the cube array (i.e., the origin point in Fig. 3.1) is located at (2, 2, 1.5) and the interelement spacing is $\delta = 4.0$ cm. The desired point source is placed at (4.0, 3.0, 1.5). A loudspeaker, which playbacks some clean speech signals (prerecorded in a quiet office environment with a sampling frequency of 16 kHz), is used to simulate a desired source. The acoustic channel impulse responses from the source to the microphones are generated with the image model method [8]. Then, the observed signals are generated by convolving the source signal with the corresponding impulse responses and adding some white noise. All the signals are sampled at 16 kHz. Let us assume that the components of the desired signal vector, $\mathbf{x}$, are the reverberant signals at the microphones. The covariance matrix, $\mathbf{\Phi}_\mathbf{x}$, of $\mathbf{x}$, is computed with 4000 snapshots, and the covariance matrix of the white noise is $\mathbf{\Phi}_\mathbf{v} = \phi_V \mathbf{I}_M$, where ϕ_V is obtained according to the specified input SNR. We set $M_0 = 3$, i.e., $M = 27$.

From (4.38), the normalized MSE (NMSE) is defined as

$$\overline{\mathcal{J}}\left(\overline{\mathbf{h}}\right) = \frac{\mathcal{J}\left(\overline{\mathbf{h}}\right)}{\mathbf{i}_1^T \mathbf{\Phi}_\mathbf{x} \mathbf{i}_1} \tag{4.93}$$

and from (4.40), the desired signal distortion index is defined as

$$\upsilon\left(\overline{\mathbf{h}}\right) = \frac{\mathcal{J}_\mathrm{d}\left(\overline{\mathbf{h}}\right)}{\mathbf{i}_1^T \mathbf{\Phi}_\mathbf{x} \mathbf{i}_1}. \tag{4.94}$$

We first study the performance of the Wiener-type filter, $\overline{\mathbf{h}}_\mathrm{W}$, given in (4.39). Figure 4.1 shows plots of the full-mode SNR of $\mathbf{z}$, $\mathrm{SNR}_{\mathbf{z},P}$, output SNR, $\mathrm{oSNR}\left(\overline{\mathbf{h}}\right)$, NMSE, $\overline{\mathcal{J}}\left(\overline{\mathbf{h}}\right)$, and desired signal distortion index, $\upsilon\left(\overline{\mathbf{h}}\right)$, of the Wiener-type filter, as a function of the input SNR, for different values of P. It can be clearly seen that PCA contributes to white noise reduction, and the smaller is the value of P, the less is the noise in $\mathbf{z}$ as compared to $\mathbf{y}$. It is also observed that the Wiener-type filter can further improve the SNR as compared to $\mathbf{z}$ except when $P = 1$. When $P = 1$, no further SNR improvement should be expected since there is only one component Z_1 and one can easily check that $\mathrm{oSNR}\left(\overline{\mathbf{h}}\right) = \mathrm{SNR}_{\mathbf{z},1}$.

Figure 4.2 shows plots of the full-mode SNR of $\mathbf{z}$, $\mathrm{SNR}_{\mathbf{z},P}$, output SNR, $\mathrm{oSNR}\left(\overline{\mathbf{h}}\right)$, NMSE, $\overline{\mathcal{J}}\left(\overline{\mathbf{h}}\right)$, and desired signal distortion index, $\upsilon\left(\overline{\mathbf{h}}\right)$, of the minimum-distortion filter, $\overline{\mathbf{h}}_\mathrm{MD}$, given in (4.41), as a function of the input SNR, for different values of P. As seen, this filter does not improve the SNR and achieves lower output SNR and lower distortion as the value of P increases. Comparing Figs. 4.1 and 4.2, one can see that the minimum-distortion filter achieves a lower desired signal distortion index (but also a lower output SNR) than the Wiener-type filter. When $P = 1$, the Wiener-type and minimum-distortion filters yield the same performance. This is not surprising since these two filters correspond to the same maximum SNR filter in this case.

We then consider a general noise environment with diffuse and white noises; then, the covariance matrix of the noise is $\mathbf{\Phi}_\mathbf{v} = \phi_\mathrm{d}\mathbf{\Gamma}_\mathrm{d} + \alpha\phi_V \mathbf{I}_M$, where $\alpha > 0$ is related to the ratio between the powers of the white and diffuse noises. We set $M_0 = 3$, $\alpha = 0.01$, and $f = 2000$ Hz for the single frequency diffuse noise, i.e., $\mathbf{\Gamma}_\mathrm{d}$. The desired signal is still the reverberant speech as in the previous example, and the variance of the noise, ϕ_V, is computed according to the specified input SNR

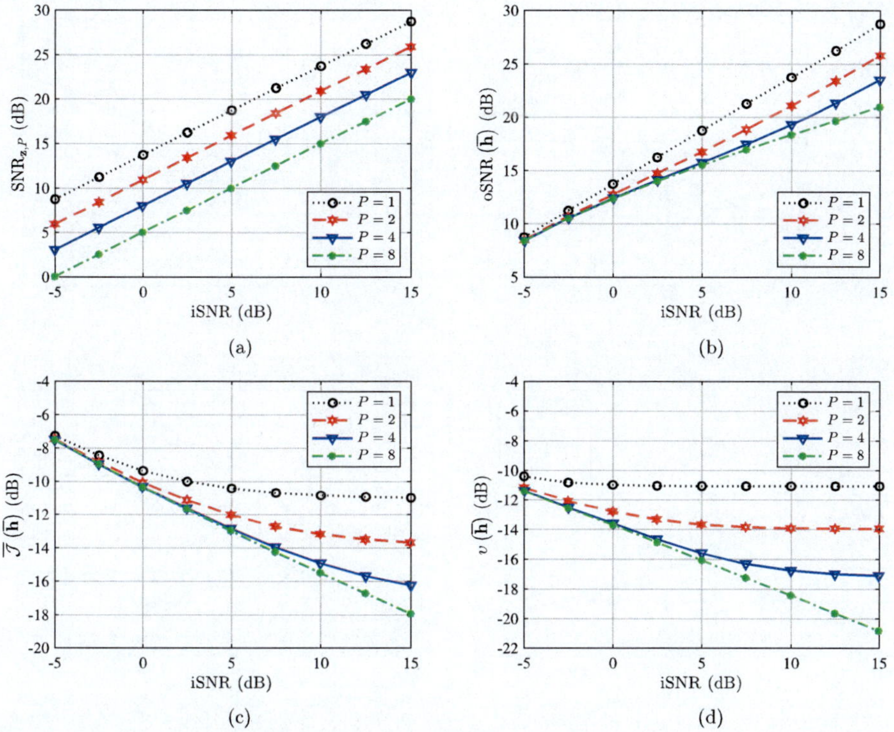

Fig. 4.1 Performance of the Wiener-type filter, $\overline{\mathbf{h}}_{\mathrm{W}}$, as a function of the input SNR, for different values of P, in the white noise case: (**a**) full-mode SNR of $\mathbf{z}$, (**b**) output SNR, (**c**) NMSE, and (**d**) desired signal distortion index. Condition: $M_0 = 3$

and value of α. The NMSE, $\overline{\mathcal{J}}\left(\overline{\mathbf{h}}'\right)$, and desired signal distortion index, $\upsilon\left(\overline{\mathbf{h}}'\right)$, are defined in a similar way to (4.93) and (4.94).

Figure 4.3 shows plots of the full-mode SNR of $\mathbf{z}'$, $\mathrm{SNR}_{\mathbf{z}',P}$, output SNR, $\mathrm{oSNR}\left(\overline{\mathbf{h}}'\right)$, NMSE, $\overline{\mathcal{J}}\left(\overline{\mathbf{h}}'\right)$, and desired signal distortion index, $\upsilon\left(\overline{\mathbf{h}}'\right)$, of the Wiener-type filter, $\overline{\mathbf{h}}'_{\mathrm{W}}$, given in (4.72), as a function of the input SNR, for different values of P. It can be observed that significant noise reduction is achieved with PCA, and the smaller is the value of P, the more is the noise reduction. It is also observed that the Wiener-type filter can sometimes further improve the SNR, and the higher is the value of P, the lower is the desired signal distortion index but the lower is the output SNR.

Figure 4.4 shows plots of the full-mode SNR of $\mathbf{z}'$, $\mathrm{SNR}_{\mathbf{z}',P}$, output SNR, $\mathrm{oSNR}\left(\overline{\mathbf{h}}'\right)$, NMSE, $\overline{\mathcal{J}}\left(\overline{\mathbf{h}}'\right)$, and desired signal distortion index, $\upsilon\left(\overline{\mathbf{h}}'\right)$, of the minimum-distortion filter, $\overline{\mathbf{h}}'_{\mathrm{MD}}$, given in (4.74), as a function of the input SNR, for different values of P. It can be observed that this filter improves the SNR, and the higher is the value of P, the lower are the output SNR and desired signal distortion

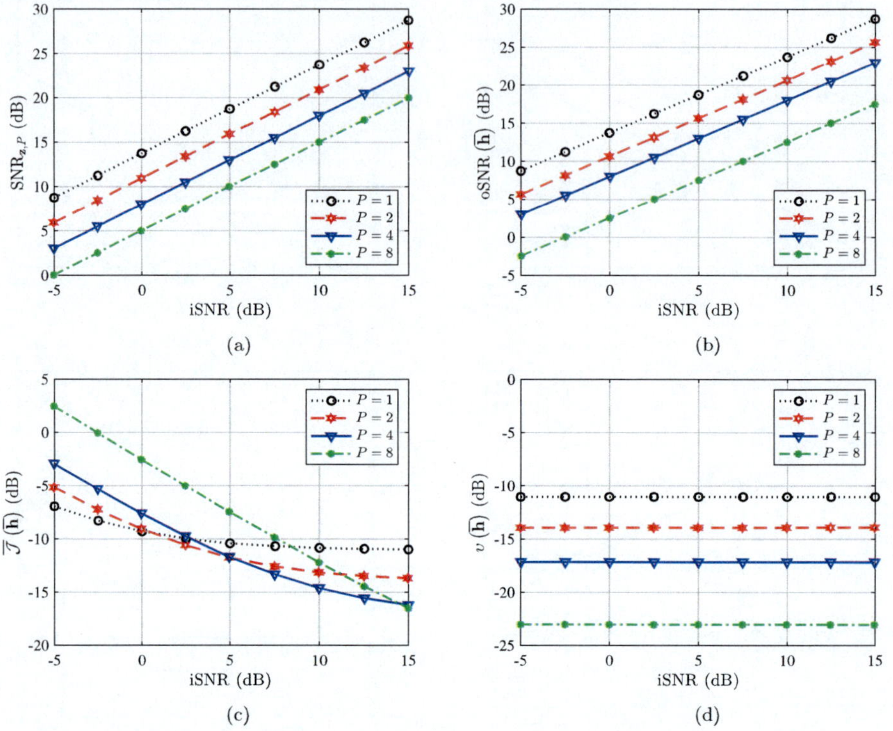

Fig. 4.2 Performance of the minimum-distortion filter, $\overline{\mathbf{h}}_{\mathrm{MD}}$, as a function of the input SNR, for different values of P, in the white noise case: (**a**) full-mode SNR of **z**, (**b**) output SNR, (**c**) NMSE, and (**d**) desired signal distortion index. Condition: $M_0 = 3$

index. It can also be observed that the minimum-distortion filter achieves a lower desired signal distortion index as compared to the Wiener-type filter.

Let us now study some examples of PCA applications to beamforming. We still consider a 3-D cube array consisting of $M = M_0^3$ omnidirectional microphones, as shown in Fig. 3.1, with $\delta = 4.0$ cm. The desired signal impinges on the array from $(\theta_s, \phi_s) = (80°, 0°)$.

We first consider a spherically isotropic (diffuse) noise field, where the input SNR of the diffuse noise is 10 dB, i.e., $\mathrm{iSNR_d} = 10$ dB. The coherence matrix of $\mathbf{y}$ is constructed according to (4.80), and the two matrices $\mathbf{\Gamma_y}$ and $\mathbf{\Gamma}_\mathrm{d}$ are jointly diagonalized according to (4.82) and (4.83). To demonstrate the performance of the AMWNG beamformer, $\overline{\mathbf{h}}'_{\mathrm{AMWNG}}$, given in (4.89), we choose $M_0 = 3$. Figures 4.5 and 4.6 show plots of the power patterns, $\left|\mathcal{B}_{\theta,\phi}(\mathbf{h})\right|^2$, of this beamformer versus the azimuth angle for $\theta = \theta_s$, and versus the elevation angle for $\phi = \phi_s$, respectively, for $f = 2000$ Hz and different values of P. It is observed that these power patterns achieve narrower main beams as the value of P decreases. Figure 4.7 shows plots of the DF, $\mathcal{D}\left(\overline{\mathbf{h}}'\right)$, and WNG, $\mathcal{W}\left(\overline{\mathbf{h}}'\right)$, of the AMWNG beamformer, as a function

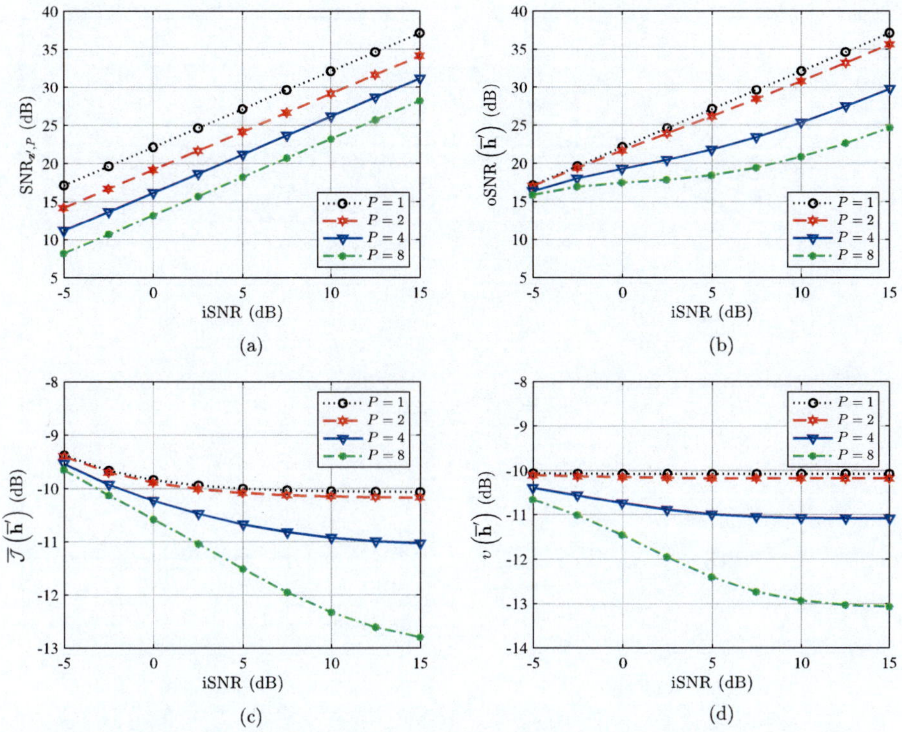

Fig. 4.3 Performance of the Wiener-type filter, $\overline{\mathbf{h}}'_{\mathrm{W}}$, as a function of the input SNR for different values of P, in the general noise case: (**a**) full-mode SNR of $\mathbf{z}'$, (**b**) output SNR, (**c**) NMSE, and (**d**) desired signal distortion index. Condition: $M_0 = 3$

of frequency, for different values of P. It is seen that the DF decreases, while the WNG increases as the value of P increases. This is also consistent with previous studies on subspace superdirective beamformers based on joint diagonalization [9, 10]. Figure 4.8 shows plots of the DF, $\mathcal{D}\left(\overline{\mathbf{h}}'\right)$, and WNG, $\mathcal{W}\left(\overline{\mathbf{h}}'\right)$, of the AMWNG beamformer, as a function of frequency, for different numbers of sensors, $M = M_0^3$, and $P = 4$. As seen, the larger is the number of sensors, the higher is the DF, while the lower is the WNG.

To demonstrate the performance of the AMDF beamformer, $\overline{\mathbf{h}}'_{\mathrm{AMDF}}$, given in (4.90), we choose $P = 1$. Figures 4.9 and 4.10 show plots of the power patterns, $\left|\mathcal{B}_{\theta,\phi}\left(\mathbf{h}\right)\right|^2$, of this beamformer versus the azimuth angle for $\theta = \theta_{\mathrm{s}}$, and versus the elevation angle for $\phi = \phi_{\mathrm{s}}$, respectively, for $f = 2000$ Hz and different numbers of sensors, $M = M_0^3$. Figure 4.11 shows plots of the DF, $\mathcal{D}\left(\overline{\mathbf{h}}'\right)$, and WNG, $\mathcal{W}\left(\overline{\mathbf{h}}'\right)$, of the AMDF beamformer, as a function of frequency, for different numbers of sensors, $M = M_0^3$. As seen, the larger is the number of sensors, the narrower are the main beams of the power patterns, and the higher is the DF but the lower is the WNG.

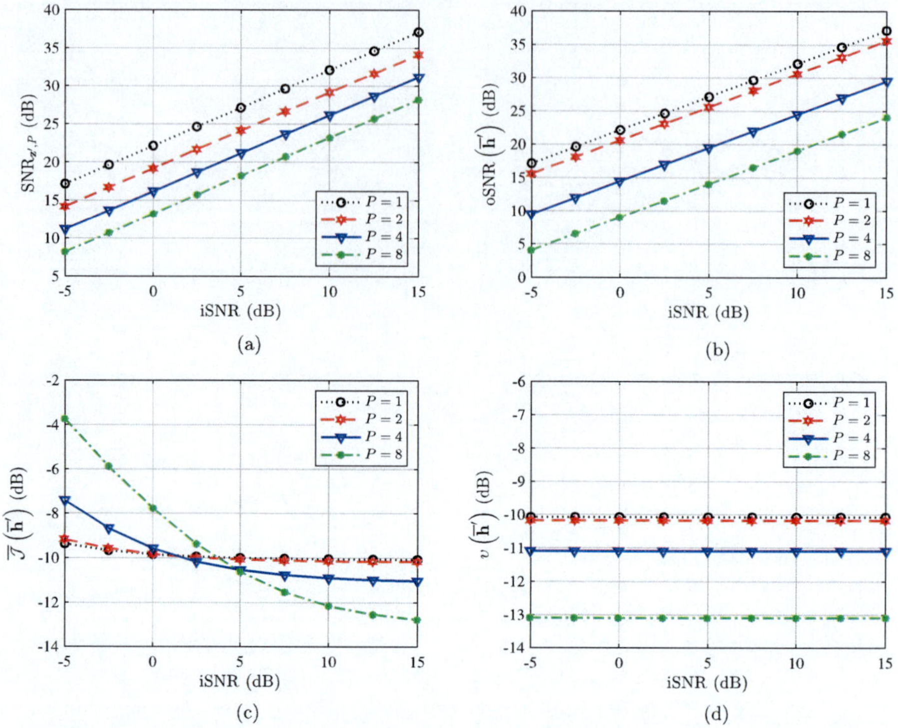

Fig. 4.4 Performance of the minimum-distortion filter, $\overline{\mathbf{h}}'_{\mathrm{MD}}$, as a function of the input SNR, for different values of P, in the general noise case: (**a**) full-mode SNR of $\mathbf{z}'$, (**b**) output SNR, (**c**) NMSE, and (**d**) desired signal distortion index. Condition: $M_0 = 3$

We also consider an environment with point source noises, diffuse noise, and white noise, where the two statistically independent interferences impinge on the cube array from two different directions $\left(\theta_{\mathrm{i},1}, \phi_{\mathrm{i},1}\right)$ and $\left(\theta_{\mathrm{i},2}, \phi_{\mathrm{i},2}\right)$. The two interferences are incoherent with the desired signal. Then, the covariance matrix of the noise signal is

$$\mathbf{\Phi}_{\mathbf{v}} = \phi_{V_{\mathrm{i},1}} \mathbf{d}_{\theta_{\mathrm{i},1},\phi_{\mathrm{i},1}} \mathbf{d}^H_{\theta_{\mathrm{i},1},\phi_{\mathrm{i},1}} + \phi_{V_{\mathrm{i},2}} \mathbf{d}_{\theta_{\mathrm{i},2},\phi_{\mathrm{i},2}} \mathbf{d}^H_{\theta_{\mathrm{i},2},\phi_{\mathrm{i},2}} + \phi_{V_{\mathrm{d}}} \mathbf{\Gamma}_{\mathrm{d}} + \phi_{V_{\mathrm{w}}} \mathbf{I}_M, \tag{4.95}$$

where $\phi_{V_{\mathrm{i},1}}$, $\phi_{V_{\mathrm{i},2}}$, $\phi_{V_{\mathrm{d}}}$, and $\phi_{V_{\mathrm{w}}}$ are the variances of the two point source noises, diffuse noise, and white noise, respectively. We choose $\left(\theta_{\mathrm{i},1}, \phi_{\mathrm{i},1}\right) = (80^\circ, 150^\circ)$, $\left(\theta_{\mathrm{i},2}, \phi_{\mathrm{i},2}\right) = (80^\circ, 210^\circ)$ and assume that $\phi_{V_{\mathrm{i},1}} = \phi_{V_{\mathrm{i},2}} = \phi_{V_{\mathrm{d}}} = \phi_V$, $\phi_{V_{\mathrm{w}}} = \alpha\phi_V$, where $\alpha > 0$ is related to the ratio between the powers of the white and diffuse noises. The overall input SNR is set to 10 dB.

To demonstrate the performance of the AMSNRG beamformer, $\overline{\mathbf{h}}'_{\mathrm{AMWNG}}$, given in (4.92), we choose $P = 1$. Figures 4.12 and 4.13 show plots of the power patterns, $\left|\mathcal{B}_{\theta,\phi}\left(\mathbf{h}\right)\right|^2$, of this beamformer, versus the azimuth angle for $\theta = \theta_{\mathrm{s}}$, and versus the

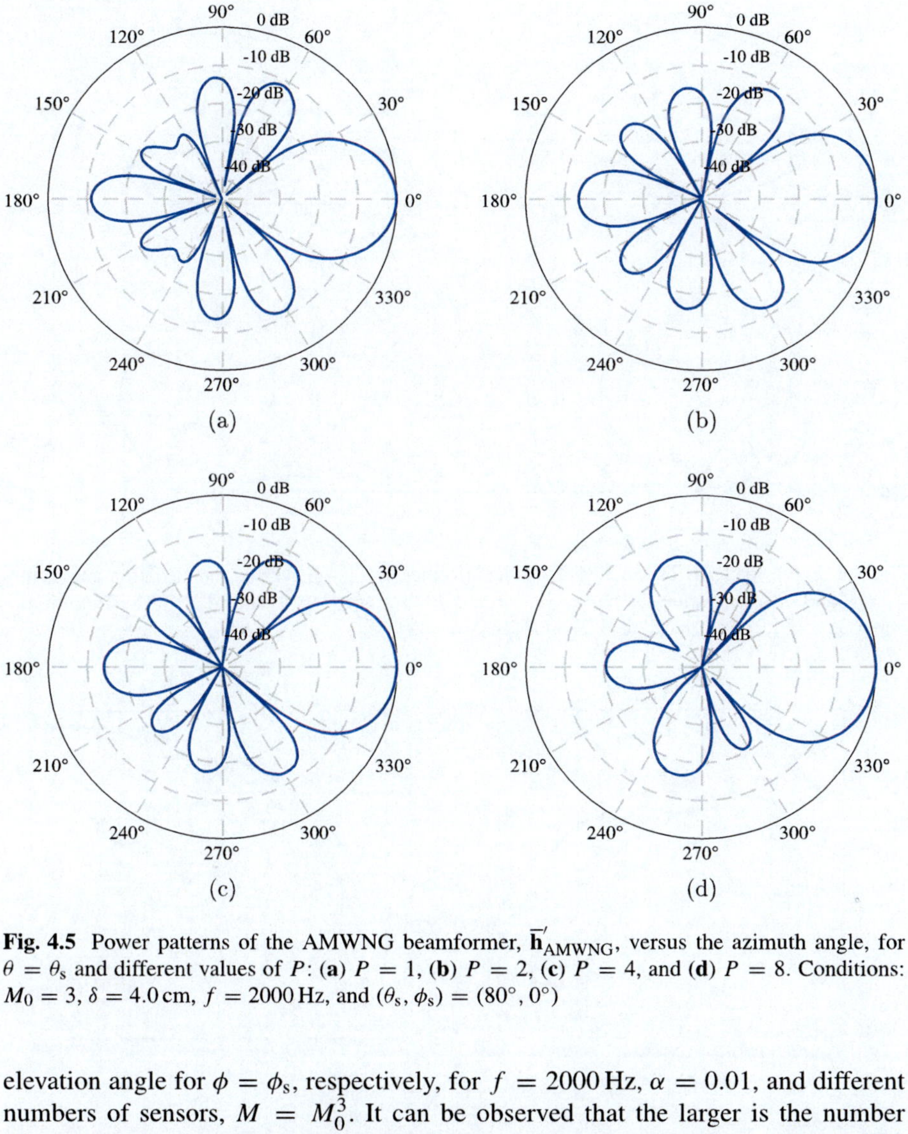

Fig. 4.5 Power patterns of the AMWNG beamformer, $\overline{\mathbf{h}}'_{\text{AMWNG}}$, versus the azimuth angle, for $\theta = \theta_s$ and different values of P: **(a)** $P = 1$, **(b)** $P = 2$, **(c)** $P = 4$, and **(d)** $P = 8$. Conditions: $M_0 = 3$, $\delta = 4.0$ cm, $f = 2000$ Hz, and $(\theta_s, \phi_s) = (80°, 0°)$

elevation angle for $\phi = \phi_s$, respectively, for $f = 2000$ Hz, $\alpha = 0.01$, and different numbers of sensors, $M = M_0^3$. It can be observed that the larger is the number of sensors, the narrower are the main beams of the power patterns. Figure 4.14 shows plots of the SNR gain, $\mathcal{G}\left(\overline{\mathbf{h}}'\right)$, as a function of frequency, of the AMSNRG beamformer, for different numbers of sensors, $M = M_0^3$, and different values of α. It is seen that for a fixed value of α, increasing the number of sensors can improve the SNR gain. It can also be observed that for the same number of sensors, the SNR gain decreases as the value of α increases. This is easy to understand as a larger value of α means that noise signals at different sensors are more white and less coherent.

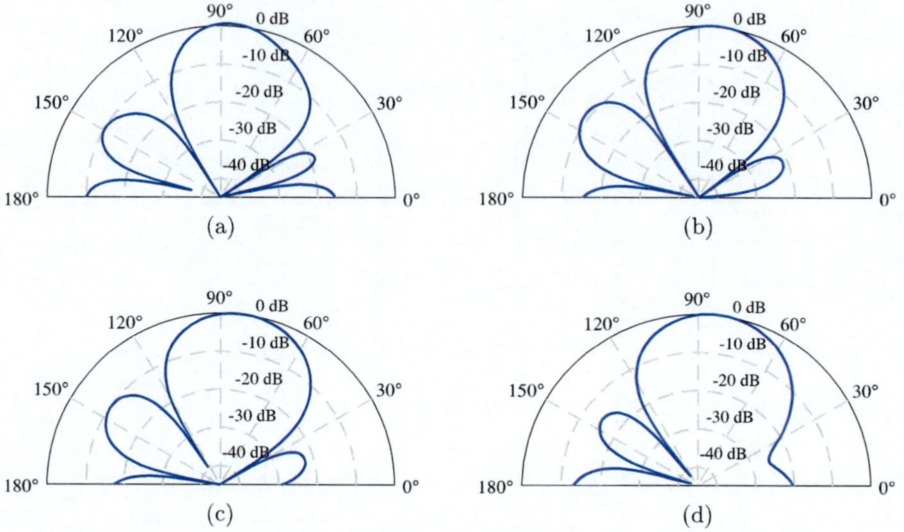

Fig. 4.6 Power patterns of the AMWNG beamformer, $\overline{\mathbf{h}}'_{\mathrm{AMWNG}}$, versus the elevation angle, for $\phi = \phi_{\mathrm{s}}$ and different values of P: (**a**) $P = 1$, (**b**) $P = 2$, (**c**) $P = 4$, and (**d**) $P = 8$. Conditions: $M_0 = 3$, $\delta = 4.0\,\mathrm{cm}$, $f = 2000\,\mathrm{Hz}$, and $(\theta_{\mathrm{s}}, \phi_{\mathrm{s}}) = (80°, 0°)$

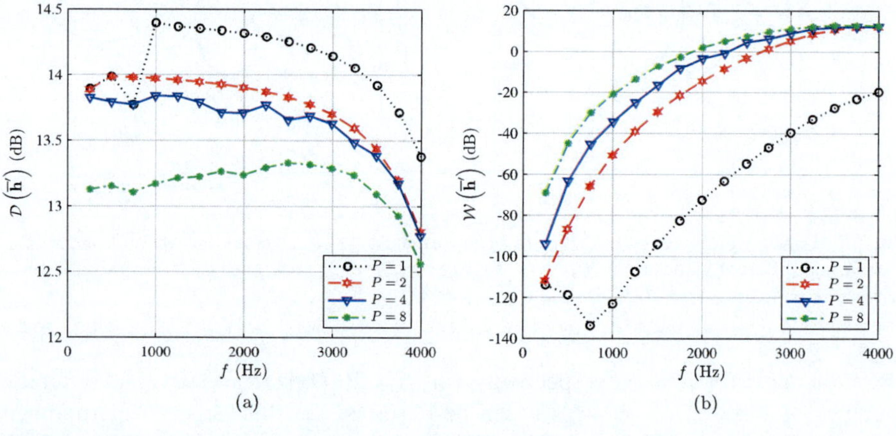

Fig. 4.7 DF and WNG of the AMWNG beamformer, $\overline{\mathbf{h}}'_{\mathrm{AMWNG}}$, as a function of frequency, for different values of P: (**a**) DF and (**b**) WNG. Conditions: $M_0 = 3$, $\delta = 4.0\,\mathrm{cm}$, and $(\theta_{\mathrm{s}}, \phi_{\mathrm{s}}) = (80°, 0°)$

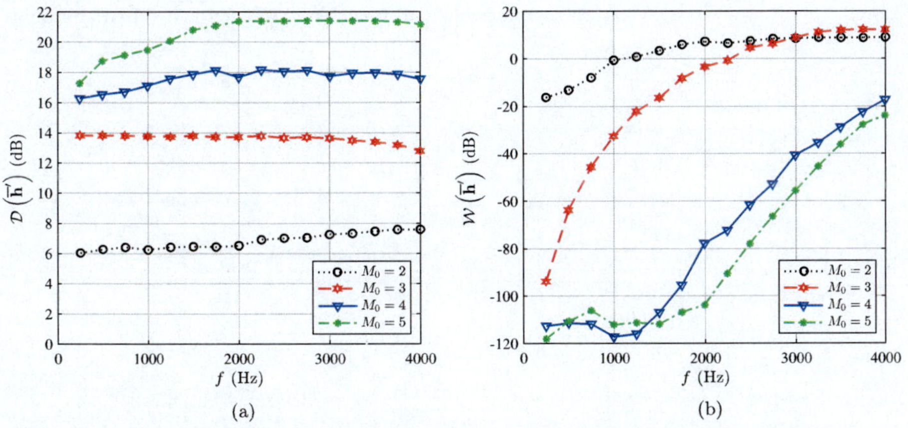

Fig. 4.8 DF and WNG of the AMWNG beamformer, $\overline{\mathbf{h}}'_{\mathrm{AMWNG}}$, as a function of frequency, for different numbers of sensors, $M = M_0^3$: (**a**) DF and (**b**) WNG. Conditions: $\delta = 4.0\,\mathrm{cm}$, $P = 4$, and $(\theta_{\mathrm{s}}, \phi_{\mathrm{s}}) = (80^\circ, 0^\circ)$

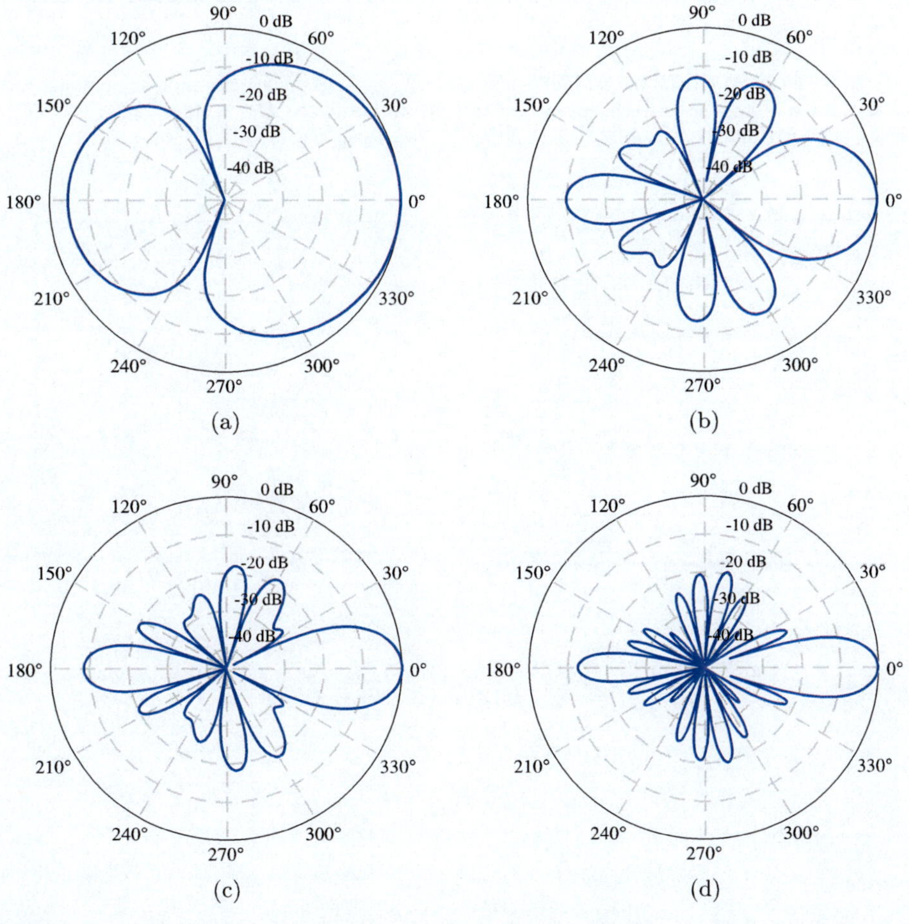

Fig. 4.9 Power patterns of the AMDF beamformer, $\overline{\mathbf{h}}'_{\mathrm{AMDF}}$, versus the azimuth angle, for $\theta = \theta_{\mathrm{s}}$ and different numbers of sensors, $M = M_0^3$: (**a**) $M_0 = 2$, (**b**) $M_0 = 3$, (**c**) $M_0 = 4$, and (**d**) $M_0 = 5$. Conditions: $\delta = 4.0\,\mathrm{cm}$, $f = 2000\,\mathrm{Hz}$, $P = 1$, and $(\theta_{\mathrm{s}}, \phi_{\mathrm{s}}) = (80^\circ, 0^\circ)$

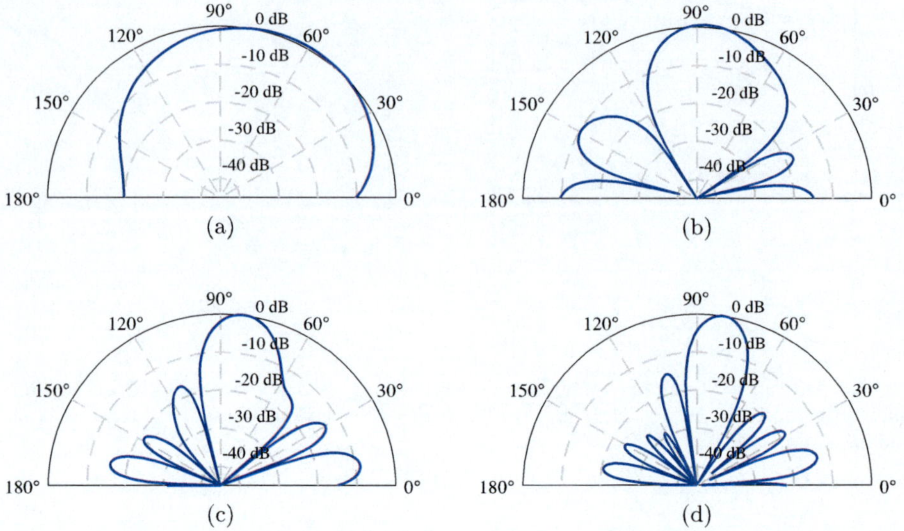

Fig. 4.10 Power patterns of the AMDF beamformer, $\overline{\mathbf{h}}'_{\text{AMDF}}$, versus the elevation angle, for $\phi = \phi_s$ and different numbers of sensors, $M = M_0^3$: (**a**) $M_0 = 2$, (**b**) $M_0 = 3$, (**c**) $M_0 = 4$, and (**d**) $M_0 = 5$. Conditions: $\delta = 4.0\,\text{cm}$, $f = 2000\,\text{Hz}$, $P = 1$, and $(\theta_s, \phi_s) = (80°, 0°)$

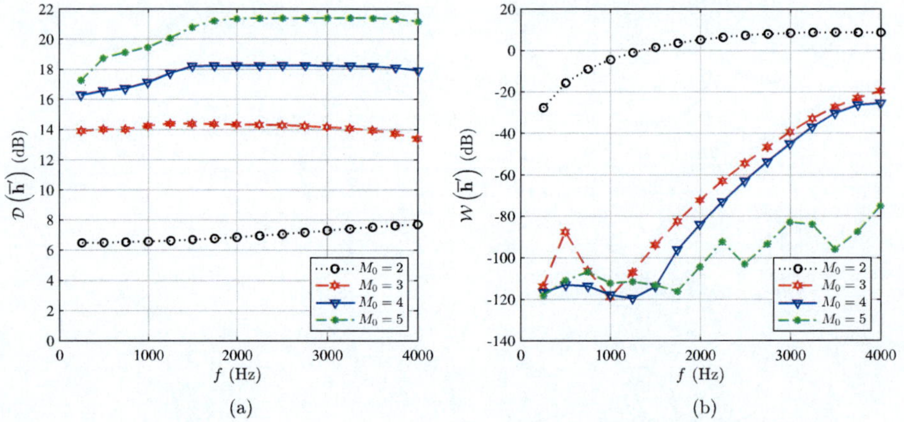

Fig. 4.11 DF and WNG of the AMDF beamformer, $\overline{\mathbf{h}}'_{\text{AMDF}}$, as a function of frequency, for different numbers of sensors, $M = M_0^3$: (**a**) DF and (**b**) WNG. Conditions: $\delta = 4.0\,\text{cm}$, $P = 1$, and $(\theta_s, \phi_s) = (80°, 0°)$

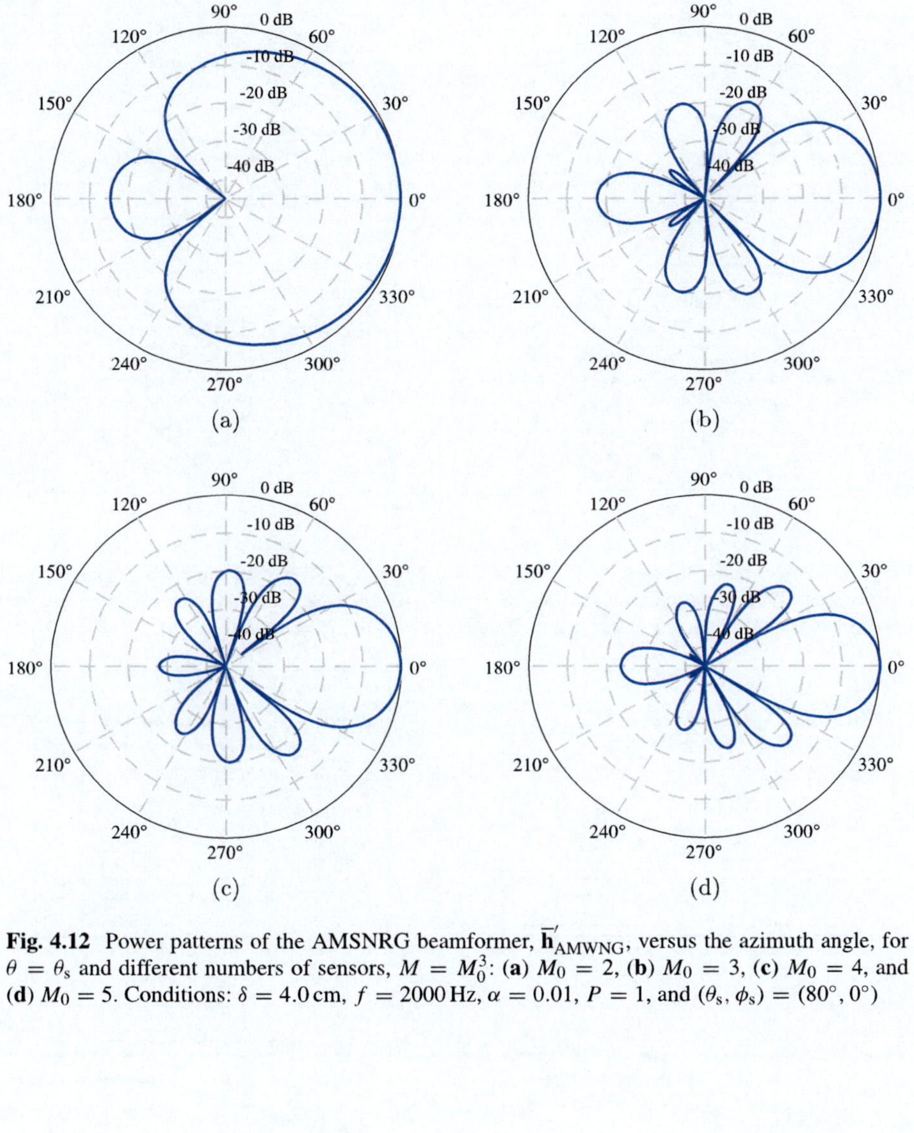

Fig. 4.12 Power patterns of the AMSNRG beamformer, $\overline{\mathbf{h}}'_{\mathrm{AMWNG}}$, versus the azimuth angle, for $\theta = \theta_s$ and different numbers of sensors, $M = M_0^3$: (**a**) $M_0 = 2$, (**b**) $M_0 = 3$, (**c**) $M_0 = 4$, and (**d**) $M_0 = 5$. Conditions: $\delta = 4.0$ cm, $f = 2000$ Hz, $\alpha = 0.01$, $P = 1$, and $(\theta_s, \phi_s) = (80°, 0°)$

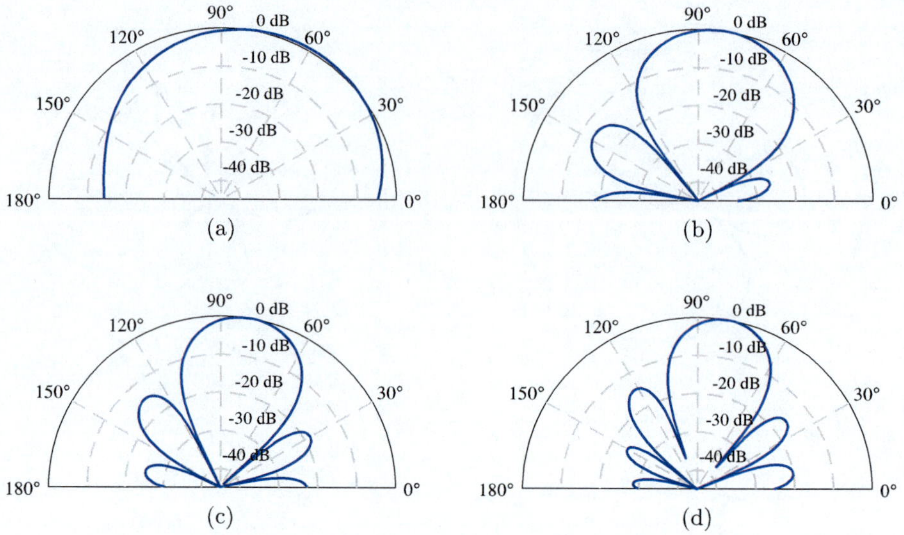

Fig. 4.13 Power patterns of the AMSNRG beamformer, $\overline{\mathbf{h}}'_{\mathrm{AMWNG}}$, versus the elevation angle, for $\phi = \phi_s$ and different numbers of sensors, $M = M_0^3$: (**a**) $M_0 = 2$, (**b**) $M_0 = 3$, (**c**) $M_0 = 4$, and (**d**) $M_0 = 5$. Conditions: $\delta = 4.0\,\mathrm{cm}$, $f = 2000\,\mathrm{Hz}$, $\alpha = 0.01$, $P = 1$, and $(\theta_s, \phi_s) = (80°, 0°)$

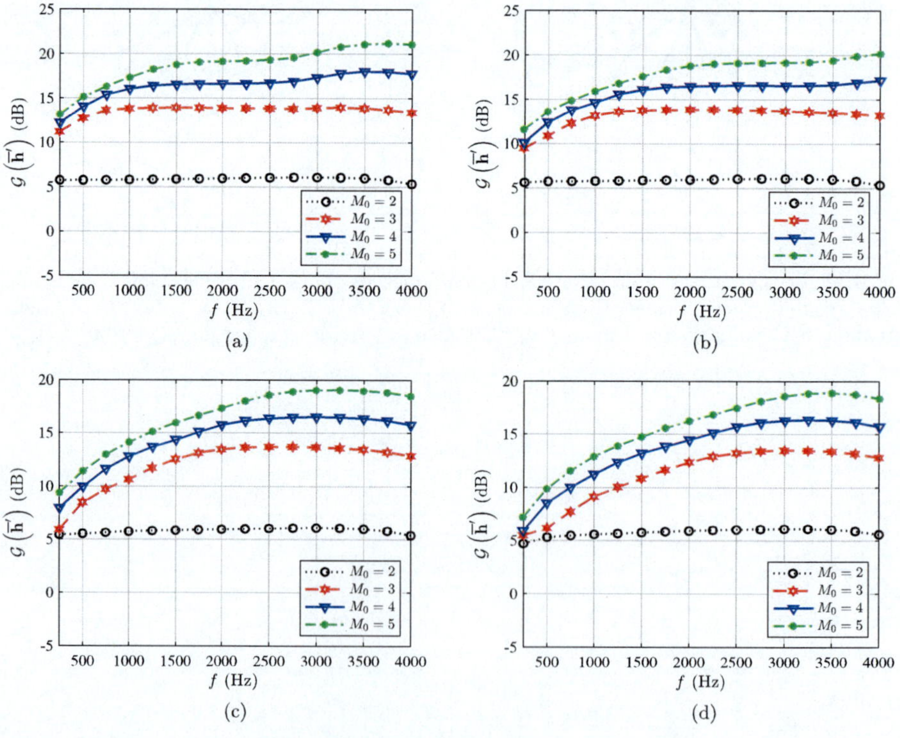

Fig. 4.14 SNR gain of the AMSNRG beamformer, $\overline{\mathbf{h}}'_{\mathrm{AMWNG}}$, as a function of frequency, for different numbers of sensors, $M = M_0^3$, and different values of α: (**a**) $\alpha = 10^{-6}$, (**b**) $\alpha = 10^{-4}$, (**c**) $\alpha = 10^{-2}$, and (**d**) $\alpha = 10^{-1}$. Conditions: $\delta = 4.0\,\mathrm{cm}$, $P = 1$, and $(\theta_s, \phi_s) = (80°, 0°)$

References

1. I.T. Jolliffe, *Principal Component Analysis*, 2nd edn. (Springer, New York, 2002)
2. J. Benesty, *Fundamentals of Speech Enhancement*. Springer Briefs in Electrical & Computer Engineering (Springer, Berlin, 2018)
3. L.L. Scharf, D.W. Tufts, Rank reduction for modeling stationary signals. IEEE Trans. Acoust. Speech Signal Process. **35**, 350–355 (1987)
4. J.N. Franklin, *Matrix Theory* (Prentice-Hall, Englewood Cliffs, 1968)
5. G. Huang, J. Benesty, T. Long, J. Chen, A family of maximum SNR filters for noise reduction. IEEE/ACM Trans. Audio Speech Lang. Process. **22**(12), 2034–2047 (2014)
6. J.R. Jensen, J. Benesty, M.G. Christensen, Noise reduction with optimal variable span linear filters. IEEE/ACM Trans. Audio Speech Lang. Process. **24**, 631–644 (2016)
7. J. Benesty, M.G. Christensen, J.R. Jensen, *Signal Enhancement with Variable Span Linear Filters* (Springer Nature, Switzerland, 2016)
8. J.B. Allen, D.A. Berkley, Image method for efficiently simulating small-room acoustics. J. Acoust. Soc. Am. **65**, 943–950 (1979)
9. C. Li, J. Benesty, G. Huang, J. Chen, Subspace superdirective beamformers based on joint diagonalization, in *2016 IEEE International Conference on Acoustics, Speech and Signal Processing (ICASSP)* (2016), pp. 400–404
10. G. Huang, J. Benesty, J. Chen, Subspace superdirective beamforming with uniform circular microphone arrays, in *2016 IEEE International Workshop on Acoustic Signal Enhancement (IWAENC)* (2016), pp. 1–5

Chapter 5
Low-Rank Beamforming

Nature is usually low rank! This means that there is nonnegligible redundancy in the observations. So, it is important to be able to translate this idea into equations in order to make things work better in practice. In this chapter, we discuss this concept and explain how it can be applied to beamforming. Then, we derive a large class of low-rank fixed and adaptive beamformers.

5.1 Fundamental Concepts

In this section, we explain the main ideas behind low-rank concepts for beamforming, which were already successfully applied in different contexts and topics [1–4].

Let $\mathbf{h}$ be a beamforming filter of length M. It is assumed that $M = M_1 M_2$. Without loss of generality, it may also be assumed that $M_1 \geq M_2$. Then, we can decompose $\mathbf{h}$ as

$$\mathbf{h} = \left[\mathbf{a}_1^T \ \mathbf{a}_2^T \ \cdots \ \mathbf{a}_{M_2}^T \right]^T, \tag{5.1}$$

where $\mathbf{a}_m, \ m = 1, 2, \ldots, M_2$ are M_2 short beamformers of length M_1 each. Equivalently, we can write $\mathbf{h}$ as a rectangular matrix of size $M_1 \times M_2$:

$$\begin{aligned} \mathbf{H} &= \left[\mathbf{a}_1 \ \mathbf{a}_2 \ \cdots \ \mathbf{a}_{M_2} \right] \\ &= \text{ivec}\left(\mathbf{h}\right), \end{aligned} \tag{5.2}$$

where ivec$(\cdot)$ is the inverse of the vectorization operator. As a result, (5.1) can also be expressed as

$$\mathbf{h} = \text{vec}\left(\mathbf{H}\right), \tag{5.3}$$

J. Benesty et al., *Microphone Arrays*, Springer Topics in Signal Processing 22,
https://doi.org/10.1007/978-3-031-36974-2_5

where $\text{vec}(\cdot)$ denotes vectorization, i.e., the operation of converting a matrix into a vector. Expression (5.3) is fundamental here since the notion of matrix rank can be explicitly exploited as explained in the rest of this section.

Indeed, thanks to the well-known singular value decomposition (SVD), $\mathbf{H}$ can be factorized as

$$\begin{aligned}\mathbf{H} &= \mathbf{U}_1 \boldsymbol{\Sigma} \mathbf{U}_2^H \\ &= \sum_{m=1}^{M_2} \sigma_m \mathbf{u}_{1,m} \mathbf{u}_{2,m}^H,\end{aligned} \tag{5.4}$$

where $\mathbf{U}_1$ and $\mathbf{U}_2$ are two unitary matrices of sizes $M_1 \times M_1$ and $M_2 \times M_2$, respectively, and $\boldsymbol{\Sigma}$ is an $M_1 \times M_2$ rectangular diagonal matrix with nonnegative real numbers on the diagonal. The columns of $\mathbf{U}_1$ (resp. $\mathbf{U}_2$) are called the left-singular (resp. right-singular) vectors of $\mathbf{H}$, while the diagonal entries $\sigma_m,\ m = 1, 2, \ldots, M_2$ of $\boldsymbol{\Sigma}$ are known as the singular values of $\mathbf{H}$ with $\sigma_1 \geq \sigma_2 \geq \cdots \geq \sigma_{M_2} \geq 0$. We can conveniently write $\mathbf{H}$ as

$$\mathbf{H} = \sum_{m=1}^{M_2} \mathbf{h}_{1,m} \mathbf{h}_{2,m}^H, \tag{5.5}$$

where

$$\mathbf{h}_{1,m} = \sigma_{1,m} \mathbf{u}_{1,m}, \tag{5.6}$$

$$\mathbf{h}_{2,m} = \sigma_{2,m} \mathbf{u}_{2,m}, \tag{5.7}$$

for $m = 1, 2, \ldots, M_2$, with $\sigma_{1,m}\sigma_{2,m} = \sigma_m$. We deduce that the beamforming filter is

$$\begin{aligned}\mathbf{h} &= \text{vec}\left(\mathbf{H}\right) \\ &= \sum_{m=1}^{M_2} \text{vec}\left(\mathbf{h}_{1,m} \mathbf{h}_{2,m}^H\right) \\ &= \sum_{m=1}^{M_2} \text{vec}\left[\mathbf{h}_{1,m} \left(\mathbf{h}_{2,m}^*\right)^T\right] \\ &= \sum_{m=1}^{M_2} \mathbf{h}_{2,m}^* \otimes \mathbf{h}_{1,m},\end{aligned} \tag{5.8}$$

where $\otimes$ is the Kronecker product.

However, in most practical scenarios, **H** is never really full rank for many good reasons, especially when the number of microphones composing the array is large. Indeed, we know by experience that when the data has some structure, the singular values of a rectangular matrix that depends on this data will decrease rapidly. Let $P \ll M_2$ and let us define the following matrix:

$$\mathbf{H}(P) = \sum_{p=1}^{P} \sigma_p \mathbf{u}_{1,p} \mathbf{u}_{2,p}^H. \tag{5.9}$$

The objective is to verify whether **H** can be well approximated by $\mathbf{H}(P)$.

Let **a** be a complex-valued vector of length M. The 2-norm or Euclidean norm of this vector is defined as

$$\|\mathbf{a}\|_2 = \sqrt{\mathbf{a}^H \mathbf{a}}. \tag{5.10}$$

Let **A** be a complex-valued rectangular matrix of size $M_1 \times M_2$. The Frobenius norm and the 2-norm of this matrix are, respectively,

$$\|\mathbf{A}\|_\mathrm{F} = \sqrt{\mathrm{tr}\left(\mathbf{A}^H \mathbf{A}\right)} \tag{5.11}$$

and

$$\|\mathbf{A}\|_2 = \max_{\|\mathbf{x}\|_2 = 1} \|\mathbf{A}\mathbf{x}\|_2 . \tag{5.12}$$

Next, we state a theorem given in [5, 6], which helps to prove that **H** can be well approximated by $\mathbf{H}(P)$; this is known as the low-rank approximation. Let $\mathrm{rank}\,(\mathbf{H}) = R \leq M_2$ and let $\mathcal{S}$ be the set of $M_1 \times M_2$ rectangular matrices of rank equal to $P < R$. Then, the solution to the following minimization problem:

$$\min_{\mathbf{G} \in \mathcal{S}} \|\mathbf{H} - \mathbf{G}\|_2 \quad \text{or} \quad \min_{\mathbf{G} \in \mathcal{S}} \|\mathbf{H} - \mathbf{G}\|_\mathrm{F} \tag{5.13}$$

is given by (5.9). Furthermore, we have

$$\min_{\mathbf{G} \in \mathcal{S}} \|\mathbf{H} - \mathbf{G}\|_2 = \|\mathbf{H} - \mathbf{H}(P)\|_2 = \sigma_{P+1} \tag{5.14}$$

and

$$\min_{\mathbf{G} \in \mathcal{S}} \|\mathbf{H} - \mathbf{G}\|_\mathrm{F} = \|\mathbf{H} - \mathbf{H}(P)\|_\mathrm{F} = \sqrt{\sum_{i=P+1}^{M_2} \sigma_i^2}. \tag{5.15}$$

Consequently, as long as the normalized distance between **H** and $\mathbf{H}(P)$, i.e.,

$$\frac{\|\mathbf{H} - \mathbf{H}(P)\|_F}{\|\mathbf{H}\|_F}, \tag{5.16}$$

remains very small, it is preferable to work with $\mathbf{H}(P)$.

In the rest of this chapter, we assume that rank $(\mathbf{H}) = P \leq M_2$. Consequently, the low-rank filter, $\mathbf{h}$, can be decomposed as

$$\mathbf{h} = \sum_{p=1}^{P} \mathbf{h}_{2,p}^* \otimes \mathbf{h}_{1,p}, \tag{5.17}$$

where the beamforming filters $\mathbf{h}_{1,p}$ and $\mathbf{h}_{2,p}$ have lengths M_1 and M_2, respectively. From the following relations:

$$\begin{aligned} \mathbf{h}_{2,p}^* \otimes \mathbf{h}_{1,p} &= \left(\mathbf{h}_{2,p}^* \times 1\right) \otimes \left(\mathbf{I}_{M_1} \times \mathbf{h}_{1,p}\right) \\ &= \left(\mathbf{h}_{2,p}^* \otimes \mathbf{I}_{M_1}\right) \mathbf{h}_{1,p} \end{aligned} \tag{5.18}$$

and

$$\begin{aligned} \mathbf{h}_{2,p}^* \otimes \mathbf{h}_{1,p} &= \left(\mathbf{I}_{M_2} \times \mathbf{h}_{2,p}^*\right) \otimes \left(\mathbf{h}_{1,p} \times 1\right) \\ &= \left(\mathbf{I}_{M_2} \otimes \mathbf{h}_{1,p}\right) \mathbf{h}_{2,p}^*, \end{aligned} \tag{5.19}$$

where $\mathbf{I}_{M_1}$ and $\mathbf{I}_{M_2}$ are the identity matrices of sizes $M_1 \times M_1$ and $M_2 \times M_2$, respectively, (5.17) may be rewritten as

$$\begin{aligned} \mathbf{h} &= \sum_{p=1}^{P} \mathbf{H}_{2,p}^* \mathbf{h}_{1,p} \\ &= \underline{\mathbf{H}}_2 \underline{\mathbf{h}}_1 \end{aligned} \tag{5.20}$$

$$\begin{aligned} &= \sum_{p=1}^{P} \mathbf{H}_{1,p} \mathbf{h}_{2,p}^* \\ &= \underline{\mathbf{H}}_1 \underline{\mathbf{h}}_2, \end{aligned} \tag{5.21}$$

where

$$\mathbf{H}_{2,p}^* = \mathbf{h}_{2,p}^* \otimes \mathbf{I}_{M_1},$$
$$\mathbf{H}_{1,p} = \mathbf{I}_{M_2} \otimes \mathbf{h}_{1,p}$$

are matrices of sizes $M \times M_1$ and $M \times M_2$, respectively;

$$\underline{\mathbf{H}}_2 = \begin{bmatrix} \mathbf{H}_{2,1}^* & \mathbf{H}_{2,2}^* & \cdots & \mathbf{H}_{2,P}^* \end{bmatrix},$$

$$\underline{\mathbf{H}}_1 = \begin{bmatrix} \mathbf{H}_{1,1} & \mathbf{H}_{1,2} & \cdots & \mathbf{H}_{1,P} \end{bmatrix}$$

are matrices of sizes $M \times PM_1$ and $M \times PM_2$, respectively; and

$$\underline{\mathbf{h}}_1 = \begin{bmatrix} \mathbf{h}_{1,1}^T & \mathbf{h}_{1,2}^T & \cdots & \mathbf{h}_{1,P}^T \end{bmatrix}^T,$$

$$\underline{\mathbf{h}}_2 = \begin{bmatrix} \mathbf{h}_{2,1}^H & \mathbf{h}_{2,2}^H & \cdots & \mathbf{h}_{2,P}^H \end{bmatrix}^T$$

are beamforming filters of lengths PM_1 and PM_2, respectively.

Therefore, we are now going to do beamforming with $\underline{\mathbf{h}}_1$ and $\underline{\mathbf{h}}_2$ instead of $\mathbf{h}$. There are at least two great advantages. First, complexity can be reduced since we have smaller matrices to invert. Second, in the context of adaptive beamforming, we need much less data to estimate the different statistics. Consequently, we should expect low-rank beamformers to be much more reliable and robust than their conventional counterparts, in general.

5.2 Signal Model and Beamforming

We consider a 3-D microphone array that picks up the sound in its environment and the signal model of Chap. 3. Therefore, the observed signal vector of length M is

$$\begin{aligned} \mathbf{y} &= \begin{bmatrix} Y_1 & Y_2 & \cdots & Y_M \end{bmatrix}^T \\ &= \mathbf{x} + \mathbf{v} \\ &= \mathbf{d}_{\theta_s,\phi_s} X + \mathbf{v}, \end{aligned} \tag{5.22}$$

where

$$\mathbf{d}_{\theta_s,\phi_s} = \begin{bmatrix} e^{\jmath 2\pi f \mathbf{a}_{\theta_s,\phi_s}^T \mathbf{r}_1/c} & e^{\jmath 2\pi f \mathbf{a}_{\theta_s,\phi_s}^T \mathbf{r}_2/c} & \cdots & e^{\jmath 2\pi f \mathbf{a}_{\theta_s,\phi_s}^T \mathbf{r}_M/c} \end{bmatrix}^T \tag{5.23}$$

is the steering vector of length M. The covariance matrix of $\mathbf{y}$ is

$$\begin{aligned} \mathbf{\Phi}_{\mathbf{y}} &= E\left(\mathbf{y}\mathbf{y}^H\right) \\ &= \mathbf{\Phi}_{\mathbf{x}} + \mathbf{\Phi}_{\mathbf{v}} \\ &= \phi_X \mathbf{d}_{\theta_s,\phi_s} \mathbf{d}_{\theta_s,\phi_s}^H + \mathbf{\Phi}_{\mathbf{v}} \end{aligned}$$

$$= \phi_X \mathbf{d}_{\theta_s,\phi_s} \mathbf{d}_{\theta_s,\phi_s}^H + \phi_V \boldsymbol{\Gamma}_\mathbf{v}. \tag{5.24}$$

In the case of the spherically isotropic (diffuse) noise field, (5.24) becomes

$$\boldsymbol{\Phi}_\mathbf{y} = \phi_X \mathbf{d}_{\theta_s,\phi_s} \mathbf{d}_{\theta_s,\phi_s}^H + \phi_d \boldsymbol{\Gamma}_d, \tag{5.25}$$

where ϕ_d is the variance of the diffuse noise and $\boldsymbol{\Gamma}_d$ is defined in Chap. 3.

In our scenario, an estimate of the desired signal, X, is

$$\begin{aligned} Z &= \mathbf{h}^H \mathbf{y} \\ &= \underline{\mathbf{h}}_1^H \underline{\mathbf{H}}_2^H \mathbf{y} \quad (5.26) \\ &= \underline{\mathbf{h}}_2^H \underline{\mathbf{H}}_1^H \mathbf{y}, \quad (5.27) \end{aligned}$$

where now $\mathbf{h}$ is a low-rank beamformer. Then, the variance of Z is

$$\begin{aligned} \phi_Z &= \mathbf{h}^H \boldsymbol{\Phi}_\mathbf{y} \mathbf{h} \\ &= \underline{\mathbf{h}}_1^H \underline{\mathbf{H}}_2^H \boldsymbol{\Phi}_\mathbf{y} \underline{\mathbf{H}}_2 \underline{\mathbf{h}}_1 \quad (5.28) \\ &= \underline{\mathbf{h}}_2^H \underline{\mathbf{H}}_1^H \boldsymbol{\Phi}_\mathbf{y} \underline{\mathbf{H}}_1 \underline{\mathbf{h}}_2. \quad (5.29) \end{aligned}$$

From the above, it is clear that the distortionless constraint is

$$\begin{aligned} 1 &= \mathbf{h}^H \mathbf{d}_{\theta_s,\phi_s} \\ &= \underline{\mathbf{h}}_1^H \underline{\mathbf{H}}_2^H \mathbf{d}_{\theta_s,\phi_s} \quad (5.30) \\ &= \underline{\mathbf{h}}_2^H \underline{\mathbf{H}}_1^H \mathbf{d}_{\theta_s,\phi_s}. \quad (5.31) \end{aligned}$$

5.3 Performance Measures

In this section, we briefly discuss the most important performance measures in the context of low-rank beamforming.

The first measure is the beampattern, which is given by

$$\begin{aligned} \mathcal{B}_{\theta,\phi}(\mathbf{h}) &= \mathbf{d}_{\theta,\phi}^H \mathbf{h} \\ &= \mathbf{d}_{\theta,\phi}^H \underline{\mathbf{H}}_2 \underline{\mathbf{h}}_1 \quad (5.32) \\ &= \mathbf{d}_{\theta,\phi}^H \underline{\mathbf{H}}_1 \underline{\mathbf{h}}_2. \quad (5.33) \end{aligned}$$

Now, all next performance measures are defined twice by assuming that one of the two beamforming filters $\underline{\mathbf{h}}_1$ and $\underline{\mathbf{h}}_2$ is fixed. This will facilitate the derivation of optimal beamformers in the next section.

From ϕ_Z and assuming that $\underline{\mathbf{h}}_2$ is fixed, we can express the output SNR as

$$\begin{aligned} \text{oSNR}\left(\underline{\mathbf{h}}_1 | \underline{\mathbf{h}}_2\right) &= \phi_X \frac{\left|\underline{\mathbf{h}}_1^H \underline{\mathbf{H}}_2^H \mathbf{d}_{\theta_s,\phi_s}\right|^2}{\underline{\mathbf{h}}_1^H \underline{\mathbf{H}}_2^H \mathbf{\Phi}_{\mathbf{v}} \underline{\mathbf{H}}_2 \underline{\mathbf{h}}_1} \\ &= \frac{\phi_X}{\phi_V} \times \frac{\left|\underline{\mathbf{h}}_1^H \underline{\mathbf{H}}_2^H \mathbf{d}_{\theta_s,\phi_s}\right|^2}{\underline{\mathbf{h}}_1^H \underline{\mathbf{H}}_2^H \mathbf{\Gamma}_{\mathbf{v}} \underline{\mathbf{H}}_2 \underline{\mathbf{h}}_1}. \end{aligned} \tag{5.34}$$

In the same way, by assuming that $\underline{\mathbf{h}}_1$ is fixed, we can define the output SNR as

$$\begin{aligned} \text{oSNR}\left(\underline{\mathbf{h}}_2 | \underline{\mathbf{h}}_1\right) &= \phi_X \frac{\left|\underline{\mathbf{h}}_2^H \underline{\mathbf{H}}_1^H \mathbf{d}_{\theta_s,\phi_s}\right|^2}{\underline{\mathbf{h}}_2^H \underline{\mathbf{H}}_1^H \mathbf{\Phi}_{\mathbf{v}} \underline{\mathbf{H}}_1 \underline{\mathbf{h}}_2} \\ &= \frac{\phi_X}{\phi_V} \times \frac{\left|\underline{\mathbf{h}}_2^H \underline{\mathbf{H}}_1^H \mathbf{d}_{\theta_s,\phi_s}\right|^2}{\underline{\mathbf{h}}_2^H \underline{\mathbf{H}}_1^H \mathbf{\Gamma}_{\mathbf{v}} \underline{\mathbf{H}}_1 \underline{\mathbf{h}}_2}. \end{aligned} \tag{5.35}$$

Then, the SNR gains when $\underline{\mathbf{h}}_2$ is fixed and when $\underline{\mathbf{h}}_1$ is fixed are, respectively,

$$\begin{aligned} \mathcal{G}\left(\underline{\mathbf{h}}_1 | \underline{\mathbf{h}}_2\right) &= \frac{\text{oSNR}\left(\underline{\mathbf{h}}_1 | \underline{\mathbf{h}}_2\right)}{\text{iSNR}} \\ &= \frac{\left|\underline{\mathbf{h}}_1^H \underline{\mathbf{H}}_2^H \mathbf{d}_{\theta_s,\phi_s}\right|^2}{\underline{\mathbf{h}}_1^H \underline{\mathbf{H}}_2^H \mathbf{\Gamma}_{\mathbf{v}} \underline{\mathbf{H}}_2 \underline{\mathbf{h}}_1} \end{aligned} \tag{5.36}$$

and

$$\begin{aligned} \mathcal{G}\left(\underline{\mathbf{h}}_2 | \underline{\mathbf{h}}_1\right) &= \frac{\text{oSNR}\left(\underline{\mathbf{h}}_2 | \underline{\mathbf{h}}_1\right)}{\text{iSNR}} \\ &= \frac{\left|\underline{\mathbf{h}}_2^H \underline{\mathbf{H}}_1^H \mathbf{d}_{\theta_s,\phi_s}\right|^2}{\underline{\mathbf{h}}_2^H \underline{\mathbf{H}}_1^H \mathbf{\Gamma}_{\mathbf{v}} \underline{\mathbf{H}}_1 \underline{\mathbf{h}}_2}. \end{aligned} \tag{5.37}$$

We deduce that the WNGs when $\underline{\mathbf{h}}_2$ is fixed and when $\underline{\mathbf{h}}_1$ is fixed are, respectively,

$$\mathcal{W}\left(\underline{\mathbf{h}}_1 | \underline{\mathbf{h}}_2\right) = \frac{\left|\underline{\mathbf{h}}_1^H \underline{\mathbf{H}}_2^H \mathbf{d}_{\theta_s,\phi_s}\right|^2}{\underline{\mathbf{h}}_1^H \underline{\mathbf{H}}_2^H \underline{\mathbf{H}}_2 \underline{\mathbf{h}}_1} \tag{5.38}$$

and

$$\mathcal{W}\left(\underline{\mathbf{h}}_2|\underline{\mathbf{h}}_1\right) = \frac{\left|\underline{\mathbf{h}}_2^H \underline{\mathbf{H}}_1^H \mathbf{d}_{\theta_s,\phi_s}\right|^2}{\underline{\mathbf{h}}_2^H \underline{\mathbf{H}}_1^H \underline{\mathbf{H}}_1 \underline{\mathbf{h}}_2}. \tag{5.39}$$

We also find that the DFs when $\underline{\mathbf{h}}_2$ is fixed and when $\underline{\mathbf{h}}_1$ is fixed are, respectively,

$$\mathcal{D}\left(\underline{\mathbf{h}}_1|\underline{\mathbf{h}}_2\right) = \frac{\left|\underline{\mathbf{h}}_1^H \underline{\mathbf{H}}_2^H \mathbf{d}_{\theta_s,\phi_s}\right|^2}{\underline{\mathbf{h}}_1^H \underline{\mathbf{H}}_2^H \mathbf{\Gamma}_{\mathrm{d}} \underline{\mathbf{H}}_2 \underline{\mathbf{h}}_1} \tag{5.40}$$

and

$$\mathcal{D}\left(\underline{\mathbf{h}}_2|\underline{\mathbf{h}}_1\right) = \frac{\left|\underline{\mathbf{h}}_2^H \underline{\mathbf{H}}_1^H \mathbf{d}_{\theta_s,\phi_s}\right|^2}{\underline{\mathbf{h}}_2^H \underline{\mathbf{H}}_1^H \mathbf{\Gamma}_{\mathrm{d}} \underline{\mathbf{H}}_1 \underline{\mathbf{h}}_2}. \tag{5.41}$$

Finally, the last measure of interest is the MSE criterion. When $\underline{\mathbf{h}}_2$ is fixed, the MSE criterion is

$$\begin{aligned}\mathcal{J}\left(\underline{\mathbf{h}}_1|\underline{\mathbf{h}}_2\right) &= E\left(\left|X - \underline{\mathbf{h}}_1^H \underline{\mathbf{H}}_2^H \mathbf{y}\right|^2\right) \\ &= \phi_X + \underline{\mathbf{h}}_1^H \underline{\mathbf{H}}_2^H \mathbf{\Phi}_{\mathbf{y}} \underline{\mathbf{H}}_2 \underline{\mathbf{h}}_1 - \phi_X \underline{\mathbf{h}}_1^H \underline{\mathbf{H}}_2^H \mathbf{d}_{\theta_s,\phi_s} - \phi_X \mathbf{d}_{\theta_s,\phi_s}^H \underline{\mathbf{H}}_2 \underline{\mathbf{h}}_1\end{aligned} \tag{5.42}$$

and when $\underline{\mathbf{h}}_1$ is fixed, the MSE criterion is

$$\begin{aligned}\mathcal{J}\left(\underline{\mathbf{h}}_2|\underline{\mathbf{h}}_1\right) &= E\left(\left|X - \underline{\mathbf{h}}_2^H \underline{\mathbf{H}}_1^H \mathbf{y}\right|^2\right) \\ &= \phi_X + \underline{\mathbf{h}}_2^H \underline{\mathbf{H}}_1^H \mathbf{\Phi}_{\mathbf{y}} \underline{\mathbf{H}}_1 \underline{\mathbf{h}}_2 - \phi_X \underline{\mathbf{h}}_2^H \underline{\mathbf{H}}_1^H \mathbf{d}_{\theta_s,\phi_s} - \phi_X \mathbf{d}_{\theta_s,\phi_s}^H \underline{\mathbf{H}}_1 \underline{\mathbf{h}}_2.\end{aligned} \tag{5.43}$$

5.4 Optimal Beamformers

In this section, we give some examples of low-rank fixed and adaptive beamformers.

The maximum WNG (MWNG) filters are obtained from the maximization of the WNGs. We get

$$\underline{\mathbf{h}}_{1,\mathrm{MWNG}} = \frac{\left(\underline{\mathbf{H}}_2^H \underline{\mathbf{H}}_2\right)^{-1} \underline{\mathbf{H}}_2^H \mathbf{d}_{\theta_s,\phi_s}}{\mathbf{d}_{\theta_s,\phi_s}^H \underline{\mathbf{H}}_2 \left(\underline{\mathbf{H}}_2^H \underline{\mathbf{H}}_2\right)^{-1} \underline{\mathbf{H}}_2^H \mathbf{d}_{\theta_s,\phi_s}}, \tag{5.44}$$

$$\underline{\mathbf{h}}_{2,\mathrm{MWNG}} = \frac{\left(\underline{\mathbf{H}}_1^H \underline{\mathbf{H}}_1\right)^{-1} \underline{\mathbf{H}}_1^H \mathbf{d}_{\theta_s,\phi_s}}{\mathbf{d}_{\theta_s,\phi_s}^H \underline{\mathbf{H}}_1 \left(\underline{\mathbf{H}}_1^H \underline{\mathbf{H}}_1\right)^{-1} \underline{\mathbf{H}}_1^H \mathbf{d}_{\theta_s,\phi_s}}. \tag{5.45}$$

Using the alternating least-squares (ALS) algorithm, i.e., by alternatively iterating between $\underline{\mathbf{h}}_{1,\mathrm{MWNG}}$ and $\underline{\mathbf{h}}_{2,\mathrm{MWNG}}$ as in [1, 2, 7], these two optimal filters will converge to the low-rank MWNG beamformer just after a couple of iterations:

$$\mathbf{h}_{\mathrm{MWNG}} = \sum_{p=1}^{P} \mathbf{h}_{2,p,\mathrm{MWNG}}^{*} \otimes \mathbf{h}_{1,p,\mathrm{MWNG}}, \tag{5.46}$$

where

$$\underline{\mathbf{h}}_{1,\mathrm{MWNG}} = \begin{bmatrix} \mathbf{h}_{1,1,\mathrm{MWNG}}^{T} & \mathbf{h}_{1,2,\mathrm{MWNG}}^{T} & \cdots & \mathbf{h}_{1,P,\mathrm{MWNG}}^{T} \end{bmatrix}^{T},$$

$$\underline{\mathbf{h}}_{2,\mathrm{MWNG}} = \begin{bmatrix} \mathbf{h}_{2,1,\mathrm{MWNG}}^{H} & \mathbf{h}_{2,2,\mathrm{MWNG}}^{H} & \cdots & \mathbf{h}_{2,P,\mathrm{MWNG}}^{H} \end{bmatrix}^{T}.$$

Analogously, the maximum DF (MDF) filters are obtained from the maximization of the DFs. We get

$$\underline{\mathbf{h}}_{1,\mathrm{MDF}} = \frac{\left(\underline{\mathbf{H}}_{2}^{H} \mathbf{\Gamma}_{\mathrm{d}} \underline{\mathbf{H}}_{2}\right)^{-1} \underline{\mathbf{H}}_{2}^{H} \mathbf{d}_{\theta_{\mathrm{s}},\phi_{\mathrm{s}}}}{\mathbf{d}_{\theta_{\mathrm{s}},\phi_{\mathrm{s}}}^{H} \underline{\mathbf{H}}_{2} \left(\underline{\mathbf{H}}_{2}^{H} \mathbf{\Gamma}_{\mathrm{d}} \underline{\mathbf{H}}_{2}\right)^{-1} \underline{\mathbf{H}}_{2}^{H} \mathbf{d}_{\theta_{\mathrm{s}},\phi_{\mathrm{s}}}}, \tag{5.47}$$

$$\underline{\mathbf{h}}_{2,\mathrm{MDF}} = \frac{\left(\underline{\mathbf{H}}_{1}^{H} \mathbf{\Gamma}_{\mathrm{d}} \underline{\mathbf{H}}_{1}\right)^{-1} \underline{\mathbf{H}}_{1}^{H} \mathbf{d}_{\theta_{\mathrm{s}},\phi_{\mathrm{s}}}}{\mathbf{d}_{\theta_{\mathrm{s}},\phi_{\mathrm{s}}}^{H} \underline{\mathbf{H}}_{1} \left(\underline{\mathbf{H}}_{1}^{H} \mathbf{\Gamma}_{\mathrm{d}} \underline{\mathbf{H}}_{1}\right)^{-1} \underline{\mathbf{H}}_{1}^{H} \mathbf{d}_{\theta_{\mathrm{s}},\phi_{\mathrm{s}}}}. \tag{5.48}$$

Thanks to the ALS algorithm, we get the low-rank MDF beamformer:

$$\mathbf{h}_{\mathrm{MDF}} = \sum_{p=1}^{P} \mathbf{h}_{2,p,\mathrm{MDF}}^{*} \otimes \mathbf{h}_{1,p,\mathrm{MDF}}, \tag{5.49}$$

where

$$\underline{\mathbf{h}}_{1,\mathrm{MDF}} = \begin{bmatrix} \mathbf{h}_{1,1,\mathrm{MDF}}^{T} & \mathbf{h}_{1,2,\mathrm{MDF}}^{T} & \cdots & \mathbf{h}_{1,P,\mathrm{MDF}}^{T} \end{bmatrix}^{T},$$

$$\underline{\mathbf{h}}_{2,\mathrm{MDF}} = \begin{bmatrix} \mathbf{h}_{2,1,\mathrm{MDF}}^{H} & \mathbf{h}_{2,2,\mathrm{MDF}}^{H} & \cdots & \mathbf{h}_{2,P,\mathrm{MDF}}^{H} \end{bmatrix}^{T}.$$

The two above low-rank beamformers are fixed. Our first proposed low-rank adaptive beamformer is the MVDR. Indeed, the two MVDR filters are obtained from the maximization of the output SNRs. We get

$$\underline{\mathbf{h}}_{1,\mathrm{MVDR}} = \frac{\left(\underline{\mathbf{H}}_{2}^{H} \mathbf{\Phi}_{\mathbf{v}} \underline{\mathbf{H}}_{2}\right)^{-1} \underline{\mathbf{H}}_{2}^{H} \mathbf{d}_{\theta_{\mathrm{s}},\phi_{\mathrm{s}}}}{\mathbf{d}_{\theta_{\mathrm{s}},\phi_{\mathrm{s}}}^{H} \underline{\mathbf{H}}_{2} \left(\underline{\mathbf{H}}_{2}^{H} \mathbf{\Phi}_{\mathbf{v}} \underline{\mathbf{H}}_{2}\right)^{-1} \underline{\mathbf{H}}_{2}^{H} \mathbf{d}_{\theta_{\mathrm{s}},\phi_{\mathrm{s}}}}, \tag{5.50}$$

$$\underline{\mathbf{h}}_{2,\text{MVDR}} = \frac{\left(\underline{\mathbf{H}}_1^H \mathbf{\Phi}_{\text{v}} \underline{\mathbf{H}}_1\right)^{-1} \underline{\mathbf{H}}_1^H \mathbf{d}_{\theta_s,\phi_s}}{\mathbf{d}_{\theta_s,\phi_s}^H \underline{\mathbf{H}}_1 \left(\underline{\mathbf{H}}_1^H \mathbf{\Phi}_{\text{v}} \underline{\mathbf{H}}_1\right)^{-1} \underline{\mathbf{H}}_1^H \mathbf{d}_{\theta_s,\phi_s}}. \tag{5.51}$$

From the ALS algorithm, we find that the low-rank MVDR beamformer is

$$\mathbf{h}_{\text{MVDR}} = \sum_{p=1}^{P} \mathbf{h}_{2,p,\text{MVDR}}^{*} \otimes \mathbf{h}_{1,p,\text{MVDR}}, \tag{5.52}$$

where

$$\underline{\mathbf{h}}_{1,\text{MVDR}} = \begin{bmatrix} \mathbf{h}_{1,1,\text{MVDR}}^T & \mathbf{h}_{1,2,\text{MVDR}}^T & \cdots & \mathbf{h}_{1,P,\text{MVDR}}^T \end{bmatrix}^T,$$

$$\underline{\mathbf{h}}_{2,\text{MVDR}} = \begin{bmatrix} \mathbf{h}_{2,1,\text{MVDR}}^H & \mathbf{h}_{2,2,\text{MVDR}}^H & \cdots & \mathbf{h}_{2,P,\text{MVDR}}^H \end{bmatrix}^T.$$

From the minimization of the MSE criteria, we find the two Wiener filters:

$$\begin{aligned} \underline{\mathbf{h}}_{1,\text{W}} &= \phi_X \left(\underline{\mathbf{H}}_2^H \mathbf{\Phi}_{\text{y}} \underline{\mathbf{H}}_2\right)^{-1} \underline{\mathbf{H}}_2^H \mathbf{d}_{\theta_s,\phi_s} \\ &= \frac{\phi_X \left(\underline{\mathbf{H}}_2^H \mathbf{\Phi}_{\text{v}} \underline{\mathbf{H}}_2\right)^{-1} \underline{\mathbf{H}}_2^H \mathbf{d}_{\theta_s,\phi_s}}{1 + \phi_X \mathbf{d}_{\theta_s,\phi_s}^H \underline{\mathbf{H}}_2 \left(\underline{\mathbf{H}}_2^H \mathbf{\Phi}_{\text{v}} \underline{\mathbf{H}}_2\right)^{-1} \underline{\mathbf{H}}_2^H \mathbf{d}_{\theta_s,\phi_s}}, \end{aligned} \tag{5.53}$$

$$\begin{aligned} \underline{\mathbf{h}}_{2,\text{W}} &= \phi_X \left(\underline{\mathbf{H}}_1^H \mathbf{\Phi}_{\text{y}} \underline{\mathbf{H}}_1\right)^{-1} \underline{\mathbf{H}}_1^H \mathbf{d}_{\theta_s,\phi_s} \\ &= \frac{\phi_X \left(\underline{\mathbf{H}}_1^H \mathbf{\Phi}_{\text{v}} \underline{\mathbf{H}}_1\right)^{-1} \underline{\mathbf{H}}_1^H \mathbf{d}_{\theta_s,\phi_s}}{1 + \phi_X \mathbf{d}_{\theta_s,\phi_s}^H \underline{\mathbf{H}}_1 \left(\underline{\mathbf{H}}_1^H \mathbf{\Phi}_{\text{v}} \underline{\mathbf{H}}_1\right)^{-1} \underline{\mathbf{H}}_1^H \mathbf{d}_{\theta_s,\phi_s}}. \end{aligned} \tag{5.54}$$

As a result, the low-rank Wiener beamformer is

$$\mathbf{h}_{\text{W}} = \sum_{p=1}^{P} \mathbf{h}_{2,p,\text{W}}^{*} \otimes \mathbf{h}_{1,p,\text{W}}, \tag{5.55}$$

where

$$\underline{\mathbf{h}}_{1,\text{W}} = \begin{bmatrix} \mathbf{h}_{1,1,\text{W}}^T & \mathbf{h}_{1,2,\text{W}}^T & \cdots & \mathbf{h}_{1,P,\text{W}}^T \end{bmatrix}^T,$$

$$\underline{\mathbf{h}}_{2,\text{W}} = \begin{bmatrix} \mathbf{h}_{2,1,\text{W}}^H & \mathbf{h}_{2,2,\text{W}}^H & \cdots & \mathbf{h}_{2,P,\text{W}}^H \end{bmatrix}^T.$$

Very often in practice, it is desirable to be able to compromise between noise reduction and speech distortion. For this purpose, we can derive the tradeoff filters:

$$\underline{\mathbf{h}}_{1,\mathrm{T},\mu} = \frac{\phi_X \left(\underline{\mathbf{H}}_2^H \mathbf{\Phi}_{\mathrm{v}} \underline{\mathbf{H}}_2\right)^{-1} \underline{\mathbf{H}}_2^H \mathbf{d}_{\theta_{\mathrm{s}},\phi_{\mathrm{s}}}}{\mu + \phi_X \mathbf{d}_{\theta_{\mathrm{s}},\phi_{\mathrm{s}}}^H \underline{\mathbf{H}}_2 \left(\underline{\mathbf{H}}_2^H \mathbf{\Phi}_{\mathrm{v}} \underline{\mathbf{H}}_2\right)^{-1} \underline{\mathbf{H}}_2^H \mathbf{d}_{\theta_{\mathrm{s}},\phi_{\mathrm{s}}}}, \tag{5.56}$$

$$\underline{\mathbf{h}}_{2,\mathrm{T},\mu} = \frac{\phi_X \left(\underline{\mathbf{H}}_1^H \mathbf{\Phi}_{\mathrm{v}} \underline{\mathbf{H}}_1\right)^{-1} \underline{\mathbf{H}}_1^H \mathbf{d}_{\theta_{\mathrm{s}},\phi_{\mathrm{s}}}}{\mu + \phi_X \mathbf{d}_{\theta_{\mathrm{s}},\phi_{\mathrm{s}}}^H \underline{\mathbf{H}}_1 \left(\underline{\mathbf{H}}_1^H \mathbf{\Phi}_{\mathrm{v}} \underline{\mathbf{H}}_1\right)^{-1} \underline{\mathbf{H}}_1^H \mathbf{d}_{\theta_{\mathrm{s}},\phi_{\mathrm{s}}}}, \tag{5.57}$$

where $\mu \geq 0$ is a tuning parameter, which can be frequency dependent. From the ALS algorithm, we find that the low-rank tradeoff beamformer is

$$\mathbf{h}_{\mathrm{T},\mu} = \sum_{p=1}^{P} \mathbf{h}_{2,p,\mathrm{T},\mu}^* \otimes \mathbf{h}_{1,p,\mathrm{T},\mu}, \tag{5.58}$$

where

$$\underline{\mathbf{h}}_{1,\mathrm{T},\mu} = \left[\mathbf{h}_{1,1,\mathrm{T},\mu}^T \;\; \mathbf{h}_{1,2,\mathrm{T},\mu}^T \;\; \cdots \;\; \mathbf{h}_{1,P,\mathrm{T},\mu}^T \right]^T,$$

$$\underline{\mathbf{h}}_{2,\mathrm{T},\mu} = \left[\mathbf{h}_{2,1,\mathrm{T},\mu}^H \;\; \mathbf{h}_{2,2,\mathrm{T},\mu}^H \;\; \cdots \;\; \mathbf{h}_{2,P,\mathrm{T},\mu}^H \right]^T.$$

One can see that for

- $\mu = 1$, $\mathbf{h}_{\mathrm{T},1} = \mathbf{h}_{\mathrm{W}}$, which is the low-rank Wiener beamformer;
- $\mu = 0$, $\mathbf{h}_{\mathrm{T},0} = \mathbf{h}_{\mathrm{MVDR}}$, which is the low-rank MVDR beamformer;
- $\mu > 1$, results in a beamformer with low residual noise (from a broadband perspective) at the expense of high desired signal distortion (as compared to Wiener); and
- $\mu < 1$, results in a beamformer with high residual noise and low desired signal distortion (as compared to Wiener).

Now, assume that we have one interference impinging on the array from the direction $(\theta_{\mathrm{i}}, \phi_{\mathrm{i}}) \neq (\theta_{\mathrm{s}}, \phi_{\mathrm{s}})$ and would like to place null in that direction with a low-rank beamformer $\mathbf{h}$, and, meanwhile, recover the desired source coming from the direction $(\theta_{\mathrm{s}}, \phi_{\mathrm{s}})$. The combination of these two constraints leads to the constraint equations:

$$\mathbf{C}_{\mathrm{s,i}}^H \underline{\mathbf{H}}_2 \underline{\mathbf{h}}_1 = \begin{bmatrix} 1 \\ 0 \end{bmatrix}, \tag{5.59}$$

$$\mathbf{C}_{\mathrm{s,i}}^H \underline{\mathbf{H}}_1 \underline{\mathbf{h}}_2 = \begin{bmatrix} 1 \\ 0 \end{bmatrix}, \tag{5.60}$$

where

$$\mathbf{C}_{\mathrm{s,i}} = \left[\mathbf{d}_{\theta_{\mathrm{s}},\phi_{\mathrm{s}}} \;\; \mathbf{d}_{\theta_{\mathrm{i}},\phi_{\mathrm{i}}} \right] \tag{5.61}$$

is the constraint matrix of size $M \times 2$, whose *two* columns are linearly independent, and $\mathbf{d}_{\theta_\mathrm{i},\phi_\mathrm{i}}$ is the steering vector in the direction $(\theta_\mathrm{i}, \phi_\mathrm{i})$. From the minimization of the residual noise subject to (5.59) and (5.60), we find the two LCMV filters:

$$\underline{\mathbf{h}}_{1,\mathrm{LCMV}} = \left(\underline{\mathbf{H}}_2^H \mathbf{\Phi}_\mathrm{v} \underline{\mathbf{H}}_2\right)^{-1} \underline{\mathbf{H}}_2^H \mathbf{C}_{\mathrm{s,i}} \times \left[\mathbf{C}_{\mathrm{s,i}}^H \underline{\mathbf{H}}_2 \left(\underline{\mathbf{H}}_2^H \mathbf{\Phi}_\mathrm{v} \underline{\mathbf{H}}_2\right)^{-1} \underline{\mathbf{H}}_2^H \mathbf{C}_{\mathrm{s,i}} \right]^{-1} \begin{bmatrix} 1 \\ 0 \end{bmatrix}, \tag{5.62}$$

$$\underline{\mathbf{h}}_{2,\mathrm{LCMV}} = \left(\underline{\mathbf{H}}_1^H \mathbf{\Phi}_\mathrm{v} \underline{\mathbf{H}}_1\right)^{-1} \underline{\mathbf{H}}_1^H \mathbf{C}_{\mathrm{s,i}} \times \left[\mathbf{C}_{\mathrm{s,i}}^H \underline{\mathbf{H}}_1 \left(\underline{\mathbf{H}}_1^H \mathbf{\Phi}_\mathrm{v} \underline{\mathbf{H}}_1\right)^{-1} \underline{\mathbf{H}}_1^H \mathbf{C}_{\mathrm{s,i}} \right]^{-1} \begin{bmatrix} 1 \\ 0 \end{bmatrix}. \tag{5.63}$$

Then, thanks to the ALS algorithm, we get the low-rank LCMV beamformer:

$$\mathbf{h}_{\mathrm{LCMV}} = \sum_{p=1}^{P} \mathbf{h}_{2,p,\mathrm{LCMV}}^{*} \otimes \mathbf{h}_{1,p,\mathrm{LCMV}}, \tag{5.64}$$

where

$$\underline{\mathbf{h}}_{1,\mathrm{LCMV}} = \left[\mathbf{h}_{1,1,\mathrm{LCMV}}^T \ \mathbf{h}_{1,2,\mathrm{LCMV}}^T \ \cdots \ \mathbf{h}_{1,P,\mathrm{LCMV}}^T \right]^T,$$

$$\underline{\mathbf{h}}_{2,\mathrm{LCMV}} = \left[\mathbf{h}_{2,1,\mathrm{LCMV}}^H \ \mathbf{h}_{2,2,\mathrm{LCMV}}^H \ \cdots \ \mathbf{h}_{2,P,\mathrm{LCMV}}^H \right]^T.$$

5.5 Illustrative Examples

After a discussion on the fundamental concepts of low-rank beamforming and the derivation of a family of fixed and adaptive low-rank beamformers, we are now ready to study some examples. For convenience, we still consider a 3-D cube array consisting of $M = M_0^3$ microphones, as shown in Fig. 3.1, with $\delta = 2.0$ cm and $(\theta_\mathrm{s}, \phi_\mathrm{s}) = (80^\circ, 0^\circ)$.

As explained in Sect. 5.1, the low-rank beamforming process decomposes a filter of length M into two series of short beamforming filters of lengths M_1 and M_2, respectively, where $M = M_1 M_2$ (and $M_1 \geq M_2$). Clearly, given an integer M, there may be different decompositions that satisfy $M = M_1 M_2$. In our examples, the values of M_1 and M_2 are chosen to be as close as possible. In the rest, the following decompositions are used: when $M_0 = 2$, i.e., $M = 8$, we choose $M_1 = 4$ and $M_2 = 2$; when $M_0 = 3$, i.e., $M = 27$, we choose $M_1 = 9$ and $M_2 = 3$; when $M_0 = 4$, i.e., $M = 64$, we choose $M_1 = 8$ and $M_2 = 8$; and when $M_0 = 5$, i.e., $M = 125$, we choose $M_1 = 25$ and $M_2 = 5$. The two optimal filters are optimized

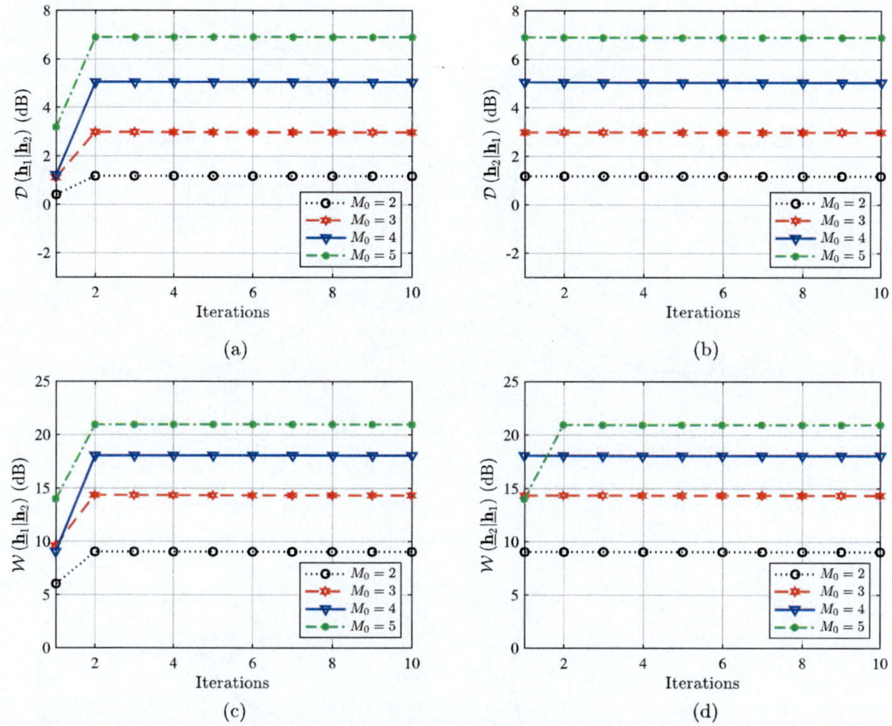

Fig. 5.1 DF and WNG of the low-rank MWNG beamformer, $\mathbf{h}_{\mathrm{MWNG}}$, as a function of iterations, for different numbers of sensors, $M = M_0^3$: (**a**) DF when $\underline{\mathbf{h}}_2$ is fixed, (**b**) DF when $\underline{\mathbf{h}}_1$ is fixed, (**c**) WNG when $\underline{\mathbf{h}}_2$ is fixed, and (**d**) WNG when $\underline{\mathbf{h}}_1$ is fixed. Conditions: $\delta = 2.0$ cm, $f = 2000$ Hz, $P = 1$, and $(\theta_s, \phi_s) = (80°, 0°)$

by using the ALS algorithm, where the filter $\underline{\mathbf{h}}_2$ is initialized as $\mathbf{h}_{2,p} = \eta \mathbf{i}_{M_2,p}$, $p = 1, 2, \ldots, P$, with η being an arbitrary constant (randomly generated), and $\mathbf{i}_{M_2,p}$ being a vector of length M_2 whose pth element is $1/M_2$ and all other elements are 0. In the implementation of a matrix inversion, a small regularization parameter, 10^{-6}, is added to the diagonal elements of the matrix.

To demonstrate the performance of the low-rank MWNG beamformer, $\mathbf{h}_{\mathrm{MWNG}}$, given in (5.46), we choose $P = 1$. We first study the performance of this beamformer as a function of iterations. Figure 5.1 shows plots of the DF when $\underline{\mathbf{h}}_2$ is fixed, $\mathcal{D}\left(\underline{\mathbf{h}}_1|\underline{\mathbf{h}}_2\right)$; DF when $\underline{\mathbf{h}}_1$ is fixed, $\mathcal{D}\left(\underline{\mathbf{h}}_2|\underline{\mathbf{h}}_1\right)$; WNG when $\underline{\mathbf{h}}_2$ is fixed, $\mathcal{W}\left(\underline{\mathbf{h}}_1|\underline{\mathbf{h}}_2\right)$; and WNG when $\underline{\mathbf{h}}_1$ is fixed, $\mathcal{W}\left(\underline{\mathbf{h}}_2|\underline{\mathbf{h}}_1\right)$, of the low-rank MWNG filters as a function of iterations, for $f = 2000$ Hz and different numbers of sensors, $M = M_0^3$. It can be observed that the two optimal filters converge just after a couple of iterations. It can also be observed from the figure that, after convergence, $\mathcal{D}\left(\underline{\mathbf{h}}_2|\underline{\mathbf{h}}_1\right)$ and $\mathcal{D}\left(\underline{\mathbf{h}}_2|\underline{\mathbf{h}}_1\right)$ are the same and $\mathcal{W}\left(\underline{\mathbf{h}}_1|\underline{\mathbf{h}}_2\right)$ and $\mathcal{W}\left(\underline{\mathbf{h}}_2|\underline{\mathbf{h}}_1\right)$ are the same, which are equal to the DF and WNG of the global low-rank MWNG beamformer. So, in the following examples, we will only show the performance of the global

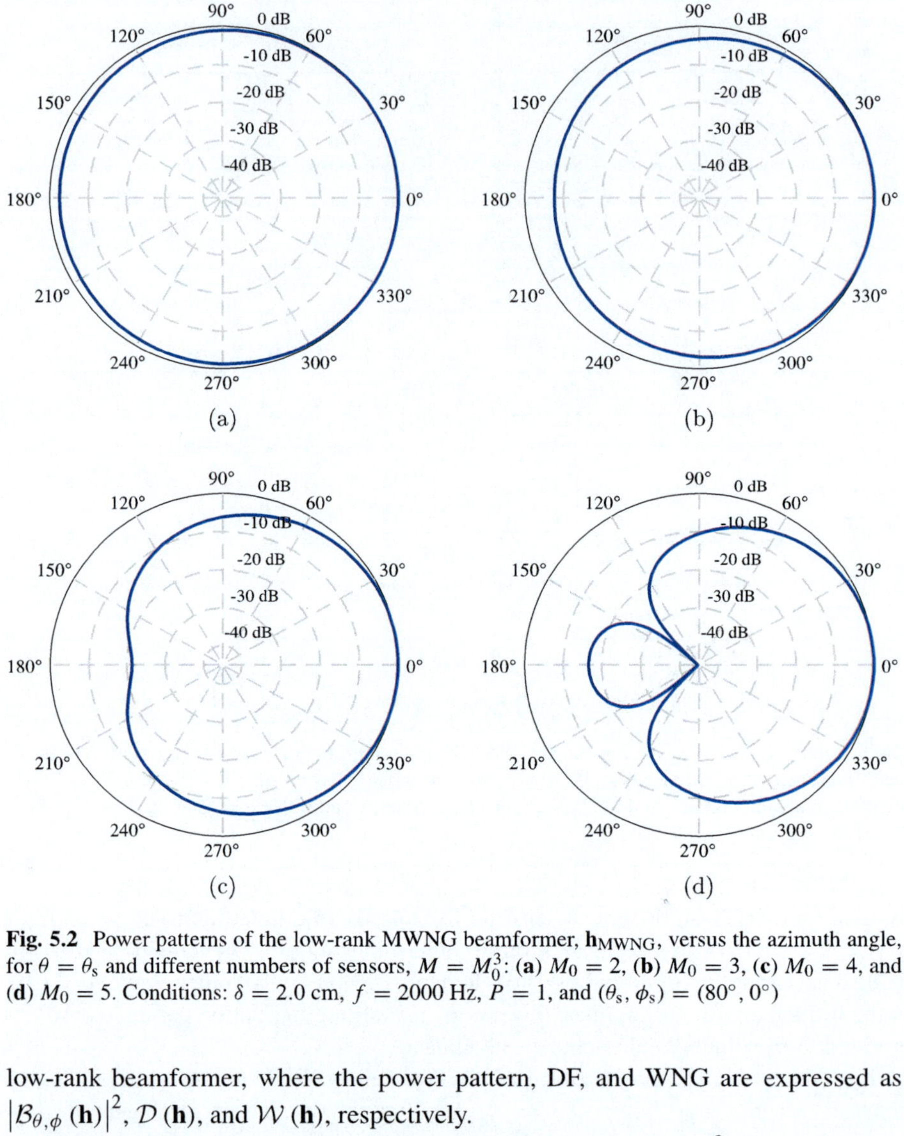

Fig. 5.2 Power patterns of the low-rank MWNG beamformer, $\mathbf{h}_{\mathrm{MWNG}}$, versus the azimuth angle, for $\theta = \theta_{\mathrm{s}}$ and different numbers of sensors, $M = M_0^3$: (**a**) $M_0 = 2$, (**b**) $M_0 = 3$, (**c**) $M_0 = 4$, and (**d**) $M_0 = 5$. Conditions: $\delta = 2.0$ cm, $f = 2000$ Hz, $P = 1$, and $(\theta_{\mathrm{s}}, \phi_{\mathrm{s}}) = (80°, 0°)$

low-rank beamformer, where the power pattern, DF, and WNG are expressed as $\left|\mathcal{B}_{\theta,\phi}(\mathbf{h})\right|^2$, $\mathcal{D}(\mathbf{h})$, and $\mathcal{W}(\mathbf{h})$, respectively.

Figures 5.2 and 5.3 show plots of the power patterns, $\left|\mathcal{B}_{\theta,\phi}(\mathbf{h})\right|^2$, of the low-rank MWNG beamformer versus the azimuth angle for $\theta = \theta_{\mathrm{s}}$, and versus the elevation angle for $\phi = \phi_{\mathrm{s}}$, respectively, for $f = 2000$ Hz and different numbers of sensors, $M = M_0^3$. As seen, the main beam of the power patterns points in the desired direction. It can also be observed that the power patterns have a wide main beam and are close to the omnidirectionality response when the number of microphones is not large. Figure 5.4 shows plots of the DF, $\mathcal{D}(\mathbf{h})$, and WNG, $\mathcal{W}(\mathbf{h})$, of the low-rank MWNG beamformer as a function of frequency for different numbers of sensors, $M = M_0^3$. As seen, the WNG of this beamformer is high, but the DF is

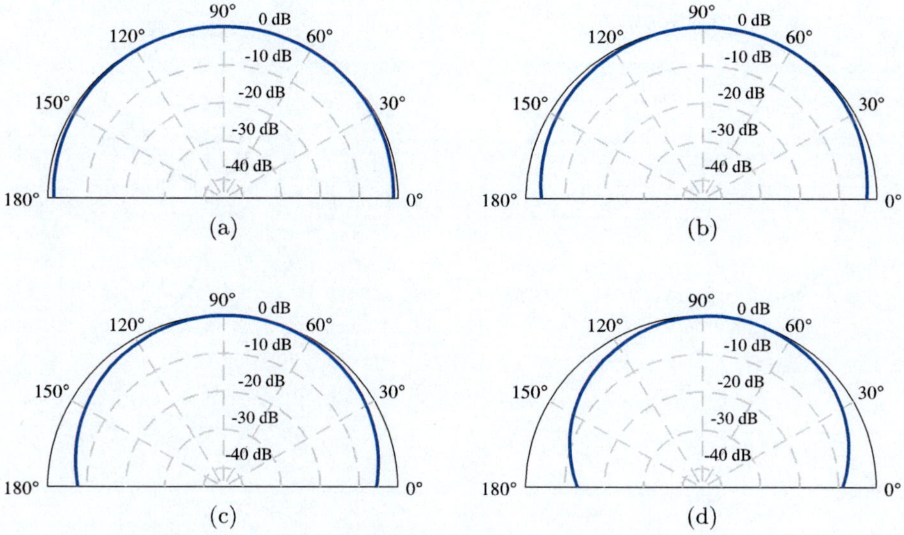

Fig. 5.3 Power patterns of the low-rank MWNG beamformer, $\mathbf{h}_{\mathrm{MWNG}}$, versus the elevation angle, for $\phi = \phi_s$ and different numbers of sensors, $M = M_0^3$: **(a)** $M_0 = 2$, **(b)** $M_0 = 3$, **(c)** $M_0 = 4$, and **(d)** $M_0 = 5$. Conditions: $\delta = 2.0$ cm, $f = 2000$ Hz, $P = 1$, and $(\theta_s, \phi_s) = (80°, 0°)$

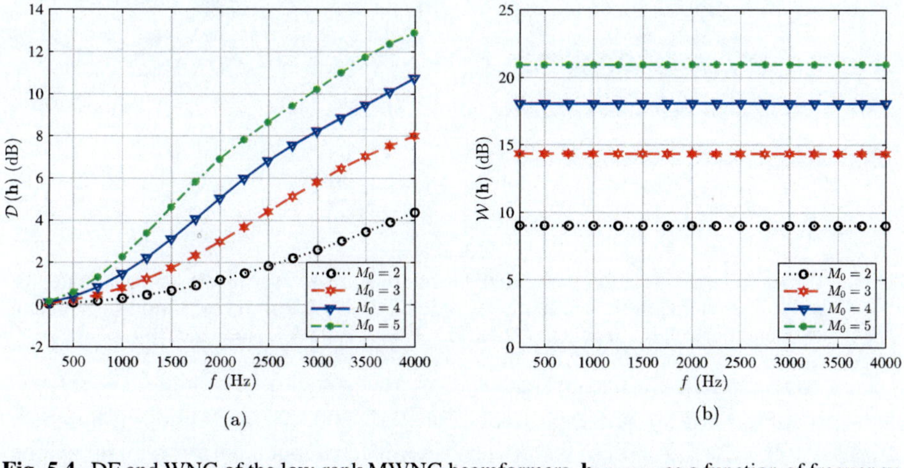

Fig. 5.4 DF and WNG of the low-rank MWNG beamformers, $\mathbf{h}_{\mathrm{MWNG}}$, as a function of frequency, for different numbers of sensors, $M = M_0^3$: **(a)** DF and **(b)** WNG. Conditions: $P = 1$, $\delta = 2.0$ cm, and $(\theta_s, \phi_s) = (80°, 0°)$

very low, particularly at low frequencies. It can also be observed from this figure that both the WNG and DF increase when M_0 increases.

To demonstrate the performance of the low-rank MDF beamformer, $\mathbf{h}_{\mathrm{MDF}}$, given in (5.49), we choose $P = 2$. Figures 5.5 and 5.6 show plots of the power patterns, $\left|\mathcal{B}_{\theta,\phi}(\mathbf{h})\right|^2$, of this beamformer versus the azimuth angle for $\theta = \theta_s$, and versus the

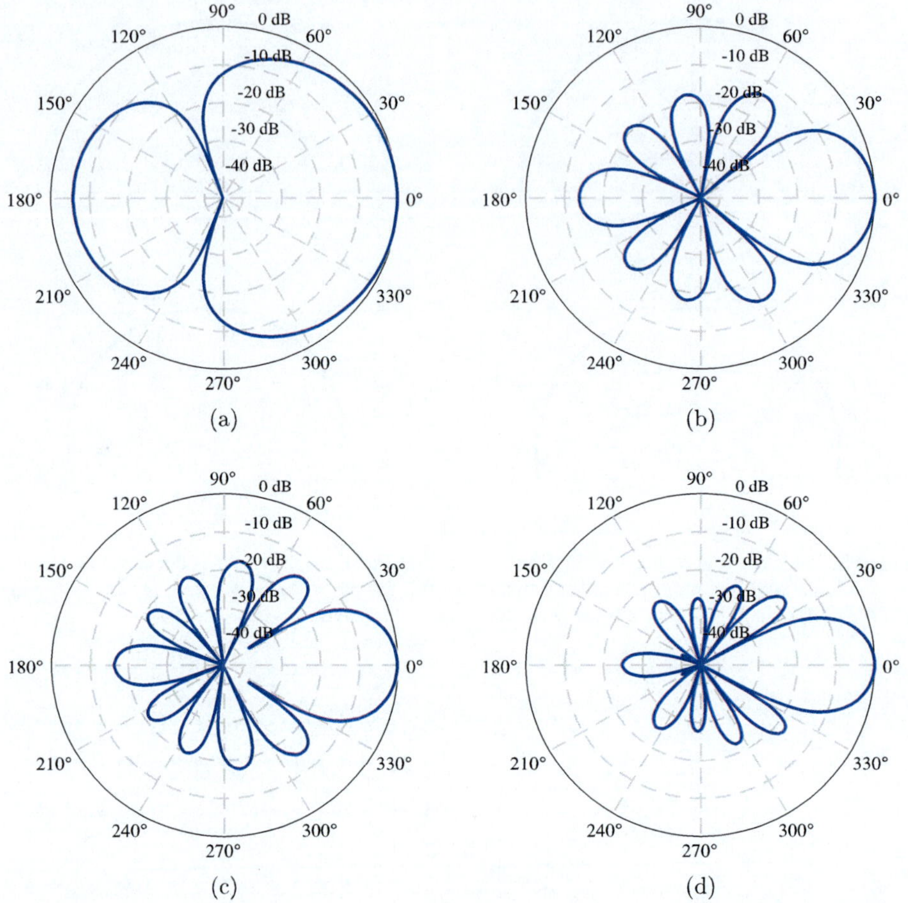

Fig. 5.5 Power patterns of the low-rank MDF beamformer, $\mathbf{h}_{\mathrm{MDF}}$, versus the azimuth angle, for $\theta = \theta_{\mathrm{s}}$ and different numbers of sensors, $M = M_0^3$: **(a)** $M_0 = 2$, **(b)** $M_0 = 3$, **(c)** $M_0 = 4$, and **(d)** $M_0 = 5$. Conditions: $\delta = 2.0$ cm, $f = 2000$ Hz, $P = 2$, and $(\theta_{\mathrm{s}}, \phi_{\mathrm{s}}) = (80^\circ, 0^\circ)$

elevation angle for $\phi = \phi_{\mathrm{s}}$, respectively, for $f = 2000$ Hz and different numbers of sensors, $M = M_0^3$. As seen, the power patterns have much narrower main beams as compared to the ones with the low-rank MWNG beamformer. It can also be observed that the power patterns achieve narrower main beams as the number of sensors increases. Figure 5.7 shows plots of the DF, $\mathcal{D}(\mathbf{h})$, and WNG, $\mathcal{W}(\mathbf{h})$, of the low-rank MDF beamformer as a function of frequency for different numbers of sensors, $M = M_0^3$. As seen, the DF of this beamformer is high and increases with the value of M, but the WNG decreases as the value of M increases. Figure 5.8 shows plots of the DF, $\mathcal{D}(\mathbf{h})$, and WNG, $\mathcal{W}(\mathbf{h})$, of this beamformer as a function of frequency for different values of P with $M_0 = 4$. It is observed that the DF increases, while the WNG decreases when P increases (note that the results in

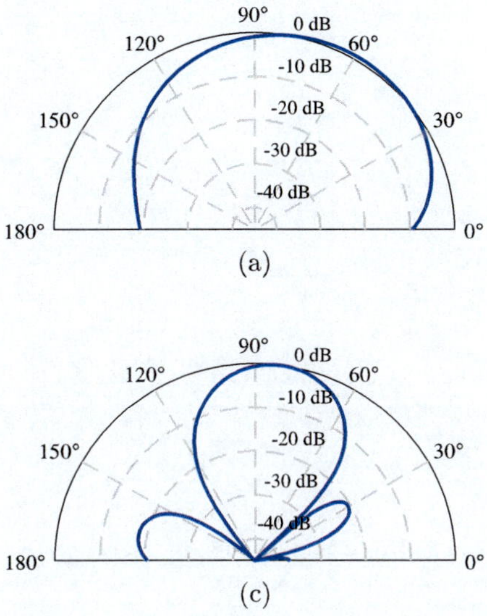

Fig. 5.6 Power patterns of the low-rank MDF beamformer, $\mathbf{h}_{\text{MDF}}$, versus the elevation angle for $\phi = \phi_s$ and different numbers of sensors, $M = M_0^3$: (**a**) $M_0 = 2$, (**b**) $M_0 = 3$, (**c**) $M_0 = 4$, and (**d**) $M_0 = 5$. Conditions: $\delta = 2.0$ cm, $f = 2000$ Hz, $P = 2$, and $(\theta_s, \phi_s) = (80°, 0°)$

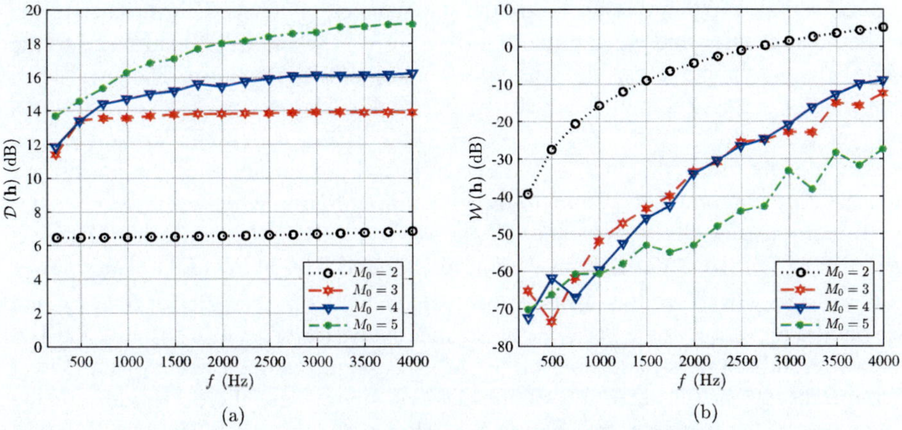

Fig. 5.7 DF and WNG of the low-rank MDF beamformer, $\mathbf{h}_{\text{MDF}}$, as a function of frequency, for different numbers of sensors, $M = M_0^3$: (**a**) DF and (**b**) WNG. Conditions: $\delta = 2.0$ cm, $P = 2$, and $(\theta_s, \phi_s) = (80°, 0°)$

Figs. 5.7 and 5.8 are based on the average of ten Monte Carlo experiments, i.e., randomly generating the constant η when initializing the filter $\underline{\mathbf{h}}_2$, to avoid small up and down fluctuations in the curves).

Now, let us study the performance of the low-rank adaptive beamformers. We consider an environment with point source noises and white noise, where the two

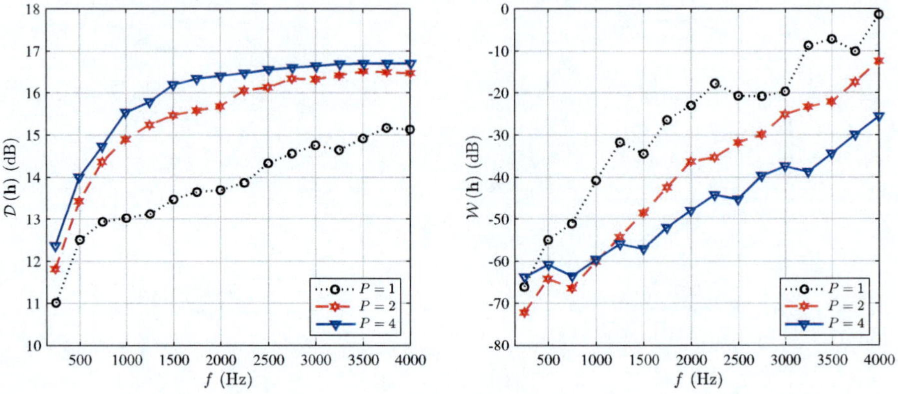

(a) (b)

Fig. 5.8 DF and WNG of the low-rank MDF beamformer, $\mathbf{h}_{\text{MDF}}$, as a function of frequency, for different ranks, P: **(a)** DF and **(b)** WNG. Conditions: $\delta = 2.0$ cm, $P = 2$, and $(\theta_s, \phi_s) = (80°, 0°)$

statistically independent interferences impinge on the cube array from two different directions $(\theta_{i,1}, \phi_{i,1})$ and $(\theta_{i,2}, \phi_{i,2})$. The two interferences are incoherent with the desired signal, and their variances are the same and equal to ϕ_V. The variance of the white noise is ϕ_{V_w}, and we assume that $\phi_{V_w} = \alpha\phi_V$, where $\alpha > 0$ is related to the ratio between the powers of the white noise and point source noise. The covariance matrix of the noise signal is given in (3.75). We choose $(\theta_s, \phi_s) = (80°, 0°)$, $(\theta_{i,1}, \phi_{i,1}) = (80°, 150°)$, $(\theta_{i,2}, \phi_{i,2}) = (80°, 210°)$, and iSNR $= 0$ dB. The variance of the desired signal, ϕ_X, is set to 1, and the variances of the noises, ϕ_V and ϕ_{V_w}, are computed according to the given input SNR and values of α.

To demonstrate the performance of the low-rank MVDR beamformer, $\mathbf{h}_{\text{MVDR}}$, given in (5.52), we choose $P = 2$. Figures 5.9 and 5.10 show plots of the power patterns, $\left|\mathcal{B}_{\theta,\phi}(\mathbf{h})\right|^2$, of this beamformer versus the azimuth angle for $\theta = \theta_s$, and versus the elevation angle for $\phi = \phi_s$, respectively, for $\alpha = 0.01$, $f = 2000$ Hz, and different numbers of sensors, $M = M_0^3$. As seen, the main beam of the power patterns points in the direction of the desired signal, and there are nulls in the directions of the interferences. Figure 5.11 shows plots of the SNR gain, $\mathcal{G}(\mathbf{h})$, of the low-rank MVDR beamformer as a function of frequency for different numbers of sensors, $M = M_0^3$, and different values of α. It is observed that the SNR gain increases as the value of M_0 increases. It can also be observed that the SNR gain decreases when the value of α increases. This is because as the value of α increases, the noise signals at the microphone arrays are less coherent; as a result, the beamformer suppresses the noise less effectively.

To demonstrate the performance of the low-rank Wiener beamformer, $\mathbf{h}_{\text{W}}$, given in (5.55), we choose $P = 2$. Figure 5.12 shows plots of the SNR gain, $\mathcal{G}(\mathbf{h})$, of this beamformer as a function of frequency for different numbers of sensors, $M = M_0^3$,

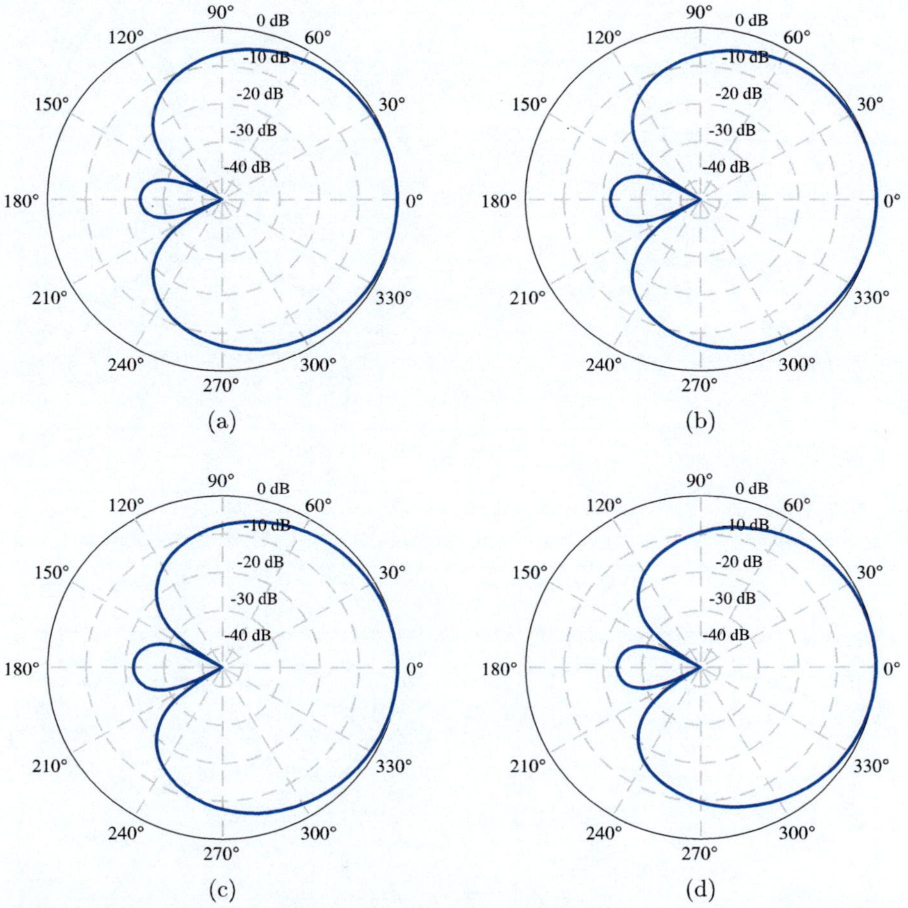

Fig. 5.9 Power patterns of the low-rank MVDR beamformer, $\mathbf{h}_{\mathrm{MVDR}}$, versus the azimuth angle, for $\theta = \theta_{\mathrm{s}}$ and different numbers of sensors, $M = M_0^3$: (**a**) $M_0 = 2$, (**b**) $M_0 = 3$, (**c**) $M_0 = 4$, and (**d**) $M_0 = 5$. Conditions: $\delta = 2.0$ cm, $f = 2000$ Hz, $\alpha = 0.01$, $P = 2$, and $(\theta_{\mathrm{s}}, \phi_{\mathrm{s}}) = (80^\circ, 0^\circ)$

and different values of α. Similarly, the SNR gain increases when M_0 increases, and the larger is the value of α, the lower is the SNR gain.

To demonstrate the performance of the low-rank tradeoff beamformer, $\mathbf{h}_{\mathrm{T},\mu}$, given in (5.58), we choose $M_0 = 3$, $\alpha = 0.01$, and $P = 2$. We use broadband performance metrics to better evaluate the performance of this beamformer, where the broadband SNR gain, noise reduction factor, desired signal reduction factor, and desired signal distortion index are, respectively, given in (3.78), (3.79), (3.80), and (3.81). Figure 5.13 shows plots of the broadband SNR gain, $\overline{\mathcal{G}(\mathbf{h})}$; broadband noise reduction factor, $\overline{\xi(\mathbf{h})}$; broadband desired signal reduction factor, $\overline{\xi_{\mathrm{d}}(\mathbf{h})}$; and broadband desired signal distortion index, $\overline{\upsilon(\mathbf{h})}$, of the tradeoff beamformer as a function of the input SNR for different values of μ. Note that for $\mu = 1$, the low-

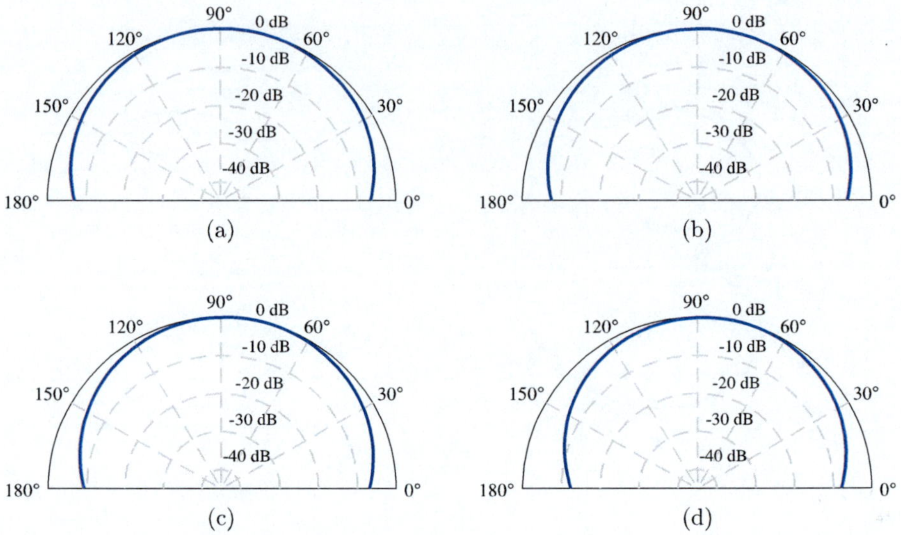

Fig. 5.10 Power patterns of the low-rank MVDR beamformer, $\mathbf{h}_{\text{MVDR}}$, versus the elevation angle for $\phi = \phi_s$ and different numbers of sensors, $M = M_0^3$: (**a**) $M_0 = 2$, (**b**) $M_0 = 3$, (**c**) $M_0 = 4$, and (**d**) $M_0 = 5$. Conditions: $\delta = 2.0$ cm, $f = 2000$ Hz, $\alpha = 0.01$, $P = 2$, and $(\theta_s, \phi_s) = (80°, 0°)$

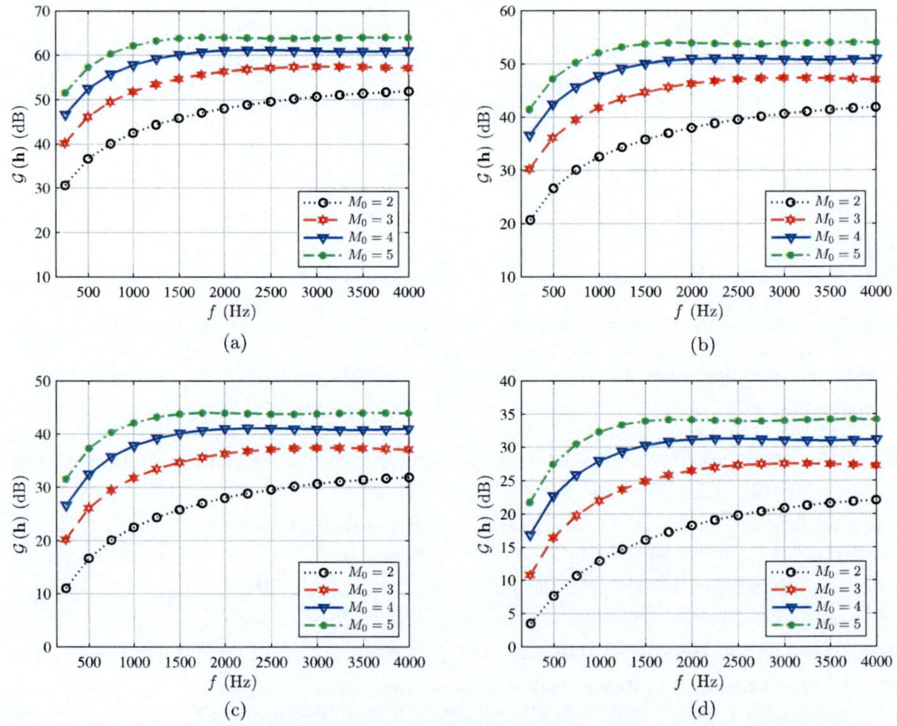

Fig. 5.11 SNR gain of the low-rank MVDR beamformer, $\mathbf{h}_{\text{MVDR}}$, as a function of frequency, for different values of α and different numbers of sensors, $M = M_0^3$: (**a**) $\alpha = 10^{-4}$, (**b**) $\alpha = 10^{-3}$, (**c**) $\alpha = 10^{-2}$, and (**d**) $\alpha = 10^{-1}$. Conditions: $\delta = 2.0$ cm, $P = 2$, and $(\theta_s, \phi_s) = (80°, 0°)$

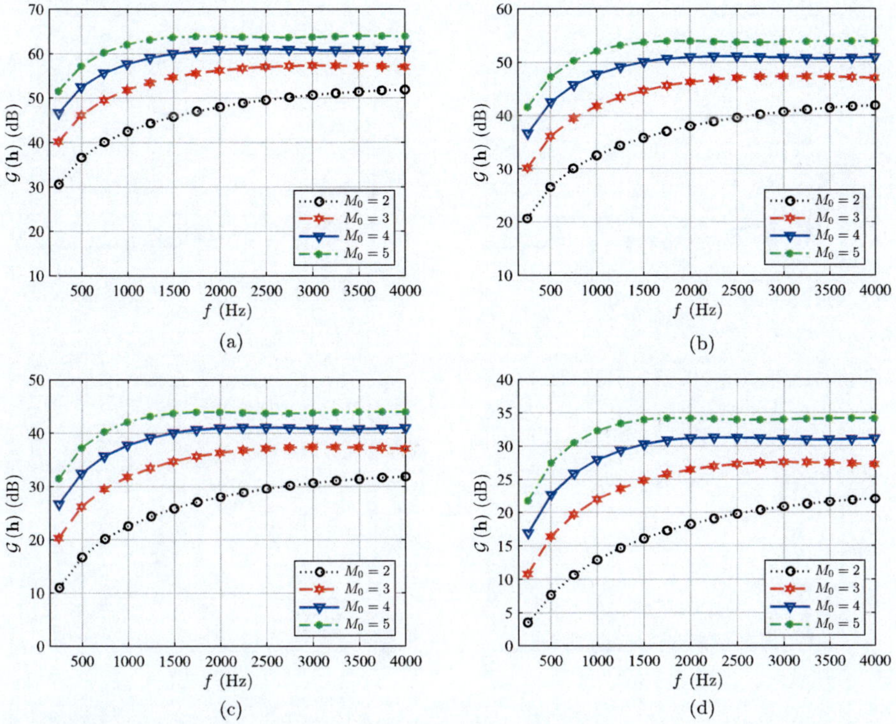

Fig. 5.12 SNR gain of the low-rank Wiener beamformer, $\mathbf{h}_W$, as a function of frequency, for different values of α and different numbers of sensors, $M = M_0^3$: (**a**) $\alpha = 10^{-4}$, (**b**) $\alpha = 10^{-3}$, (**c**) $\alpha = 10^{-2}$, and (**d**) $\alpha = 10^{-1}$. Conditions: $\delta = 2.0$ cm, $P = 2$, and $(\theta_s, \phi_s) = (80^\circ, 0^\circ)$

rank tradeoff and low-rank Wiener beamformers are identical. It can be seen that the SNR gain, noise reduction factor, desired noise reduction factor, and desired signal distortion index all increase as the value of μ increases, which indicates that the low-rank tradeoff beamformer achieves a good compromise between noise reduction and desired signal distortion from a broadband perspective.

To demonstrate the performance of the low-rank LCMV beamformer, $\mathbf{h}_{\text{LCMV}}$, given in (5.64), we choose $P = 2$ and $\alpha = 0.01$, and set two null constraints in the direction of the interferences, i.e., $\left(\theta_{i,1}, \phi_{i,1}\right) = (80^\circ, 150^\circ)$ and $\left(\theta_{i,2}, \phi_{i,2}\right) = (80^\circ, 210^\circ)$. Figures 5.14 and 5.15 show plots of the power patterns, $\left|\mathcal{B}_{\theta,\phi}(\mathbf{h})\right|^2$, of this beamformer versus the azimuth angle for $\theta = \theta_s$, and versus the elevation angle for $\phi = \phi_s$, respectively, for $f = 2000$ Hz and different numbers of sensors, $M = M_0^3$. As seen, the main beam of the power patterns points in the direction of the desired signal, and there are nulls in the directions of the interferences. Figure 5.16 shows plots of the SNR gain, $\mathcal{G}(\mathbf{h})$, of this beamformer as a function of frequency for different numbers of sensors, $M = M_0^3$, and different values of α. As seen, the SNR gain increases when M_0 increases, and the larger is the value of α, the lower is the SNR gain.

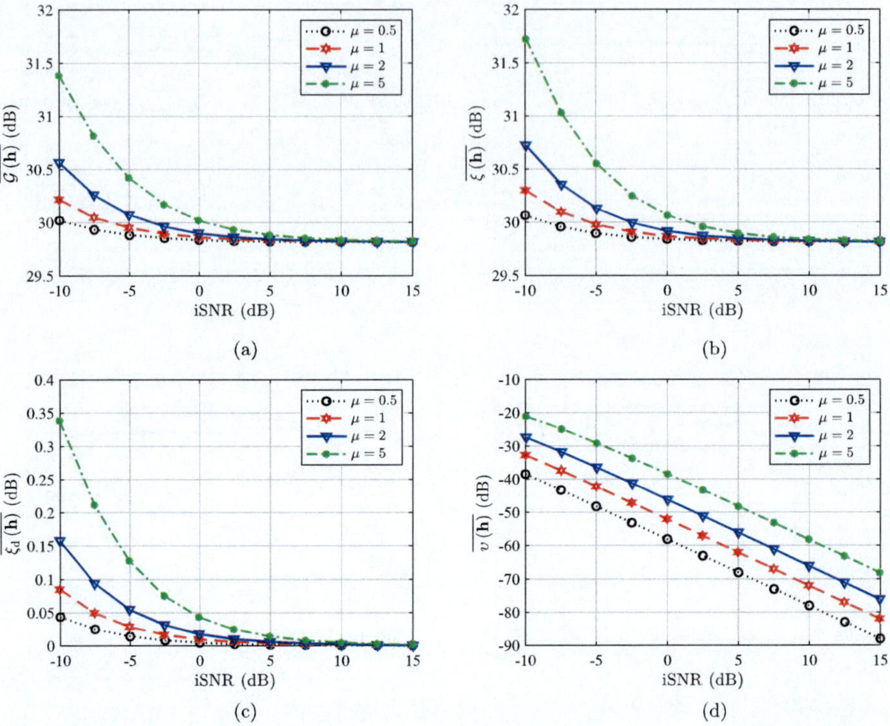

Fig. 5.13 Performance of the low-rank tradeoff beamformer, $\mathbf{h}_{\mathrm{T},\mu}$, as a function of the input SNR, for different values of μ: (**a**) broadband SNR gain, (**b**) broadband noise reduction factor, (**c**) broadband desired signal reduction factor, and (**d**) broadband desired signal distortion index. Conditions: $M_0 = 3$, $\delta = 2.0$ cm, $\alpha = 0.01$, $P = 2$, and $(\theta_{\mathrm{s}}, \phi_{\mathrm{s}}) = (80^\circ, 0^\circ)$

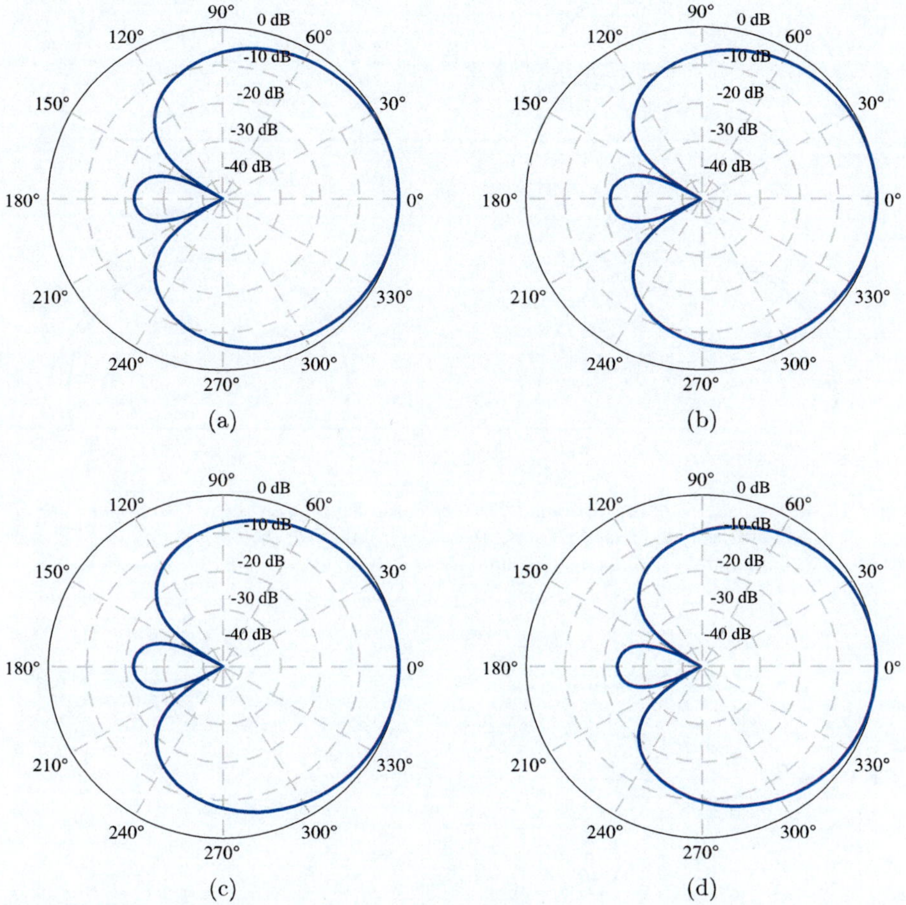

Fig. 5.14 Power patterns of the low-rank LCMV beamformer, $\mathbf{h}_{\mathrm{LCMV}}$, versus the azimuth angle, for $\theta = \theta_s$ and different numbers of sensors, $M = M_0^3$: **(a)** $M_0 = 2$, **(b)** $M_0 = 3$, **(c)** $M_0 = 4$, and **(d)** $M_0 = 5$. Conditions: $\delta = 2.0$ cm, $f = 2000$ Hz, $\alpha = 0.01$, $P = 2$, and $(\theta_s, \phi_s) = (80^\circ, 0^\circ)$

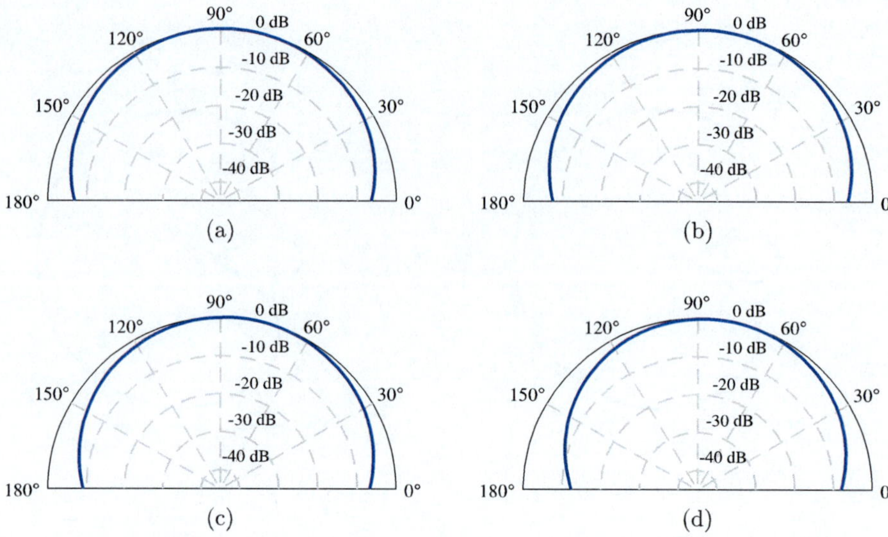

Fig. 5.15 Power patterns of the low-rank LCMV beamformer, $\mathbf{h}_{\mathrm{LCMV}}$, versus the elevation angle, for $\phi = \phi_{\mathrm{s}}$ and different numbers of sensors, $M = M_0^3$: (**a**) $M_0 = 2$, (**b**) $M_0 = 3$, (**c**) $M_0 = 4$, and (**d**) $M_0 = 5$. Conditions: $\delta = 2.0$ cm, $f = 2000$ Hz, $\alpha = 0.01$, $P = 2$, and $(\theta_{\mathrm{s}}, \phi_{\mathrm{s}}) = (80^\circ, 0^\circ)$

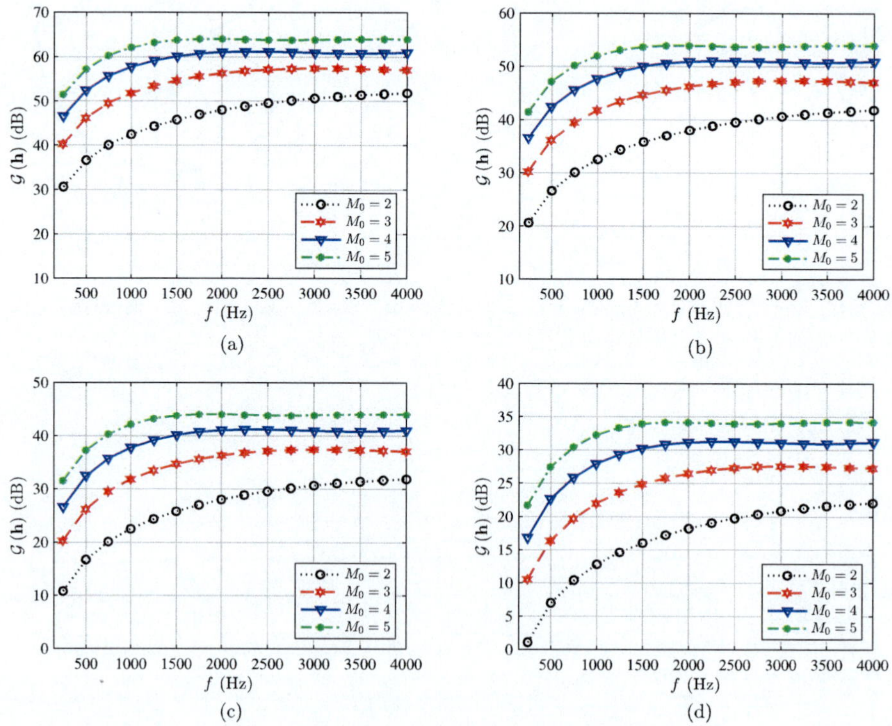

Fig. 5.16 SNR gain of the robust low-rank LCMV beamformer, $\mathbf{h}_{\mathrm{LCMV}}$, as a function of frequency, for different values of α and different numbers of sensors, $M = M_0^3$: (**a**) $\alpha = 10^{-4}$, (**b**) $\alpha = 10^{-3}$, (**c**) $\alpha = 10^{-2}$, and (**d**) $\alpha = 10^{-1}$. Conditions: $\delta = 2.0$ cm, $P = 2$, and $(\theta_{\mathrm{s}}, \phi_{\mathrm{s}}) = (80^\circ, 0^\circ)$

References

1. J. Benesty, C. Paleologu, L.-M. Dogariu, S. Ciochină, Identification of linear and bilinear systems: a unified study. Electronics **10**(15), 33 (2021)
2. C. Paleologu, J. Benesty, S. Ciochină, Linear system identification based on a Kronecker product decomposition. IEEE/ACM Trans. Audio Speech Lang. Process. **26**, 1793–1808 (2018)
3. J. Benesty, I Cohen, J. Chen, *Array Processing–Kronecker Product Beamforming* (Springer, Cham, Switzerland, 2019)
4. G. Huang, J. Benesty, I. Cohen, J. Chen, Kronecker product multichannel linear filtering for adaptive weighted prediction error-based speech dereverberation. IEEE/ACM Trans. Audio Speech Lang. Process. **30**, 1277–1289 (2022)
5. G.H. Golub, C.F. Van Loan, *Matrix Computations, 3rd Edition* (The John Hopkins University Press, Baltimore, MD, 1996)
6. W. Gander, M.J. Gander, F. Kwok, *Scientific Computing—An Introduction Using Maple and MATLAB* (Springer, Berlin, 2014)
7. J. Benesty, C. Paleologu, S. Ciochină, On the identification of bilinear forms with the Wiener filter. IEEE Signal Process. Lett **24**, 653–657 (2017)

Chapter 6
Distortionless Beamforming

Distortionless beamforming plays a huge role in microphone array processing, in particular, and array processing, in general. Indeed, the most interesting and practical (fixed or adaptive) beamformers are distortionless. Therefore, it is of interest to study this important family of beamformers and understand how they really work. This is the objective of this chapter. A byproduct of this investigation is the derivation of optimal robust beamformers, which are so much useful in sensor arrays.

6.1 Signal Model, Performance Measures, and Distortionless Beamformers

We consider the signal model of Chap. 3 (with a 3-D array of M microphones), i.e.,

$$\begin{aligned}\mathbf{y} &= \begin{bmatrix} Y_1 & Y_2 & \cdots & Y_M \end{bmatrix}^T \\ &= \mathbf{x} + \mathbf{v} \\ &= \mathbf{d}_{\theta_s,\phi_s} X + \mathbf{v},\end{aligned} \tag{6.1}$$

where

$$\mathbf{d}_{\theta_s,\phi_s} = \begin{bmatrix} e^{\jmath 2\pi f \mathbf{a}_{\theta_s,\phi_s}^T \mathbf{r}_1/c} & e^{\jmath 2\pi f \mathbf{a}_{\theta_s,\phi_s}^T \mathbf{r}_2/c} & \cdots & e^{\jmath 2\pi f \mathbf{a}_{\theta_s,\phi_s}^T \mathbf{r}_M/c} \end{bmatrix}^T \tag{6.2}$$

is the steering vector of length M. The covariance matrix of $\mathbf{y}$ is

$$\mathbf{\Phi}_\mathbf{y} = E\left(\mathbf{y}\mathbf{y}^H\right)$$

J. Benesty et al., *Microphone Arrays*, Springer Topics in Signal Processing 22,
https://doi.org/10.1007/978-3-031-36974-2_6

$$= \mathbf{\Phi}_{\mathbf{x}} + \mathbf{\Phi}_{\mathbf{v}}, \tag{6.3}$$

which is also

$$\begin{aligned} \mathbf{\Phi}_{\mathbf{y}} &= \phi_X \mathbf{d}_{\theta_s,\phi_s} \mathbf{d}_{\theta_s,\phi_s}^H + \mathbf{\Phi}_{\mathbf{v}} \\ &= \phi_X \mathbf{d}_{\theta_s,\phi_s} \mathbf{d}_{\theta_s,\phi_s}^H + \phi_V \mathbf{\Gamma}_{\mathbf{v}} \end{aligned} \tag{6.4}$$

and, in the case of the spherically isotropic (diffuse) noise field, (6.4) becomes

$$\mathbf{\Phi}_{\mathbf{y}} = \phi_X \mathbf{d}_{\theta_s,\phi_s} \mathbf{d}_{\theta_s,\phi_s}^H + \phi_{\mathrm{d}} \mathbf{\Gamma}_{\mathrm{d}}, \tag{6.5}$$

where ϕ_{d} is the variance of the diffuse noise and $\mathbf{\Gamma}_{\mathrm{d}}$ is defined in Chap. 3.

Then, the beamformer output signal is

$$\begin{aligned} Z &= \mathbf{h}^H \mathbf{y} \\ &= X \mathbf{h}^H \mathbf{d}_{\theta_s,\phi_s} + \mathbf{h}^H \mathbf{v}, \end{aligned} \tag{6.6}$$

where $\mathbf{h}$ is the beamforming filter of length M. We clearly see from (6.6) that the distortionless constraint is

$$\mathbf{h}^H \mathbf{d}_{\theta_s,\phi_s} = 1. \tag{6.7}$$

In this chapter, it is assumed that (6.7) is always satisfied.

In this context, we recall three important measures; they are the SNR gain:

$$\mathcal{G}(\mathbf{h}) = \frac{\left|\mathbf{h}^H \mathbf{d}_{\theta_s,\phi_s}\right|^2}{\mathbf{h}^H \mathbf{\Gamma}_{\mathbf{v}} \mathbf{h}}, \tag{6.8}$$

WNG:

$$\mathcal{W}(\mathbf{h}) = \frac{\left|\mathbf{h}^H \mathbf{d}_{\theta_s,\phi_s}\right|^2}{\mathbf{h}^H \mathbf{h}}, \tag{6.9}$$

and DF:

$$\mathcal{D}(\mathbf{h}) = \frac{\left|\mathbf{h}^H \mathbf{d}_{\theta_s,\phi_s}\right|^2}{\mathbf{h}^H \mathbf{\Gamma}_{\mathrm{d}} \mathbf{h}}, \tag{6.10}$$

from which we can derive, respectively, the MVDR, DS (or MWNG), and MDF beamformers (see Chap. 3):

$$\mathbf{h}_{\mathrm{MVDR}} = \frac{\mathbf{\Gamma}_{\mathbf{v}}^{-1} \mathbf{d}_{\theta_s,\phi_s}}{\mathbf{d}_{\theta_s,\phi_s}^H \mathbf{\Gamma}_{\mathbf{v}}^{-1} \mathbf{d}_{\theta_s,\phi_s}}, \tag{6.11}$$

$$\mathbf{h}_{\mathrm{DS}} = \frac{\mathbf{d}_{\theta_s,\phi_s}}{M}, \tag{6.12}$$

and

$$\mathbf{h}_{\mathrm{MDF}} = \frac{\boldsymbol{\Gamma}_{\mathrm{d}}^{-1}\mathbf{d}_{\theta_s,\phi_s}}{\mathbf{d}_{\theta_s,\phi_s}^H \boldsymbol{\Gamma}_{\mathrm{d}}^{-1}\mathbf{d}_{\theta_s,\phi_s}}. \tag{6.13}$$

One can easily check that the three above beamformers are distortionless. Let $\boldsymbol{\Gamma} \in \{\boldsymbol{\Gamma}_{\mathbf{v}}, \boldsymbol{\Gamma}_{\mathrm{d}}\}$. In the rest, we may write $\mathbf{h}_{\mathrm{M}} = \mathbf{h}_{\mathrm{MVDR}}$ if $\boldsymbol{\Gamma} = \boldsymbol{\Gamma}_{\mathbf{v}}$ (which is an adaptive beamformer), $\mathbf{h}_{\mathrm{M}} = \mathbf{h}_{\mathrm{MDF}}$ if $\boldsymbol{\Gamma} = \boldsymbol{\Gamma}_{\mathrm{d}}$ (which is a fixed beamformer), and call

$$\mathbf{h}_{\mathrm{M}} = \frac{\boldsymbol{\Gamma}^{-1}\mathbf{d}_{\theta_s,\phi_s}}{\mathbf{d}_{\theta_s,\phi_s}^H \boldsymbol{\Gamma}^{-1}\mathbf{d}_{\theta_s,\phi_s}} \tag{6.14}$$

the M beamformer.

Next, our objective is to show how distortionless beamformers work and are connected. We also show how to produce compromises among different measures for robust beamforming.

6.2 Principle of Distortionless Beamforming

Since we are only interested in distortionless (fixed and adaptive) beamformers satisfying (6.7), we will then only consider this class of beamformers. Next, we give an important property, which shows how a distortionless beamformer can be decomposed. Obviously, this decomposition is not unique but it reveals some inherent structure due to the distortionless constraint.

Property 6.1 Any distortionless fixed or adaptive beamformer can be written as [1]

$$\mathbf{h} = \mathbf{h}_{\mathrm{DS}} + \overline{\mathbf{U}}\,\overline{\mathbf{h}}, \tag{6.15}$$

where $\mathbf{h}_{\mathrm{DS}}$ is the DS beamformer given in (6.12), $\overline{\mathbf{U}}$ (defined below) is a semi-unitary matrix of size $M \times (M-1)$, $\overline{\mathbf{h}}$ is a beamforming filter of length $M-1$, and $\mathbf{h}_{\mathrm{DS}}^H\overline{\mathbf{U}}\,\overline{\mathbf{h}} = 0$. In other words, any distortionless beamformer can be written as the sum of two orthogonal filters: the DS beamformer, which is fixed, and another reduced-rank beamformer, i.e., $\overline{\mathbf{U}}\,\overline{\mathbf{h}}$, which can be fixed or adaptive, depending on $\overline{\mathbf{h}}$.

Proof Using the well-known eigenvalue decomposition, the rank-1 matrix $\mathbf{d}_{\theta_s,\phi_s}\mathbf{d}_{\theta_s,\phi_s}^H$ can be diagonalized as

$$\mathbf{U}^H\mathbf{d}_{\theta_s,\phi_s}\mathbf{d}_{\theta_s,\phi_s}^H\mathbf{U} = \mathrm{diag}\left(M, 0, \ldots, 0\right), \tag{6.16}$$

where

$$\mathbf{U} = \begin{bmatrix} \frac{\mathbf{d}_{\theta_s,\phi_s}}{\sqrt{M}} & \mathbf{u}_2 & \cdots & \mathbf{u}_M \end{bmatrix} = \begin{bmatrix} \frac{\mathbf{d}_{\theta_s,\phi_s}}{\sqrt{M}} & \overline{\mathbf{U}} \end{bmatrix} \tag{6.17}$$

is a unitary matrix, i.e., $\mathbf{U}^H\mathbf{U} = \mathbf{U}\mathbf{U}^H = \mathbf{I}_M$. In practice, the semi-unitary matrix $\overline{\mathbf{U}}$ can be easily obtained from the Gram-Schmidt orthonormalization process. Since $\mathbf{U}$ has full rank, we can always write $\mathbf{h}$ in the new basis:

$$\mathbf{h} = \mathbf{U}\begin{bmatrix} \widetilde{H} \\ \overline{\mathbf{h}} \end{bmatrix} = \frac{\mathbf{d}_{\theta_s,\phi_s}}{\sqrt{M}}\widetilde{H} + \overline{\mathbf{U}}\,\overline{\mathbf{h}}. \tag{6.18}$$

Applying the distortionless constraint to the previous expression and using the fact that $\mathbf{d}_{\theta_s,\phi_s}^H\overline{\mathbf{U}} = \mathbf{0}$ (which also implies that $\mathbf{h}_{\mathrm{DS}}^H\overline{\mathbf{U}}\,\overline{\mathbf{h}} = 0$), we get

$$\widetilde{H} = \frac{1}{\sqrt{M}}. \tag{6.19}$$

Finally, substituting (6.19) into (6.18), we obtain the proposed decomposition in (6.15). ■

With the distortionless beamformer, $\mathbf{h}$, as given in (6.15), the WNG can be expressed as

$$\mathcal{W}(\mathbf{h}) = \frac{1}{\mathbf{h}_{\mathrm{DS}}^H\mathbf{h}_{\mathrm{DS}} + \overline{\mathbf{h}}^H\overline{\mathbf{h}}} = \frac{\mathcal{W}(\mathbf{h}_{\mathrm{DS}})}{1 + \mathcal{W}(\mathbf{h}_{\mathrm{DS}})\mathcal{A}_{\mathrm{w}}\left(\overline{\mathbf{h}}\right)}, \tag{6.20}$$

where

$$\mathcal{W}(\mathbf{h}_{\mathrm{DS}}) = M, \tag{6.21}$$

$$\mathcal{A}_{\mathrm{w}}\left(\overline{\mathbf{h}}\right) = \overline{\mathbf{h}}^H\overline{\mathbf{h}}. \tag{6.22}$$

This shows how the WNG of any distortionless beamformer can be written as a function of the WNG of $\mathbf{h}_{\mathrm{DS}}$. As we already know, since $\mathcal{A}_{\mathrm{w}}\left(\overline{\mathbf{h}}\right) \geq 0$, we have

$$\mathcal{W}(\mathbf{h}) \leq \mathcal{W}(\mathbf{h}_{\mathrm{DS}}), \ \forall \mathbf{h} \tag{6.23}$$

and, obviously, $\mathcal{W}(\mathbf{h})$ is maximized when $\overline{\mathbf{h}} = \mathbf{0}$.

The DF of the distortionless beamformer, $\mathbf{h}$, in (6.15) is

$$\mathcal{D}(\mathbf{h}) = \frac{\mathcal{D}(\mathbf{h}_{\mathrm{DS}})}{1 + \mathcal{D}(\mathbf{h}_{\mathrm{DS}}) \mathcal{A}_{\mathrm{d}}\left(\overline{\mathbf{h}}\right)}, \tag{6.24}$$

where

$$\mathcal{D}(\mathbf{h}_{\mathrm{DS}}) = \frac{M^2}{\mathbf{d}_{\theta_{\mathrm{s}},\phi_{\mathrm{s}}}^H \mathbf{\Gamma}_{\mathrm{d}} \mathbf{d}_{\theta_{\mathrm{s}},\phi_{\mathrm{s}}}}, \tag{6.25}$$

$$\mathcal{A}_{\mathrm{d}}\left(\overline{\mathbf{h}}\right) = \mathbf{h}_{\mathrm{DS}}^H \mathbf{\Gamma}_{\mathrm{d}} \overline{\mathbf{U}}\, \overline{\mathbf{h}} + \overline{\mathbf{h}}^H \overline{\mathbf{U}}^H \mathbf{\Gamma}_{\mathrm{d}} \mathbf{h}_{\mathrm{DS}} + \overline{\mathbf{h}}^H \overline{\mathbf{U}}^H \mathbf{\Gamma}_{\mathrm{d}} \overline{\mathbf{U}}\, \overline{\mathbf{h}}. \tag{6.26}$$

This shows how the DF of any distortionless beamformer can be written as a function of the DF of $\mathbf{h}_{\mathrm{DS}}$. The MDF beamformer is deduced by maximizing the DF or, equivalently, by minimizing $\mathcal{A}_{\mathrm{d}}\left(\overline{\mathbf{h}}\right)$. We obtain

$$\overline{\mathbf{h}}_{\mathrm{MDF}} = -\left(\overline{\mathbf{U}}^H \mathbf{\Gamma}_{\mathrm{d}} \overline{\mathbf{U}}\right)^{-1} \overline{\mathbf{U}}^H \mathbf{\Gamma}_{\mathrm{d}} \mathbf{h}_{\mathrm{DS}}. \tag{6.27}$$

As a result, another interesting formulation of the MDF beamformer is

$$\begin{aligned} \mathbf{h}_{\mathrm{MDF}} &= \mathbf{h}_{\mathrm{DS}} - \overline{\mathbf{U}}\left(\overline{\mathbf{U}}^H \mathbf{\Gamma}_{\mathrm{d}} \overline{\mathbf{U}}\right)^{-1} \overline{\mathbf{U}}^H \mathbf{\Gamma}_{\mathrm{d}} \mathbf{h}_{\mathrm{DS}} \\ &= \mathbf{\Gamma}_{\mathrm{d}}^{-1/2} \mathbf{P}_{\mathrm{d}} \mathbf{\Gamma}_{\mathrm{d}}^{1/2} \mathbf{h}_{\mathrm{DS}}, \end{aligned} \tag{6.28}$$

where

$$\mathbf{P}_{\mathrm{d}} = \mathbf{I}_M - \mathbf{\Gamma}_{\mathrm{d}}^{1/2} \overline{\mathbf{U}}\left(\overline{\mathbf{U}}^H \mathbf{\Gamma}_{\mathrm{d}} \overline{\mathbf{U}}\right)^{-1} \overline{\mathbf{U}}^H \mathbf{\Gamma}_{\mathrm{d}}^{1/2} \tag{6.29}$$

is an orthogonal projection matrix. With $\mathbf{h}_{\mathrm{MDF}}$ in (6.28), the WNG and DF are, respectively,

$$\mathcal{W}(\mathbf{h}_{\mathrm{MDF}}) = \frac{1}{\mathbf{h}_{\mathrm{DS}}^H \mathbf{\Gamma}_{\mathrm{d}}^{1/2} \mathbf{P}_{\mathrm{d}} \mathbf{\Gamma}_{\mathrm{d}}^{-1} \mathbf{P}_{\mathrm{d}} \mathbf{\Gamma}_{\mathrm{d}}^{1/2} \mathbf{h}_{\mathrm{DS}}} \tag{6.30}$$

and

$$\mathcal{D}(\mathbf{h}_{\mathrm{MDF}}) = \frac{1}{\mathbf{h}_{\mathrm{DS}}^H \mathbf{\Gamma}_{\mathrm{d}}^{1/2} \mathbf{P}_{\mathrm{d}} \mathbf{\Gamma}_{\mathrm{d}}^{1/2} \mathbf{h}_{\mathrm{DS}}}. \tag{6.31}$$

In the same way, the SNR gain of the distortionless beamformer, $\mathbf{h}$, in (6.15) is

$$\mathcal{G}(\mathbf{h}) = \frac{\mathcal{G}(\mathbf{h}_{\mathrm{DS}})}{1 + \mathcal{G}(\mathbf{h}_{\mathrm{DS}})\mathcal{A}_{\mathrm{g}}\left(\overline{\mathbf{h}}\right)}, \tag{6.32}$$

where

$$\mathcal{G}(\mathbf{h}_{\mathrm{DS}}) = \frac{M^2}{\mathbf{d}^H_{\theta_{\mathrm{s}},\phi_{\mathrm{s}}}\mathbf{\Gamma}_{\mathbf{v}}\mathbf{d}_{\theta_{\mathrm{s}},\phi_{\mathrm{s}}}}, \tag{6.33}$$

$$\mathcal{A}_{\mathrm{g}}\left(\overline{\mathbf{h}}\right) = \mathbf{h}^H_{\mathrm{DS}}\mathbf{\Gamma}_{\mathbf{v}}\overline{\mathbf{U}}\,\overline{\mathbf{h}} + \overline{\mathbf{h}}^H\overline{\mathbf{U}}^H\mathbf{\Gamma}_{\mathbf{v}}\mathbf{h}_{\mathrm{DS}} + \overline{\mathbf{h}}^H\overline{\mathbf{U}}^H\mathbf{\Gamma}_{\mathbf{v}}\overline{\mathbf{U}}\,\overline{\mathbf{h}}. \tag{6.34}$$

This shows how the SNR gain of any distortionless beamformer can be written as a function of the SNR gain of $\mathbf{h}_{\mathrm{DS}}$. The MVDR beamformer is obtained by maximizing the SNR gain or, equivalently, by minimizing $\mathcal{A}_{\mathrm{g}}\left(\overline{\mathbf{h}}\right)$. We get

$$\overline{\mathbf{h}}_{\mathrm{MVDR}} = -\left(\overline{\mathbf{U}}^H\mathbf{\Gamma}_{\mathbf{v}}\overline{\mathbf{U}}\right)^{-1}\overline{\mathbf{U}}^H\mathbf{\Gamma}_{\mathbf{v}}\mathbf{h}_{\mathrm{DS}}. \tag{6.35}$$

As a result, another interesting formulation of the MVDR beamformer is

$$\begin{aligned}\mathbf{h}_{\mathrm{MVDR}} &= \mathbf{h}_{\mathrm{DS}} - \overline{\mathbf{U}}\left(\overline{\mathbf{U}}^H\mathbf{\Gamma}_{\mathbf{v}}\overline{\mathbf{U}}\right)^{-1}\overline{\mathbf{U}}^H\mathbf{\Gamma}_{\mathbf{v}}\mathbf{h}_{\mathrm{DS}} \\ &= \mathbf{\Gamma}_{\mathbf{v}}^{-1/2}\mathbf{P}_{\mathbf{v}}\mathbf{\Gamma}_{\mathbf{v}}^{1/2}\mathbf{h}_{\mathrm{DS}},\end{aligned} \tag{6.36}$$

where

$$\mathbf{P}_{\mathbf{v}} = \mathbf{I}_M - \mathbf{\Gamma}_{\mathbf{v}}^{1/2}\overline{\mathbf{U}}\left(\overline{\mathbf{U}}^H\mathbf{\Gamma}_{\mathbf{v}}\overline{\mathbf{U}}\right)^{-1}\overline{\mathbf{U}}^H\mathbf{\Gamma}_{\mathbf{v}}^{1/2} \tag{6.37}$$

is an orthogonal projection matrix. With $\mathbf{h}_{\mathrm{MVDR}}$ in (6.36), the SNR gain is

$$\mathcal{G}(\mathbf{h}_{\mathrm{MVDR}}) = \frac{1}{\mathbf{h}^H_{\mathrm{DS}}\mathbf{\Gamma}_{\mathbf{v}}^{1/2}\mathbf{P}_{\mathbf{v}}\mathbf{\Gamma}_{\mathbf{v}}^{1/2}\mathbf{h}_{\mathrm{DS}}}. \tag{6.38}$$

From the above, we can deduce a very general class of distortionless beamformers:

$$\mathbf{h} = \mathbf{\Gamma}^{-1/2}\mathbf{P}\mathbf{\Gamma}^{1/2}\mathbf{h}_{\mathrm{DS}}, \tag{6.39}$$

where $\mathbf{\Gamma}$ is the coherence matrix of any noise we have in mind and

$$\mathbf{P} = \mathbf{I}_M - \mathbf{\Gamma}^{1/2}\overline{\mathbf{U}}\left(\overline{\mathbf{U}}^H\mathbf{\Gamma}\overline{\mathbf{U}}\right)^{-1}\overline{\mathbf{U}}^H\mathbf{\Gamma}^{1/2} \tag{6.40}$$

is an orthogonal projection matrix.

6.3 Robust Beamforming

In practice, it is important to be able to compromise between WNG and DF/SNR gain in order to have a robust beamformer. In this section, we show how to do this in an optimal way, thanks in part to Property 6.1.

Assume that we want to find a beamformer whose WNG is equal to $\mathcal{W}_0$. A robust beamformer can be derived from the criterion [1]:

$$\min_{\mathbf{h}} \mathbf{h}^H \boldsymbol{\Gamma} \mathbf{h} \quad \text{subject to} \quad \begin{cases} \mathbf{h}^H \mathbf{h} = \mathcal{W}_0^{-1} \\ \mathbf{h}^H \mathbf{d}_{\theta_s, \phi_s} = 1 \end{cases}. \tag{6.41}$$

Using (6.15), where the distortionless constraint is already built in, it is easy to verify that

$$\mathbf{h}^H \boldsymbol{\Gamma} \mathbf{h} = \mathbf{h}_{\text{DS}}^H \boldsymbol{\Gamma} \mathbf{h}_{\text{DS}} - \overline{\mathbf{h}}^H \mathbf{b} - \mathbf{b}^H \overline{\mathbf{h}} + \overline{\mathbf{h}}^H \boldsymbol{\Upsilon} \overline{\mathbf{h}} \tag{6.42}$$

and

$$\mathbf{h}^H \mathbf{h} = \frac{1}{M} + \overline{\mathbf{h}}^H \overline{\mathbf{h}} = \frac{1}{\mathcal{W}_0}, \tag{6.43}$$

where

$$\mathbf{b} = -\overline{\mathbf{U}}^H \boldsymbol{\Gamma} \mathbf{h}_{\text{DS}},$$

$$\boldsymbol{\Upsilon} = \overline{\mathbf{U}}^H \boldsymbol{\Gamma} \overline{\mathbf{U}}.$$

As a result, the criterion in (6.41) is equivalent to

$$\min_{\overline{\mathbf{h}}} \left(\overline{\mathbf{h}}^H \boldsymbol{\Upsilon} \overline{\mathbf{h}} - \overline{\mathbf{h}}^H \mathbf{b} - \mathbf{b}^H \overline{\mathbf{h}} \right) \quad \text{subject to} \quad \overline{\mathbf{h}}^H \overline{\mathbf{h}} = \epsilon_+, \tag{6.44}$$

where $\epsilon_+ = \mathcal{W}_0^{-1} - M^{-1}$. From (6.44), we can write the Lagrange function:

$$\mathcal{L}\left(\overline{\mathbf{h}}, \lambda\right) = \overline{\mathbf{h}}^H \boldsymbol{\Upsilon} \overline{\mathbf{h}} - \overline{\mathbf{h}}^H \mathbf{b} - \mathbf{b}^H \overline{\mathbf{h}} - \lambda \left(\overline{\mathbf{h}}^H \overline{\mathbf{h}} - \epsilon_+ \right), \tag{6.45}$$

where λ is a Lagrange multiplier. Differentiating $\mathcal{L}\left(\overline{\mathbf{h}}, \lambda\right)$ with respect to $\overline{\mathbf{h}}$ and λ yields the equations:

$$\boldsymbol{\Upsilon} \overline{\mathbf{h}} - \lambda \overline{\mathbf{h}} = \mathbf{b}, \tag{6.46}$$

$$\overline{\mathbf{h}}^H \overline{\mathbf{h}} = \epsilon_+. \tag{6.47}$$

It can be shown [2, 3] that the smallest solution λ of the two previous equations is needed to solve (6.44).

Assume that λ is not an eigenvalue[1] of $\mathbf{\Upsilon}$ and let

$$\begin{aligned} \overline{\overline{\mathbf{h}}} &= \left(\mathbf{\Upsilon} - \lambda\mathbf{I}_{M-1}\right)^{-2}\mathbf{b} \\ &= \left(\mathbf{\Upsilon} - \lambda\mathbf{I}_{M-1}\right)^{-1}\overline{\mathbf{h}}. \end{aligned} \quad (6.48)$$

We can write (6.46)–(6.47) in terms of $\overline{\overline{\mathbf{h}}}$ [3, 4]:

$$\left(\mathbf{\Upsilon} - \lambda\mathbf{I}_{M-1}\right)^2 \overline{\overline{\mathbf{h}}} = \mathbf{b}, \quad (6.49)$$

$$\begin{aligned} \overline{\overline{\mathbf{h}}}^H \left(\mathbf{\Upsilon} - \lambda\mathbf{I}_{M-1}\right)^2 \overline{\overline{\mathbf{h}}} &= \epsilon_+ \\ &= \overline{\overline{\mathbf{h}}}^H \mathbf{b}. \end{aligned} \quad (6.50)$$

From this second equation, we easily see that

$$\frac{\mathbf{b}^H \overline{\overline{\mathbf{h}}}}{\epsilon_+} = 1. \quad (6.51)$$

Developing (6.49) and exploiting the previous result, we get

$$\begin{aligned} \left(\mathbf{\Upsilon}^2 - 2\lambda\mathbf{\Upsilon} + \lambda^2\mathbf{I}_{M-1}\right)\overline{\overline{\mathbf{h}}} &= \mathbf{b} \\ &= \frac{\mathbf{b}\mathbf{b}^H}{\epsilon_+}\overline{\overline{\mathbf{h}}}. \end{aligned} \quad (6.52)$$

As a consequence, (6.52) can be rewritten as

$$\left[\lambda^2\mathbf{I}_{M-1} - 2\lambda\mathbf{\Upsilon} + \left(\mathbf{\Upsilon}^2 - \frac{\mathbf{b}\mathbf{b}^H}{\epsilon_+}\right)\right]\overline{\overline{\mathbf{h}}} = \mathbf{0} \quad (6.53)$$

or, in a more standard form:

$$\left(\lambda^2\mathbf{A}_2 + \lambda\mathbf{A}_1 + \mathbf{A}_0\right)\overline{\overline{\mathbf{h}}} = \mathbf{Q}(\lambda)\,\overline{\overline{\mathbf{h}}} = \mathbf{0}, \quad (6.54)$$

where

[1] In fact, since $\mathbf{\Upsilon} > 0$, all eigenvalues of $\mathbf{\Upsilon}$ are real and strictly positive. In order to be consistent with the regularization technique (or robustness), $\mathbf{\Upsilon} - \lambda\mathbf{I}_{M-1}$ must also be positive definite. Therefore, λ must always be negative.

$$\mathbf{A}_2 = \mathbf{I}_{M-1},$$

$$\mathbf{A}_1 = -2\boldsymbol{\Upsilon},$$

$$\mathbf{A}_0 = \boldsymbol{\Upsilon}^2 - \frac{\mathbf{b}\mathbf{b}^H}{\epsilon_+},$$

and $\mathbf{Q}(\lambda)$ is called the λ-matrix, which is an $(M-1) \times (M-1)$ matrix polynomial of degree 2. In (6.54), we recognize the quadratic eigenvalue problem (QEP) [4], where $\overline{\overline{\mathbf{h}}}$ is the right eigenvector corresponding to the eigenvalue λ. Therefore, the solution of (6.44) is

$$\overline{\mathbf{h}} = (\boldsymbol{\Upsilon} - \lambda\mathbf{I}_{M-1})^{-1}\,\mathbf{b}, \tag{6.55}$$

where λ is the smallest eigenvalue of (6.54).

The QEP is a generalization of the standard eigenvalue problem (SEP), $\mathbf{A}\mathbf{x} = \lambda\mathbf{x}$, and the generalized eigenvalue problem (GEP), $\mathbf{A}\mathbf{x} = \lambda\mathbf{B}\mathbf{x}$. With a λ-matrix of size $(M-1) \times (M-1)$, the QEP has $2(M-1)$ eigenvalues (finite or infinite) with up to $2(M-1)$ right and $2(M-1)$ left eigenvectors, and if there are more than $M-1$ eigenvectors, they are not, of course, linearly independent [4]. In our context, since $\mathbf{A}_2 = \mathbf{I}_{M-1}$ is nonsingular, the QEP has $2(M-1)$ finite eigenvalues. Since $\mathbf{A}_0$, $\mathbf{A}_1$, and $\mathbf{A}_2$ are Hermitian, the eigenvalues are real or come in pairs (λ, λ^*) [4].

One important aspect of the QEP is the linearization process, which is not unique. We say that a $2(M-1) \times 2(M-1)$ matrix $\mathbf{A} - \lambda\mathbf{B}$ is a linearization of the $(M-1) \times (M-1)$ λ-matrix $\mathbf{Q}(\lambda)$ if the eigenvalues of these matrices coincide. Thanks to this process, we can solve the QEP with the GEP. One possibility is to take

$$\mathbf{A} = \begin{bmatrix} \mathbf{0} & \mathbf{A}_0 \\ \mathbf{A}_0 & \mathbf{A}_1 \end{bmatrix}, \tag{6.56}$$

$$\mathbf{B} = \begin{bmatrix} \mathbf{A}_0 & \mathbf{0} \\ \mathbf{0} & -\mathbf{A}_2 \end{bmatrix}. \tag{6.57}$$

It can be verified that if $\mathbf{x}$ is an eigenvector of $\mathbf{Q}(\lambda)$, then

$$\mathbf{z} = \begin{bmatrix} \mathbf{x} \\ \lambda\mathbf{x} \end{bmatrix} \tag{6.58}$$

is an eigenvector of $\mathbf{A} - \lambda\mathbf{B}$. This process corresponds to the symmetric linearization since both $\mathbf{A}$ and $\mathbf{B}$ are Hermitian. The accuracy of this approach will depend on the conditioning of $\mathbf{A}_0$. Another possibility is to take

$$\mathbf{A} = \begin{bmatrix} \mathbf{0} & -\mathbf{I}_{M-1} \\ \mathbf{A}_0 & \mathbf{A}_1 \end{bmatrix}, \tag{6.59}$$

$$\mathbf{B} = \begin{bmatrix} -\mathbf{I}_{M-1} & \mathbf{0} \\ \mathbf{0} & -\mathbf{A}_2 \end{bmatrix}$$
$$= \begin{bmatrix} -\mathbf{I}_{M-1} & \mathbf{0} \\ \mathbf{0} & -\mathbf{I}_{M-1} \end{bmatrix}. \tag{6.60}$$

This linearization is simpler, since we only need to solve the SEP, but is not symmetric. Different linearizations may give different results, so it is important to choose the right one.

Let us order the eigenvalues of $\mathbf{Q}(\lambda)$:

$$\Re(\lambda_1) \leq \Re(\lambda_2) \leq \cdots \leq \Re(\lambda_{2M-2}). \tag{6.61}$$

Then, the optimal robust beamformer that will give exactly a WNG equal to $\mathcal{W}_0$ is

$$\mathbf{h}_{\mathrm{R}} = \mathbf{h}_{\mathrm{DS}} + \overline{\mathbf{U}} \left[\boldsymbol{\Upsilon} - \Re(\lambda_1) \mathbf{I}_{M-1} \right]^{-1} \mathbf{b}. \tag{6.62}$$

Now, if we consider the eigenvalues with the P smallest real and negative parts, we can get another robust beamformer:

$$\mathbf{h}_{\mathrm{R},P} = \mathbf{h}_{\mathrm{DS}} + \overline{\mathbf{U}} \left(\boldsymbol{\Upsilon} - \overline{\lambda}_P \mathbf{I}_{M-1} \right)^{-1} \mathbf{b}, \tag{6.63}$$

where

$$\overline{\lambda}_P = \frac{1}{P} \sum_{p=1}^{P} \Re(\lambda_p). \tag{6.64}$$

In general, $\mathcal{W}\left(\mathbf{h}_{\mathrm{R},P}\right) \neq \mathcal{W}_0$, but $\mathbf{h}_{\mathrm{R},P}$ may be more reliable than $\mathbf{h}_{\mathrm{R}}$ in practice.

6.4 Another Perspective

In the previous sections, we explained the main concepts from the WNG and DS beamformer perspective. In this section, we show the same concepts but from the DF or SNR gain and M beamformer perspective.

Definition 6.1 Let $\mathbf{h}$ be a distortionless beamformer. The transformed beamformer corresponding to $\mathbf{h}$ is defined as

$$\mathbf{g} = \boldsymbol{\Gamma}^{1/2}\mathbf{h}. \tag{6.65}$$

Therefore, we find that the transformed M beamformer is

$$\mathbf{g}_{\mathrm{M}} = \mathbf{\Gamma}^{1/2}\mathbf{h}_{\mathrm{M}} \tag{6.66}$$
$$= \frac{\mathbf{\Gamma}^{-1/2}\mathbf{d}_{\theta_s,\phi_s}}{\mathbf{d}_{\theta_s,\phi_s}^H \mathbf{\Gamma}^{-1}\mathbf{d}_{\theta_s,\phi_s}}.$$

Property 6.2 Any transformed beamformer, $\mathbf{g} = \mathbf{\Gamma}^{1/2}\mathbf{h}$, where $\mathbf{h}$ is a distortionless beamformer, can be decomposed as [5]

$$\mathbf{g} = \mathbf{g}_{\mathrm{M}} + \overline{\mathbf{T}}\overline{\mathbf{g}}, \tag{6.67}$$

where $\mathbf{g}_{\mathrm{M}}$ is the transformed M beamformer given in (6.66), $\overline{\mathbf{T}}$ (defined below) is a semi-unitary matrix of size $M \times (M-1)$, $\overline{\mathbf{g}}$ is a filter of length $M-1$, and $\mathbf{g}_{\mathrm{M}}^H\overline{\mathbf{T}}\overline{\mathbf{g}} = 0$. In other words, any $\mathbf{g} = \mathbf{\Gamma}^{1/2}\mathbf{h}$, with $\mathbf{h}$ being distortionless, can be written as the sum of two orthogonal filters: the transformed M beamformer and another reduced-rank beamformer, i.e., $\overline{\mathbf{T}}\overline{\mathbf{g}}$.

Proof Thanks to the eigenvalue decomposition, the matrix $\mathbf{\Gamma}^{-1/2}\mathbf{d}_{\theta_s,\phi_s}\mathbf{d}_{\theta_s,\phi_s}^H\mathbf{\Gamma}^{-1/2}$ of rank 1 can be diagonalized as

$$\mathbf{T}^H\mathbf{\Gamma}^{-1/2}\mathbf{d}_{\theta_s,\phi_s}\mathbf{d}_{\theta_s,\phi_s}^H\mathbf{\Gamma}^{-1/2}\mathbf{T} = \mathrm{diag}\left(\nu_1, 0, \ldots, 0\right), \tag{6.68}$$

where $\nu_1 = \mathbf{d}_{\theta_s,\phi_s}^H\mathbf{\Gamma}^{-1}\mathbf{d}_{\theta_s,\phi_s}$,

$$\mathbf{T} = \begin{bmatrix} \mathbf{t}_1 & \mathbf{t}_2 & \cdots & \mathbf{t}_M \end{bmatrix} \tag{6.69}$$
$$= \begin{bmatrix} \mathbf{t}_1 & \overline{\mathbf{T}} \end{bmatrix}$$

is a unitary matrix, i.e., $\mathbf{T}^H\mathbf{T} = \mathbf{T}\mathbf{T}^H = \mathbf{I}_M$,

$$\mathbf{t}_1 = \frac{\mathbf{\Gamma}^{-1/2}\mathbf{d}_{\theta_s,\phi_s}}{\sqrt{\mathbf{d}_{\theta_s,\phi_s}^H\mathbf{\Gamma}^{-1}\mathbf{d}_{\theta_s,\phi_s}}}, \tag{6.70}$$

and the matrix $\overline{\mathbf{T}}$ contains the eigenvectors of the null eigenvalues of $\mathbf{\Gamma}^{-1/2}\mathbf{d}_{\theta_s,\phi_s}\mathbf{d}_{\theta_s,\phi_s}^H\mathbf{\Gamma}^{-1/2}$. From (6.68), it is easy to see that $\mathbf{d}_{\theta_s,\phi_s}^H\mathbf{\Gamma}^{-1/2}\overline{\mathbf{T}} = \mathbf{0}$ (which implies that $\mathbf{g}_{\mathrm{M}}^H\overline{\mathbf{T}}\overline{\mathbf{g}} = 0$). Since $\mathbf{T}$ has full rank, we can always write $\mathbf{g}$ in the new basis:

$$\mathbf{g} = \mathbf{T}\begin{bmatrix} \widetilde{G} \\ \overline{\mathbf{g}} \end{bmatrix} \tag{6.71}$$
$$= \mathbf{t}_1\widetilde{G} + \overline{\mathbf{T}}\overline{\mathbf{g}}.$$

Left multiplying both sides of the previous expression by $\mathbf{d}_{\theta_s,\phi_s}^H\mathbf{\Gamma}^{-1/2}$, we get

$$\mathbf{d}_{\theta_s,\phi_s}^H \mathbf{\Gamma}^{-1/2}\mathbf{g} = 1 = \mathbf{d}_{\theta_s,\phi_s}^H \mathbf{\Gamma}^{-1/2}\mathbf{t}_1 \widetilde{G}. \tag{6.72}$$

As a result,

$$\widetilde{G} = \frac{1}{\mathbf{d}_{\theta_s,\phi_s}^H \mathbf{\Gamma}^{-1/2}\mathbf{t}_1} = \frac{1}{\sqrt{\mathbf{d}_{\theta_s,\phi_s}^H \mathbf{\Gamma}^{-1}\mathbf{d}_{\theta_s,\phi_s}}} \tag{6.73}$$

and (6.71) becomes

$$\mathbf{g} = \frac{\mathbf{\Gamma}^{-1/2}\mathbf{d}_{\theta_s,\phi_s}}{\mathbf{d}_{\theta_s,\phi_s}^H \mathbf{\Gamma}^{-1}\mathbf{d}_{\theta_s,\phi_s}} + \mathbf{T}\overline{\mathbf{g}} = \mathbf{g}_{\mathrm{M}} + \mathbf{T}\overline{\mathbf{g}}. \quad \blacksquare \tag{6.74}$$

Using the transformed beamformer and Property 6.2, the DF of any distortionless filter, $\mathbf{h}$, can be written as

$$\begin{aligned} \mathcal{D}(\mathbf{h}) &= \frac{1}{\mathbf{g}^H\mathbf{g}} \\ &= \frac{1}{\mathbf{g}_{\mathrm{MDF}}^H \mathbf{g}_{\mathrm{MDF}} + \overline{\mathbf{g}}^H\overline{\mathbf{g}}} \\ &= \frac{\mathcal{D}(\mathbf{h}_{\mathrm{MDF}})}{1 + \mathcal{D}(\mathbf{h}_{\mathrm{MDF}})\mathcal{A}(\overline{\mathbf{g}})}, \end{aligned} \tag{6.75}$$

where

$$\mathcal{D}(\mathbf{h}_{\mathrm{MDF}}) = \mathbf{d}_{\theta_s,\phi_s}^H \mathbf{\Gamma}_{\mathrm{d}}^{-1}\mathbf{d}_{\theta_s,\phi_s}, \tag{6.76}$$

$$\mathcal{A}(\overline{\mathbf{g}}) = \overline{\mathbf{g}}^H\overline{\mathbf{g}}. \tag{6.77}$$

This shows how the DF of any distortionless beamformer, $\mathbf{h}$, can be written as a function of the DF of $\mathbf{h}_{\mathrm{MDF}}$. As we already know, since $\mathcal{A}(\overline{\mathbf{g}}) \geq 0$, we have

$$\mathcal{D}(\mathbf{h}) \leq \mathcal{D}(\mathbf{h}_{\mathrm{MDF}}), \ \forall \mathbf{h} \tag{6.78}$$

and, obviously, $\mathcal{D}(\mathbf{h})$ is maximized when $\overline{\mathbf{g}} = \mathbf{0}$. As we have just seen, the best filter $\mathbf{h}$ as far as the DF is concerned is the MDF beamformer. However, depending on how we choose $\overline{\mathbf{g}}$, there is no limit to the damage we can do to the DF of $\mathbf{h}$.

In the same way, we have

$$\mathcal{G}(\mathbf{h}) = \frac{\mathcal{G}(\mathbf{h}_{\mathrm{MVDR}})}{1 + \mathcal{G}(\mathbf{h}_{\mathrm{MVDR}})\,\mathcal{A}(\overline{\mathbf{g}})}, \tag{6.79}$$

where

$$\mathcal{G}(\mathbf{h}_{\mathrm{MVDR}}) = \mathbf{d}_{\theta_s,\phi_s}^H \boldsymbol{\Gamma}_{\mathbf{v}}^{-1} \mathbf{d}_{\theta_s,\phi_s}. \tag{6.80}$$

This shows how the SNR gain of any distortionless beamformer, **h**, can be written as a function of the SNR gain of $\mathbf{h}_{\mathrm{MVDR}}$.

Again, thanks to the transformed beamformer and Property 6.2, the WNG of any distortionless filter, **h**, can be expressed as

$$\begin{aligned} \mathcal{W}(\mathbf{h}) &= \frac{1}{\mathbf{g}^H \boldsymbol{\Gamma}^{-1} \mathbf{g}} \qquad (6.81) \\ &= \frac{\mathcal{W}(\mathbf{h}_{\mathrm{M}})}{1 + \mathcal{W}(\mathbf{h}_{\mathrm{M}})\,\mathcal{A}_{\mathrm{w}}(\overline{\mathbf{g}})}, \end{aligned}$$

where

$$\mathcal{A}_{\mathrm{w}}(\overline{\mathbf{g}}) = \mathbf{g}_{\mathrm{M}}^H \boldsymbol{\Gamma}^{-1} \overline{\mathbf{T}} \overline{\mathbf{g}} + \overline{\mathbf{g}}^H \overline{\mathbf{T}}^H \boldsymbol{\Gamma}^{-1} \mathbf{g}_{\mathrm{M}} + \overline{\mathbf{g}}^H \overline{\mathbf{T}}^H \boldsymbol{\Gamma}^{-1} \overline{\mathbf{T}}\, \overline{\mathbf{g}}. \tag{6.82}$$

This shows how the WNG of any distortionless beamformer, **h**, can be written as a function of the WNG of $\mathbf{h}_{\mathrm{M}}$. The DS beamformer is obtained by maximizing the WNG or, equivalently, by minimizing $\mathcal{A}_{\mathrm{w}}(\overline{\mathbf{g}})$. We obtain

$$\overline{\mathbf{g}}_{\mathrm{DS}} = -\left(\overline{\mathbf{T}}^H \boldsymbol{\Gamma}^{-1} \overline{\mathbf{T}}\right)^{-1} \overline{\mathbf{T}}^H \boldsymbol{\Gamma}^{-1} \mathbf{g}_{\mathrm{M}}. \tag{6.83}$$

As a result, another interesting formulation of the DS beamformer is

$$\begin{aligned} \mathbf{h}_{\mathrm{DS}} &= \mathbf{h}_{\mathrm{M}} - \boldsymbol{\Gamma}^{-1/2} \overline{\mathbf{T}} \left(\overline{\mathbf{T}}^H \boldsymbol{\Gamma}^{-1} \overline{\mathbf{T}}\right)^{-1} \overline{\mathbf{T}}^H \boldsymbol{\Gamma}^{-1/2} \mathbf{h}_{\mathrm{M}} \qquad (6.84) \\ &= \mathbf{Q}\mathbf{h}_{\mathrm{M}}, \end{aligned}$$

where

$$\mathbf{Q} = \mathbf{I}_M - \boldsymbol{\Gamma}^{-1/2} \overline{\mathbf{T}} \left(\overline{\mathbf{T}}^H \boldsymbol{\Gamma}^{-1} \overline{\mathbf{T}}\right)^{-1} \overline{\mathbf{T}}^H \boldsymbol{\Gamma}^{-1/2} \tag{6.85}$$

is an orthogonal projection matrix. With $\mathbf{h}_{\mathrm{DS}}$ in (6.84), the WNG, DF, and SNR gain are, respectively,

$$\mathcal{W}(\mathbf{h}_{\mathrm{DS}}) = \frac{1}{\mathbf{h}_{\mathrm{M}}^H \mathbf{Q} \mathbf{h}_{\mathrm{M}}} \tag{6.86}$$

and

$$\mathcal{D}\left(\mathbf{h}_{\mathrm{DS}}\right)=\frac{1}{\mathbf{h}_{\mathrm{M}}^{H}\mathbf{Q}\boldsymbol{\Gamma}_{\mathrm{d}}\mathbf{Q}\mathbf{h}_{\mathrm{M}}},\tag{6.87}$$

and

$$\mathcal{G}\left(\mathbf{h}_{\mathrm{DS}}\right)=\frac{1}{\mathbf{h}_{\mathrm{M}}^{H}\mathbf{Q}\boldsymbol{\Gamma}_{\mathbf{v}}\mathbf{Q}\mathbf{h}_{\mathrm{M}}}.\tag{6.88}$$

Also, by simple identification, one can verify that

$$\boldsymbol{\Gamma}^{-1/2}\mathbf{P}\boldsymbol{\Gamma}^{1/2}\mathbf{Q}=\mathbf{Q}\boldsymbol{\Gamma}^{-1/2}\mathbf{P}\boldsymbol{\Gamma}^{1/2}=\mathbf{I}_{M},\tag{6.89}$$

where $\mathbf{Q}$ can be seen as the pseudo-inverse of $\boldsymbol{\Gamma}^{-1/2}\mathbf{P}\boldsymbol{\Gamma}^{1/2}$.

Now, we would like to find a distortionless beamformer, $\mathbf{h}$, that maximizes the WNG and whose DF/SNR gain is equal to $\mathcal{V}_0$. We deduce that the criterion corresponding to this problem is

$$\min_{\mathbf{h}}\mathbf{h}^{H}\mathbf{h}\quad\text{subject to}\quad\begin{cases}\mathbf{h}^{H}\boldsymbol{\Gamma}\mathbf{h}=\mathcal{V}_0^{-1}\\ \mathbf{h}^{H}\mathbf{d}_{\theta_{\mathrm{s}},\phi_{\mathrm{s}}}=1\end{cases}.\tag{6.90}$$

Using Property 6.2, it is easy to verify that

$$\mathbf{h}^{H}\mathbf{h}=\mathbf{h}_{\mathrm{M}}^{H}\mathbf{h}_{\mathrm{M}}-\overline{\mathbf{g}}^{H}\mathbf{b}'-\mathbf{b}'^{H}\overline{\mathbf{g}}+\overline{\mathbf{g}}^{H}\boldsymbol{\Upsilon}'\overline{\mathbf{g}}\tag{6.91}$$

and

$$\mathbf{h}^{H}\boldsymbol{\Gamma}\mathbf{h}=\mathcal{V}_{\max}^{-1}+\overline{\mathbf{g}}^{H}\overline{\mathbf{g}}=\mathcal{V}_0^{-1},\tag{6.92}$$

where

$$\mathcal{V}_{\max}=\mathbf{d}_{\theta_{\mathrm{s}},\phi_{\mathrm{s}}}^{H}\boldsymbol{\Gamma}^{-1}\mathbf{d}_{\theta_{\mathrm{s}},\phi_{\mathrm{s}}},$$
$$\mathbf{b}'=-\overline{\mathbf{T}}^{H}\boldsymbol{\Gamma}^{-1}\mathbf{g}_{\mathrm{M}},$$
$$\boldsymbol{\Upsilon}'=\overline{\mathbf{T}}^{H}\boldsymbol{\Gamma}^{-1}\overline{\mathbf{T}}.$$

As a result, solving (6.90) is equivalent to solving

$$\min_{\overline{\mathbf{g}}}\left(\overline{\mathbf{g}}^{H}\boldsymbol{\Upsilon}'\overline{\mathbf{g}}-\overline{\mathbf{g}}^{H}\mathbf{b}'-\mathbf{b}'^{H}\overline{\mathbf{g}}\right)\quad\text{subject to}\quad\overline{\mathbf{g}}^{H}\overline{\mathbf{g}}=\epsilon'_{+},\tag{6.93}$$

where

$$\epsilon'_+ = \mathcal{V}_0^{-1} - \mathcal{V}_{\max}^{-1}. \tag{6.94}$$

The Lagrange function associated with (6.93) is

$$\mathcal{L}(\overline{\mathbf{g}}, \lambda) = \overline{\mathbf{g}}^H \boldsymbol{\Upsilon}' \overline{\mathbf{g}} - \overline{\mathbf{g}}^H \mathbf{b}' - \mathbf{b}'^H \overline{\mathbf{g}} - \lambda \left(\overline{\mathbf{g}}^H \overline{\mathbf{g}} - \epsilon'_+ \right), \tag{6.95}$$

where λ is a Lagrange multiplier. Define the vector:

$$\begin{aligned} \overline{\overline{\mathbf{g}}} &= \left(\boldsymbol{\Upsilon}' - \lambda \mathbf{I}_{M-1} \right)^{-2} \mathbf{b}' \\ &= \left(\boldsymbol{\Upsilon}' - \lambda \mathbf{I}_{M-1} \right)^{-1} \overline{\mathbf{g}}. \end{aligned} \tag{6.96}$$

Following the same approach as in Sect. 6.3, it can be shown that the QEP to solve is

$$\left[\lambda^2 \mathbf{I}_{M-1} - 2\lambda \boldsymbol{\Upsilon}' + \left(\boldsymbol{\Upsilon}'^2 - \frac{\mathbf{b}'\mathbf{b}'^H}{\epsilon'_+} \right) \right] \overline{\overline{\mathbf{g}}} = \mathbf{0} \tag{6.97}$$

or, in a more concise way:

$$\left(\lambda^2 \mathbf{A}'_2 + \lambda \mathbf{A}'_1 + \mathbf{A}'_0 \right) \overline{\overline{\mathbf{g}}} = \mathbf{Q}'(\lambda) \overline{\overline{\mathbf{g}}} = \mathbf{0}, \tag{6.98}$$

where

$$\begin{aligned} \mathbf{A}'_2 &= \mathbf{I}_{M-1}, \\ \mathbf{A}'_1 &= -2\boldsymbol{\Upsilon}', \\ \mathbf{A}'_0 &= \boldsymbol{\Upsilon}'^2 - \frac{\mathbf{b}'\mathbf{b}'^H}{\epsilon'_+}. \end{aligned}$$

As a consequence, the solution of (6.93) is

$$\overline{\mathbf{g}} = \left(\boldsymbol{\Upsilon}' - \lambda \mathbf{I}_{M-1} \right)^{-1} \mathbf{b}', \tag{6.99}$$

where λ is the smallest eigenvalue of (6.98). By using the appropriate linearization, we can find the desired eigenvalue.

Let us order the eigenvalues of $\mathbf{Q}'(\lambda)$:

$$\Re(\lambda_1) \le \Re(\lambda_2) \le \cdots \le \Re(\lambda_{2M-2}). \tag{6.100}$$

Then, the optimal beamformer that will give exactly a DF/SNR gain equal to $\mathcal{V}_0$ is

$$\mathbf{h}'_{\mathrm{O}} = \mathbf{h}_{\mathrm{M}} + \boldsymbol{\Gamma}^{-1/2} \overline{\mathbf{T}} \left[\boldsymbol{\Upsilon}' - \Re(\lambda_1) \mathbf{I}_{M-1} \right]^{-1} \mathbf{b}'. \tag{6.101}$$

Now, if we consider the eigenvalues with the P smallest real and negative parts, we can get a slightly suboptimal beamformer:

$$\mathbf{h}'_{\mathrm{SO},P} = \mathbf{h}_{\mathrm{M}} + \mathbf{\Gamma}^{-1/2}\overline{\mathbf{T}}\left(\mathbf{\Upsilon}' - \overline{\lambda}_P \mathbf{I}_{M-1}\right)^{-1}\mathbf{b}', \tag{6.102}$$

where

$$\overline{\lambda}_P = \frac{1}{P}\sum_{p=1}^{P} \Re\left(\lambda_p\right). \tag{6.103}$$

In practice, $\mathbf{h}'_{\mathrm{SO},P}$ may be more reliable than $\mathbf{h}'_{\mathrm{O}}$.

6.5 Illustrative Examples

In this section, we study some examples of distortionless beamformers. For convenience, we still consider a 3-D cube array consisting of $M = M_0^3$ microphones, as shown in Fig. 3.1, with $\delta = 2.0$ cm and $(\theta_{\mathrm{s}}, \phi_{\mathrm{s}}) = (80^\circ, 0^\circ)$.

In the first set of simulations, we evaluate the performance of the robust beamformer derived from the WNG and DS beamformer perspective, $\mathbf{h}_{\mathrm{R}}$, given in (6.62). Generally, the desired level of WNG, $\mathcal{W}_0$, should be smaller than the WNG of the DS beamformer, i.e., $\mathcal{W}_0$ should be smaller than $\mathcal{W}(\mathbf{h}_{\mathrm{DS}}) = M$; otherwise, we will just obtain the DS beamformer.

We first consider a spherically isotropic noise field, so that the noise coherence matrix is $\mathbf{\Gamma} = \mathbf{\Gamma}_{\mathrm{d}}$, and set $\mathcal{W}_0 = -20$ dB. Figures 6.1 and 6.2 show plots of the power patterns, $\left|\mathcal{B}_{\theta,\phi}(\mathbf{h})\right|^2$, of the robust beamformer versus the azimuth angle for $\theta = \theta_{\mathrm{s}}$, and versus the elevation angle for $\phi = \phi_{\mathrm{s}}$, respectively, for $f = 2000$ Hz and different numbers of sensors, $M = M_0^3$. As seen, with the same desired level of WNG, the power patterns have narrower main beams and more nulls when the number of sensors increases. Figure 6.3 shows plots of the DF, $\mathcal{D}(\mathbf{h})$, and WNG, $\mathcal{W}(\mathbf{h})$, of this beamformer as a function of frequency for $\mathcal{W}_0 = -20$ dB and different numbers of sensors, $M = M_0^3$. It is observed that the designed beamformer maintains the preset level of WNG, i.e., $\mathcal{W}_0$, over the studied frequency band of interest while achieving the maximum possible DF. It is also observed that the robust beamformer achieves higher DFs as the value of M_0 increases since there are more degrees of freedom that can be used for DF maximization. Figure 6.4 shows plots of the DF, $\mathcal{D}(\mathbf{h})$, and WNG, $\mathcal{W}(\mathbf{h})$, of this beamformer as a function of frequency for $M = 4^3$ and different desired levels of WNG, $\mathcal{W}_0$. As seen, the beamformer maintains the minimum required WNG level over the studied frequency band of interest. Also, it can be clearly observed that the DF increases when the value of the required WNG level decreases.

We then consider an environment with point source noises, diffuse noise, and white noise, where the two statistically independent interferences impinge

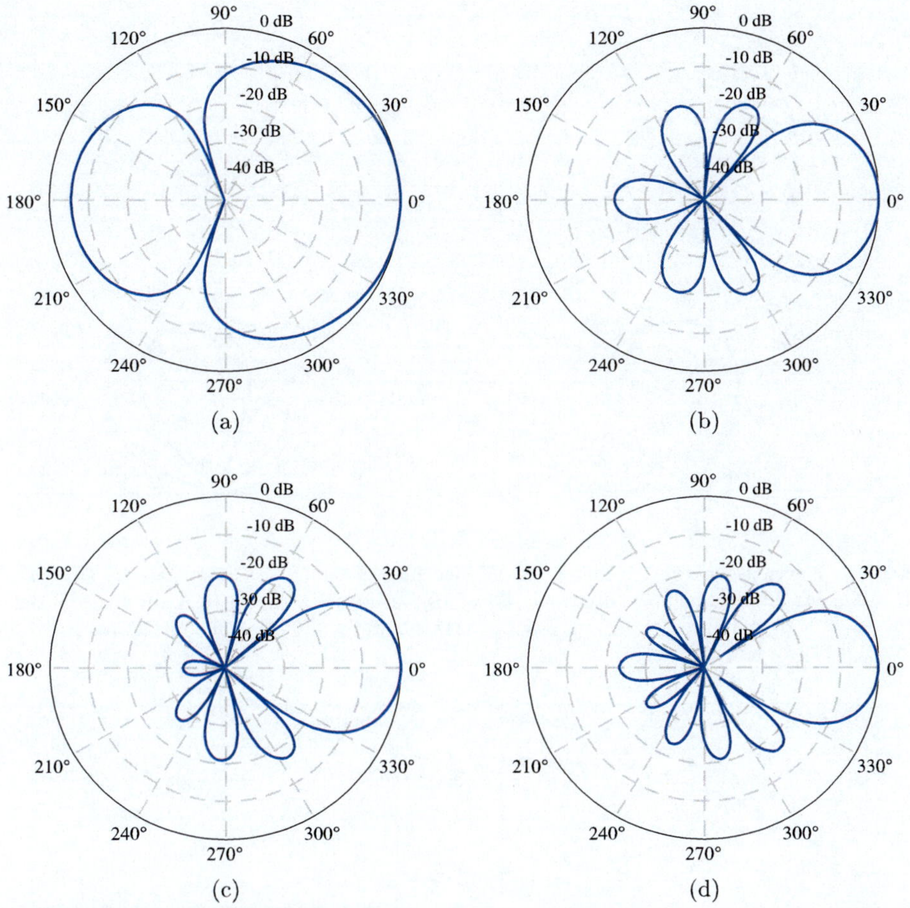

Fig. 6.1 Power patterns of the robust beamformer, $\mathbf{h}_R$, versus the azimuth angle for $\theta = \theta_s$, $\boldsymbol{\Gamma} = \boldsymbol{\Gamma}_d$, and different numbers of sensors, $M = M_0^3$: **(a)** $M_0 = 2$, **(b)** $M_0 = 3$, **(c)** $M_0 = 4$, and **(d)** $M_0 = 5$. Conditions: $\delta = 2.0$ cm, $f = 2000$ Hz, $(\theta_s, \phi_s) = (80°, 0°)$, and the desired level of WNG is $\mathcal{W}_0 = -20$ dB

on the cube array from two different directions $(\theta_{i,1}, \phi_{i,1})$ and $(\theta_{i,2}, \phi_{i,2})$. The two interferences are incoherent with the desired signal. We choose $M_0 = 4$, $(\theta_{i,1}, \phi_{i,1}) = (80°, 150°)$, and $(\theta_{i,2}, \phi_{i,2}) = (80°, 210°)$. The covariance matrix of the noise signal, $\boldsymbol{\Phi}_\mathbf{v}$, is given in (4.95), with $\alpha = 10^{-6}$ ($\alpha > 0$ is related to the ratio between the powers of the white and diffuse noises), and the overall input SNR is set to 0 dB. The variance of the desired signal, ϕ_X, is set to 1, and the variance of the noise is computed according to the specified input SNR and value of α. The matrix $\boldsymbol{\Gamma}_v$ is obtained as $\boldsymbol{\Gamma}_v = \boldsymbol{\Phi}_\mathbf{v}/\mathbf{i}_1^T \boldsymbol{\Phi}_\mathbf{v}\mathbf{i}_1$. Figures 6.5 and 6.6 show plots of the power patterns, $\left|\mathcal{B}_{\theta,\phi}(\mathbf{h})\right|^2$, of the robust beamformer versus the azimuth angle for $\theta = \theta_s$, and versus the elevation angle for $\phi = \phi_s$, respectively, for $\boldsymbol{\Gamma} = \boldsymbol{\Gamma}_v$,

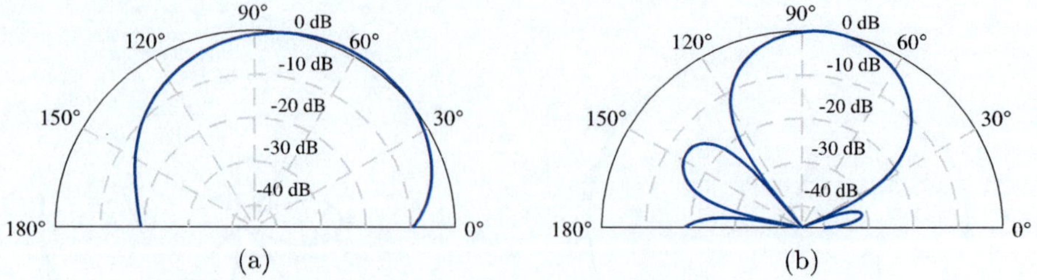

Fig. 6.2 Power patterns of the robust beamformer, $\mathbf{h}_{\text{R}}$, versus the elevation angle for $\phi = \phi_{\text{s}}$, $\mathbf{\Gamma} = \mathbf{\Gamma}_{\text{d}}$, and different numbers of sensors, $M = M_0^3$: (**a**) $M_0 = 2$, (**b**) $M_0 = 3$, (**c**) $M_0 = 4$, and (**d**) $M_0 = 5$. Conditions: $\delta = 2.0$ cm, $f = 2000$ Hz, $(\theta_{\text{s}}, \phi_{\text{s}}) = (80°, 0°)$, and the desired level of WNG is $\mathcal{W}_0 = -20$ dB

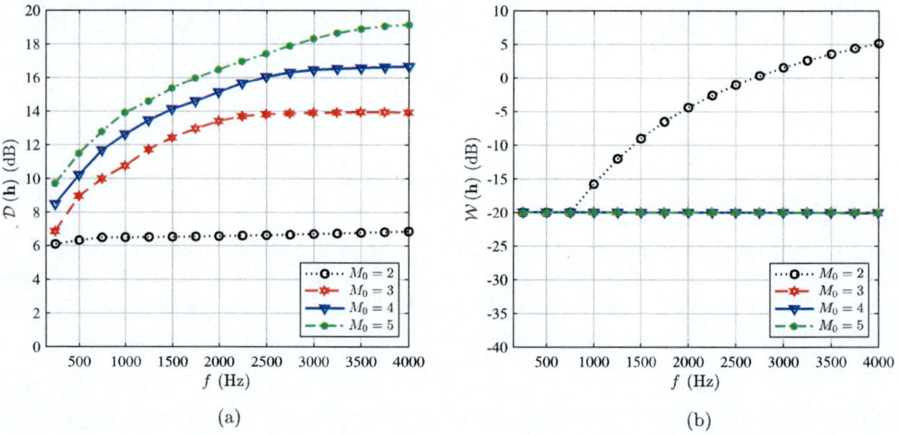

Fig. 6.3 DF and WNG of the robust beamformer, $\mathbf{h}_{\text{R}}$, as a function of frequency for $\mathbf{\Gamma} = \mathbf{\Gamma}_{\text{d}}$ and different numbers of sensors, $M = M_0^3$: (**a**) DF and (**b**) WNG. Conditions: $\delta = 2.0$ cm, $(\theta_{\text{s}}, \phi_{\text{s}}) = (80°, 0°)$, and the desired level of WNG is $\mathcal{W}_0 = -20$ dB

$f = 2000$ Hz, and different desired levels of WNG, $\mathcal{W}_0$. As seen, with the same number of microphones $M = 4^3$, the power patterns have narrower main beams and more nulls with a lower value of $\mathcal{W}_0$. Figure 6.7 shows plots of the SNR gain, $\mathcal{G}(\mathbf{h})$, and WNG, $\mathcal{W}(\mathbf{h})$, of the robust beamformer as a function of frequency for different

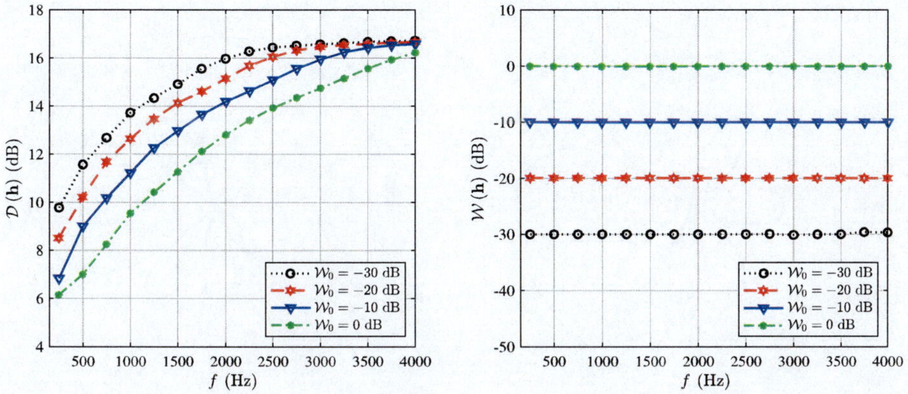

(a) (b)

Fig. 6.4 DF and WNG of the robust beamformer, $\mathbf{h}_{\mathrm{R}}$, as a function of frequency for $\mathbf{\Gamma} = \mathbf{\Gamma}_{\mathrm{d}}$ and different desired levels of WNG, $\mathcal{W}_0$: **(a)** DF and **(b)** WNG. Conditions: $M_0 = 4$, $\delta = 2.0$ cm, and $(\theta_{\mathrm{s}}, \phi_{\mathrm{s}}) = (80^\circ, 0^\circ)$

desired levels of WNG, $\mathcal{W}_0$. It can be observed that the designed beamformer maintains the preset level of WNG while achieving the maximum possible SNR gain.

Next, we study the beamformer derived from the DF (or SNR gain) and M beamformer perspective, $\mathbf{h}'_{\mathrm{O}}$, given in (6.101). Similarly, the desired DF (or SNR gain) should be smaller than the DF (or SNR gain) of the M beamformer; otherwise, the beamformer degenerates to the M beamformer. Figures 6.8 and 6.9 show plots of the power patterns, $\left|\mathcal{B}_{\theta,\phi}(\mathbf{h})\right|^2$, of the robust beamformer versus the azimuth angle for $\theta = \theta_{\mathrm{s}}$, and versus the elevation angle for $\phi = \phi_{\mathrm{s}}$, respectively, for $\mathbf{\Gamma} = \mathbf{\Gamma}_{\mathrm{d}}$, $f = 2000$ Hz, $M_0 = 4$, and different desired levels of DF, $\mathcal{V}_0$. As seen, a higher value of $\mathcal{V}_0$ corresponds to a narrower main beam and more nulls in the beampattern, which obviously means higher directivity. Figure 6.10 shows plots of the DF, $\mathcal{D}(\mathbf{h})$, and WNG, $\mathcal{W}(\mathbf{h})$, of the robust beamformer as a function of frequency for $M_0 = 4$ and different desired levels of DF, $\mathcal{V}_0$. It can be clearly observed that the optimal beamformer achieved the desired level of DF while using the additional flexibility to maximize the WNG; as a result, the value of WNG increases when the value of $\mathcal{V}_0$ decreases.

Finally, we study the performance of the beamformer $\mathbf{h}'_{\mathrm{O}}$ in the noise field consisting of point source noises, diffuse noise, and white noise, where the covariance matrix of the noise signal is given by (4.95) and with $\alpha = 10^{-4}$. The other parameters are the same as the ones in Figs. 6.5 and 6.6. Figures 6.11 and 6.12 show plots of the power patterns, $\left|\mathcal{B}_{\theta,\phi}(\mathbf{h})\right|^2$, of the robust beamformer versus the azimuth angle for $\theta = \theta_{\mathrm{s}}$, and versus the elevation angle for $\phi = \phi_{\mathrm{s}}$, respectively, for $\mathbf{\Gamma} = \mathbf{\Gamma}_{\mathrm{v}}$, $f = 2000$ Hz, and different desired levels of the SNR gain, $\mathcal{V}_0$. As seen, a lower value of $\mathcal{V}_0$ corresponds to a wider main beam and fewer nulls in the beampattern. Figure 6.13 shows plots of the SNR gain, $\mathcal{G}(\mathbf{h})$, and WNG, $\mathcal{W}(\mathbf{h})$,

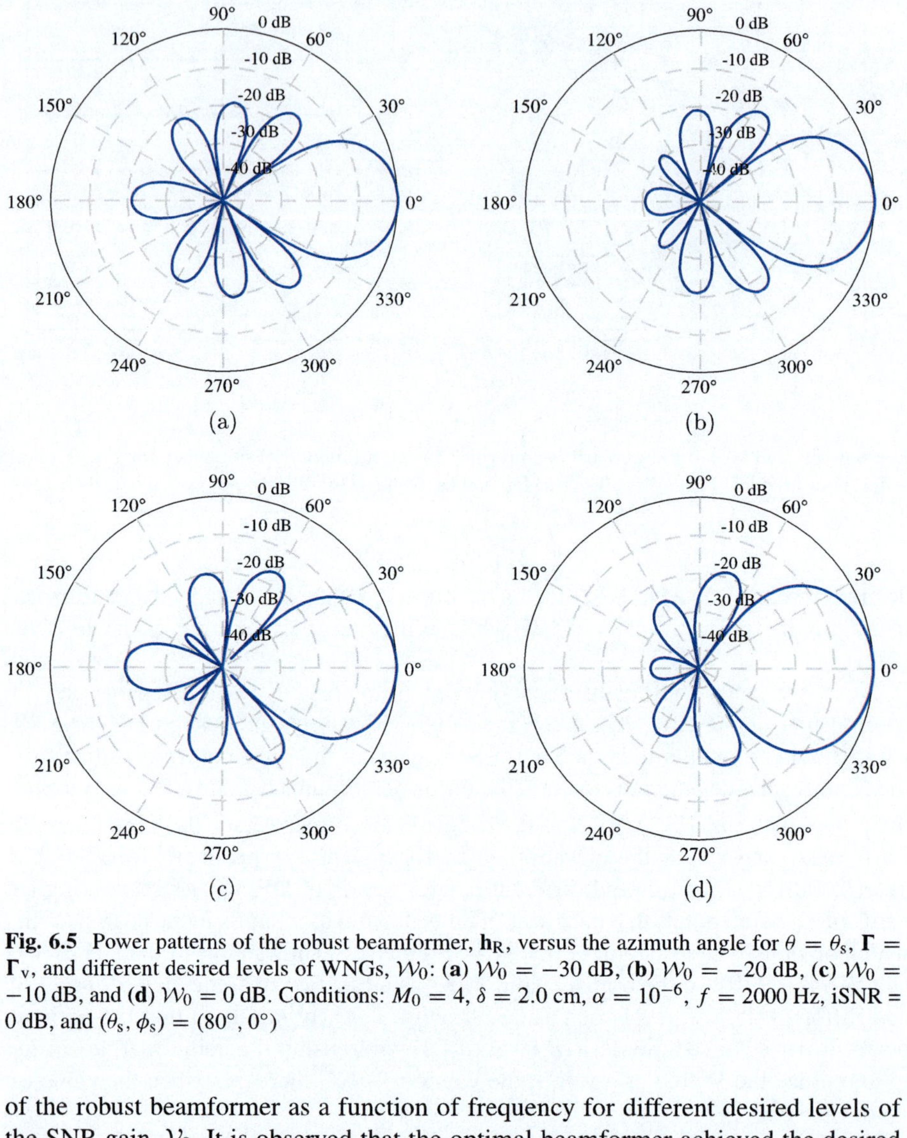

Fig. 6.5 Power patterns of the robust beamformer, $\mathbf{h}_{\mathrm{R}}$, versus the azimuth angle for $\theta = \theta_{\mathrm{s}}$, $\mathbf{\Gamma} = \mathbf{\Gamma}_{\mathrm{v}}$, and different desired levels of WNGs, $\mathcal{W}_0$: **(a)** $\mathcal{W}_0 = -30$ dB, **(b)** $\mathcal{W}_0 = -20$ dB, **(c)** $\mathcal{W}_0 = -10$ dB, and **(d)** $\mathcal{W}_0 = 0$ dB. Conditions: $M_0 = 4$, $\delta = 2.0$ cm, $\alpha = 10^{-6}$, $f = 2000$ Hz, iSNR = 0 dB, and $(\theta_{\mathrm{s}}, \phi_{\mathrm{s}}) = (80^\circ, 0^\circ)$

of the robust beamformer as a function of frequency for different desired levels of the SNR gain, $\mathcal{V}_0$. It is observed that the optimal beamformer achieved the desired level of SNR gains while using the additional flexibility to maximize the WNG; as a result, the value of WNG increases when the value of $\mathcal{V}_0$ decreases.

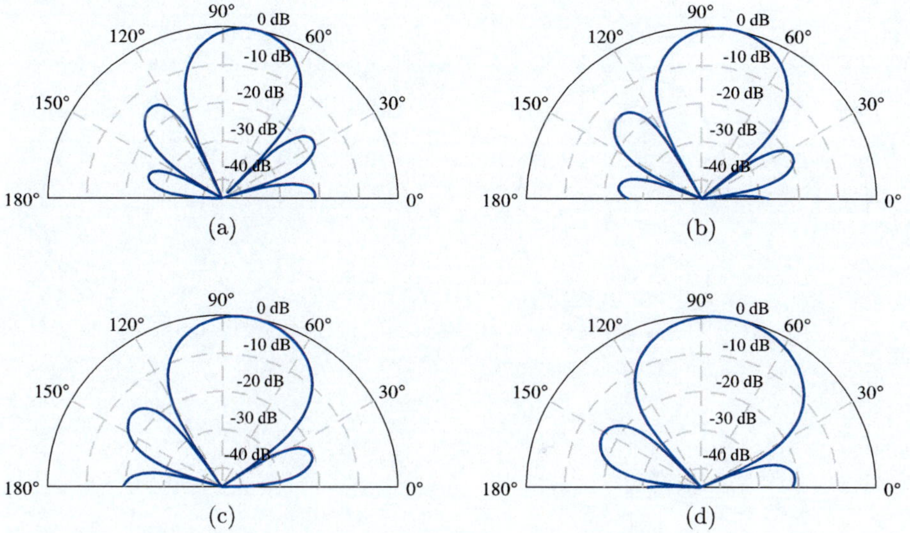

Fig. 6.6 Power patterns of the robust beamformer, $\mathbf{h}_R$, versus the elevation angle for $\phi = \phi_s$, $\mathbf{\Gamma} = \mathbf{\Gamma}_v$, and different desired levels of WNGs, $\mathcal{W}_0$: (**a**) $\mathcal{W}_0 = -30$ dB, (**b**) $\mathcal{W}_0 = -20$ dB, (**c**) $\mathcal{W}_0 = -10$ dB, and (**d**) $\mathcal{W}_0 = 0$ dB. Conditions: $M_0 = 4$, $\delta = 2.0$ cm, $\alpha = 10^{-6}$, $f = 2000$ Hz, iSNR = 0 dB, and $(\theta_s, \phi_s) = (80^\circ, 0^\circ)$

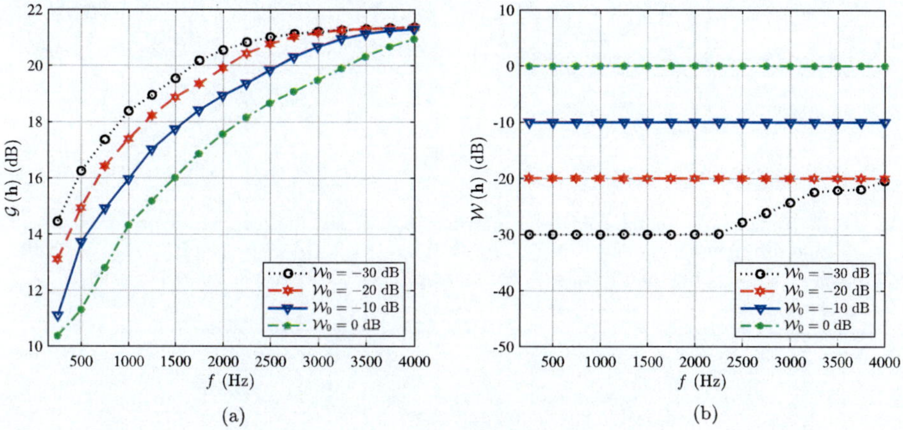

Fig. 6.7 SNR gain and WNG of the robust beamformer, $\mathbf{h}_R$, as a function of frequency for $\mathbf{\Gamma} = \mathbf{\Gamma}_v$ and different desired levels of WNG, $\mathcal{W}_0$: (**a**) SNR gain and (**b**) WNG. Conditions: $M_0 = 4$, $\delta = 2.0$ cm, $\alpha = 10^{-6}$, iSNR = 0 dB, and $(\theta_s, \phi_s) = (80^\circ, 0^\circ)$

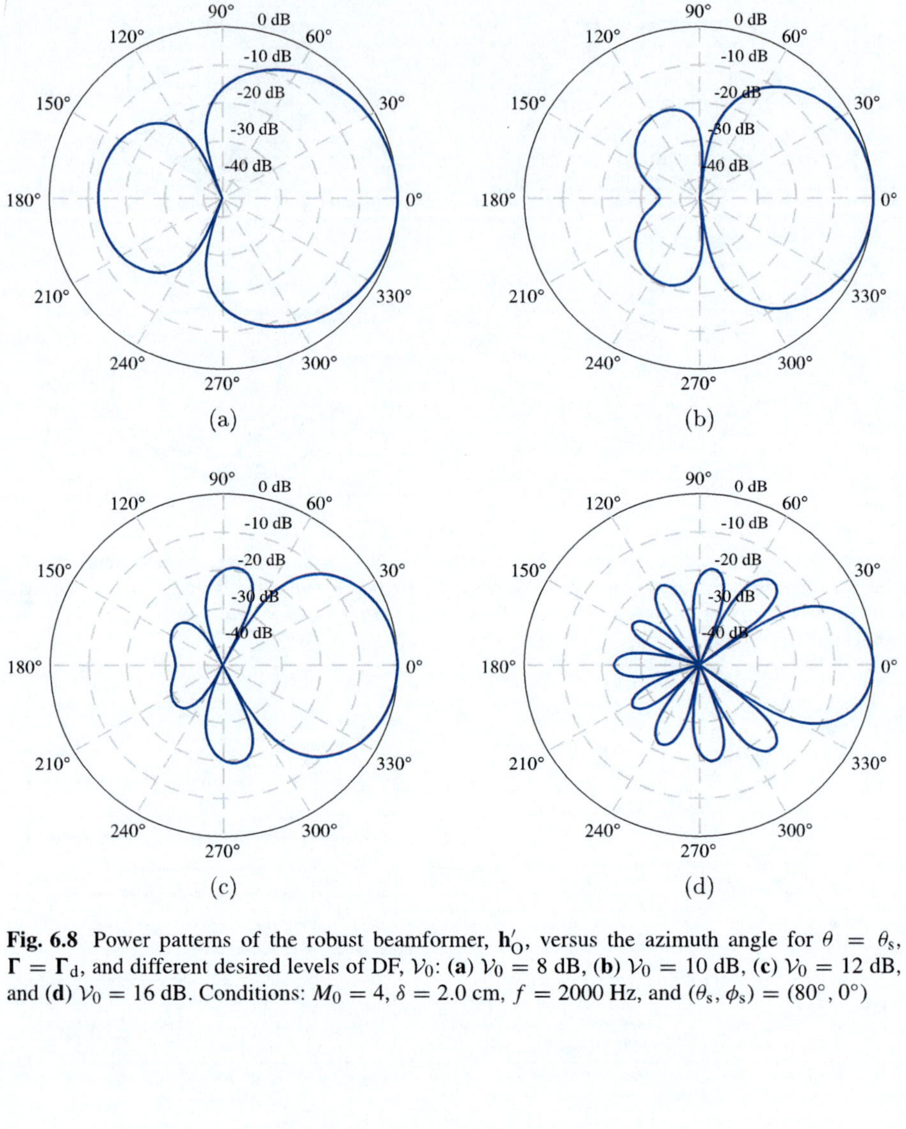

Fig. 6.8 Power patterns of the robust beamformer, $\mathbf{h}'_{\mathrm{O}}$, versus the azimuth angle for $\theta = \theta_s$, $\mathbf{\Gamma} = \mathbf{\Gamma}_d$, and different desired levels of DF, $\mathcal{V}_0$: **(a)** $\mathcal{V}_0 = 8$ dB, **(b)** $\mathcal{V}_0 = 10$ dB, **(c)** $\mathcal{V}_0 = 12$ dB, and **(d)** $\mathcal{V}_0 = 16$ dB. Conditions: $M_0 = 4$, $\delta = 2.0$ cm, $f = 2000$ Hz, and $(\theta_s, \phi_s) = (80°, 0°)$

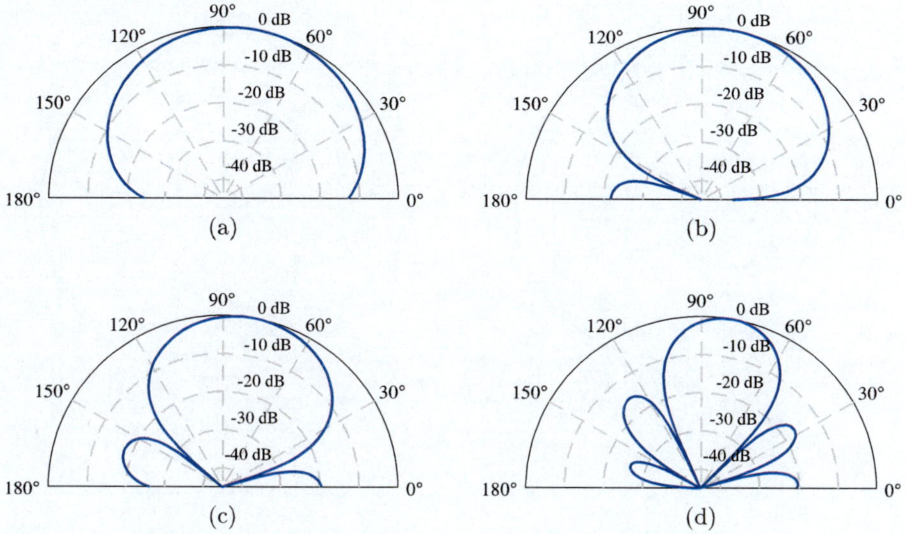

Fig. 6.9 Power patterns of the robust beamformer, $\mathbf{h}'_{\mathrm{O}}$, versus the elevation angle for $\phi = \phi_{\mathrm{s}}$, $\mathbf{\Gamma} = \mathbf{\Gamma}_{\mathrm{d}}$, and different desired levels of DF, $\mathcal{V}_{\mathrm{O}}$: (**a**) $\mathcal{V}_{\mathrm{O}} = 8$ dB, (**b**) $\mathcal{V}_{\mathrm{O}} = 10$ dB, (**c**) $\mathcal{V}_{\mathrm{O}} = 12$ dB, and (**d**) $\mathcal{V}_{\mathrm{O}} = 16$ dB. Conditions: $M_{\mathrm{O}} = 4$, $\delta = 2.0$ cm, $f = 2000$ Hz, and $(\theta_{\mathrm{s}}, \phi_{\mathrm{s}}) = (80^\circ, 0^\circ)$

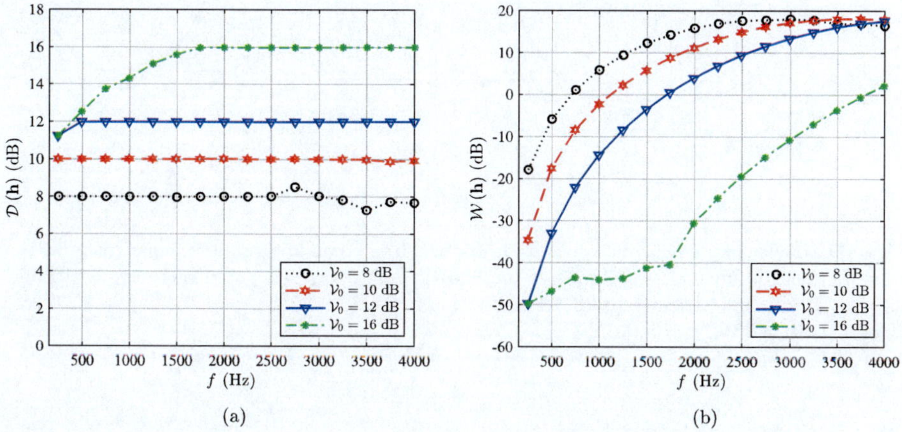

Fig. 6.10 DF and WNG of the robust beamformer, $\mathbf{h}'_{\mathrm{O}}$, as a function of frequency for $\mathbf{\Gamma} = \mathbf{\Gamma}_{\mathrm{d}}$ and different desired levels of DF, $\mathcal{V}_{\mathrm{O}}$: (**a**) DF and (**b**) WNG. Conditions: $M_{\mathrm{O}} = 4$, $\delta = 2.0$ cm, and $(\theta_{\mathrm{s}}, \phi_{\mathrm{s}}) = (80^\circ, 0^\circ)$

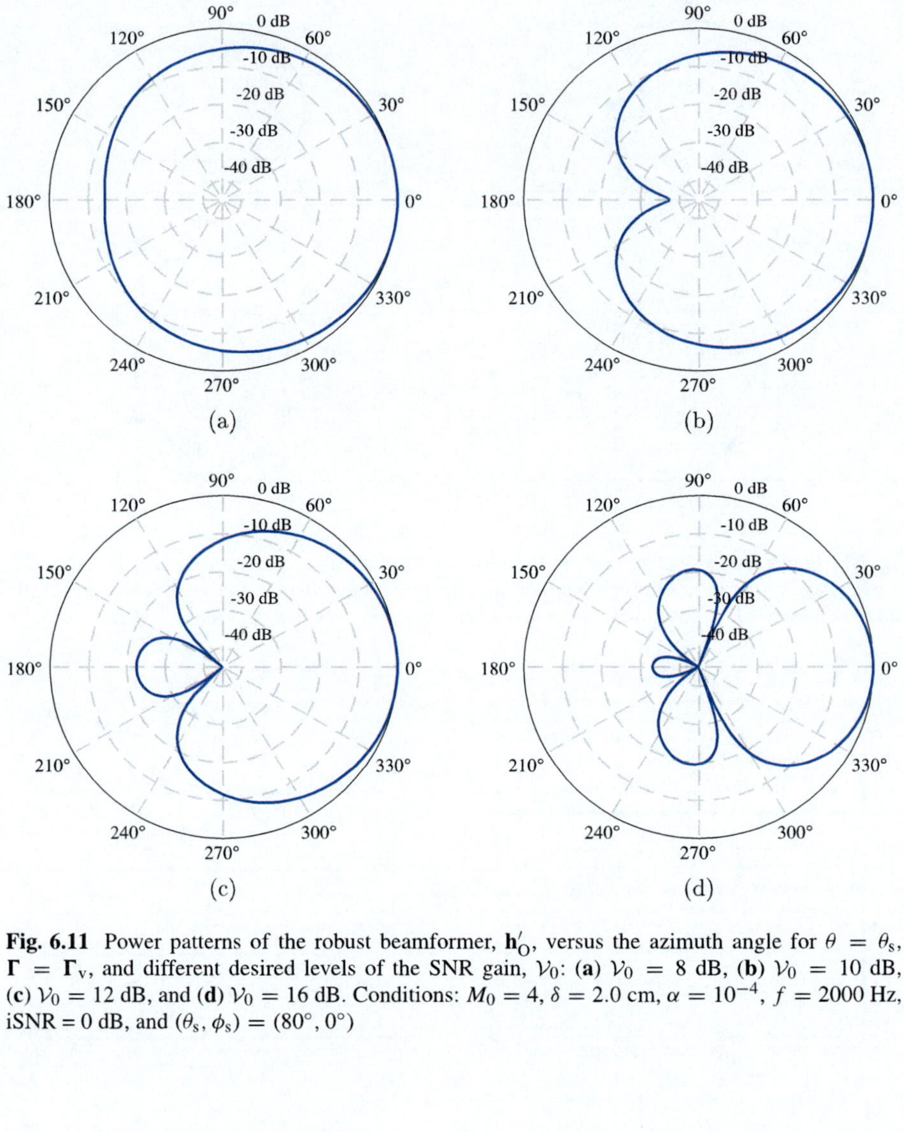

Fig. 6.11 Power patterns of the robust beamformer, $\mathbf{h}'_{\mathrm{O}}$, versus the azimuth angle for $\theta = \theta_s$, $\mathbf{\Gamma} = \mathbf{\Gamma}_v$, and different desired levels of the SNR gain, $\mathcal{V}_0$: (**a**) $\mathcal{V}_0 = 8$ dB, (**b**) $\mathcal{V}_0 = 10$ dB, (**c**) $\mathcal{V}_0 = 12$ dB, and (**d**) $\mathcal{V}_0 = 16$ dB. Conditions: $M_0 = 4$, $\delta = 2.0$ cm, $\alpha = 10^{-4}$, $f = 2000$ Hz, iSNR = 0 dB, and $(\theta_s, \phi_s) = (80^\circ, 0^\circ)$

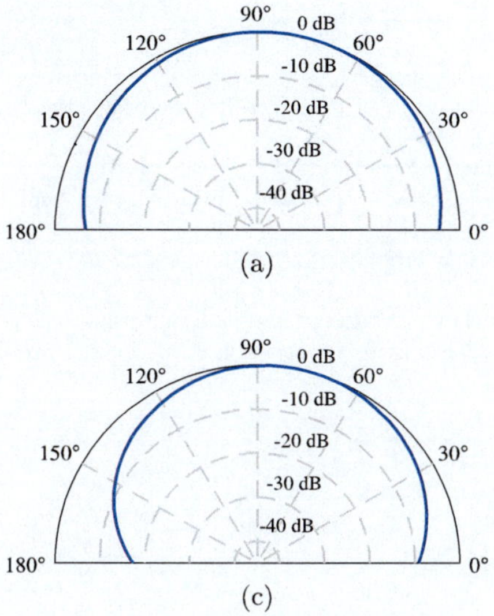

Fig. 6.12 Power patterns of the robust beamformer, $\mathbf{h}'_{\text{O}}$, versus the elevation angle for $\phi = \phi_s$, $\mathbf{\Gamma} = \mathbf{\Gamma}_v$, and different desired levels of the SNR gain, $\mathcal{V}_0$: **(a)** $\mathcal{V}_0 = 8$ dB, **(b)** $\mathcal{V}_0 = 10$ dB, **(c)** $\mathcal{V}_0 = 12$ dB, and **(d)** $\mathcal{V}_0 = 16$ dB. Conditions: $M_0 = 4$, $\delta = 2.0$ cm, $\alpha = 10^{-4}$, $f = 2000$ Hz, iSNR = 0 dB, and $(\theta_s, \phi_s) = (80°, 0°)$

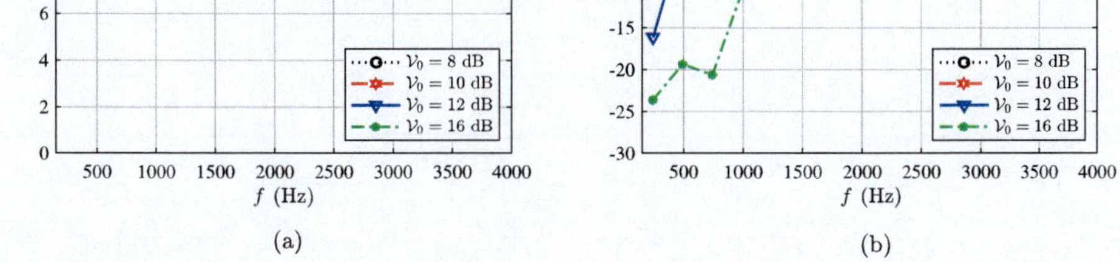

Fig. 6.13 SNR gain and WNG of the robust beamformer, $\mathbf{h}'_{\text{O}}$, as a function of frequency for $\mathbf{\Gamma} = \mathbf{\Gamma}_v$ and different desired levels of the SNR gain, $\mathcal{V}_0$: **(a)** SNR gain and **(b)** WNG. Conditions: $M_0 = 4$, $\delta = 2.0$ cm, $\alpha = 10^{-4}$, iSNR = 0 dB, and $(\theta_s, \phi_s) = (80°, 0°)$

References

1. X. Chen, J. Benesty, G. Huang, J. Chen, On the robustness of the superdirective beamformer. IEEE/ACM Trans. Audio Speech Lang. Process. **29**, 838–849 (2021)
2. W. Gander, Least squares with a quadratic constraint. Numer. Math. **36**, 291–307 (1981)
3. W. Gander, G.H. Golub, U. von Matt, A constrained eigenvalue problem. Linear Algebra Appl. **114/115**, 815–839 (1989)
4. F. Tisseur, K. Meerbergen, The quadratic eigenvalue problem. SIAM Rev. **43**(2), 235–286 (2001)
5. G. Huang, J. Benesty, J. Chen, Fundamental approaches to robust differential beamforming with high directivity factors. IEEE/ACM Trans. Audio Speech Lang. Process. **30**, 3074–3088 (2022)

Chapter 7
Differential Beamforming

Another fundamental way to perform beamforming is by considering pressure differences among microphones instead of direct pressure as in conventional beamforming. This leads to the co-called differential beamforming with at least two great advantages: frequency-invariant beampatterns and high directional gains. Although there are different strategies to derive differential beamformers, none of the developed and available approaches do it in an elegant and systematic way. This is what we attempt to do in this chapter by starting with the simplest case: first-order linear difference equations.

7.1 Signal Model

We consider the signal model of Chap. 3 (with a 3-D array of M microphones), i.e.,

$$\begin{aligned}\mathbf{y} &= \begin{bmatrix} Y_1 & Y_2 & \cdots & Y_M \end{bmatrix}^T \\ &= \mathbf{x} + \mathbf{v} \\ &= \mathbf{d}_{\theta_s,\phi_s} X + \mathbf{v},\end{aligned} \tag{7.1}$$

where

$$\begin{aligned}\mathbf{d}_{\theta_s,\phi_s} &= \left[e^{\jmath 2\pi f \mathbf{a}_{\theta_s,\phi_s}^T \mathbf{r}_1/c} \ e^{\jmath 2\pi f \mathbf{a}_{\theta_s,\phi_s}^T \mathbf{r}_2/c} \ \cdots \ e^{\jmath 2\pi f \mathbf{a}_{\theta_s,\phi_s}^T \mathbf{r}_M/c} \right]^T \\ &= \begin{bmatrix} D_{\theta_s,\phi_s,1} & D_{\theta_s,\phi_s,2} & \cdots & D_{\theta_s,\phi_s,M} \end{bmatrix}^T\end{aligned} \tag{7.2}$$

is the steering vector of length M. The covariance matrix of $\mathbf{y}$ is

J. Benesty et al., *Microphone Arrays*, Springer Topics in Signal Processing 22,
https://doi.org/10.1007/978-3-031-36974-2_7

$$\begin{aligned}\mathbf{\Phi_y} &= E\left(\mathbf{y}\mathbf{y}^H\right)\\ &= \mathbf{\Phi_x} + \mathbf{\Phi_v}\\ &= \phi_X \mathbf{d}_{\theta_s,\phi_s}\mathbf{d}_{\theta_s,\phi_s}^H + \mathbf{\Phi_v}\\ &= \phi_X \mathbf{d}_{\theta_s,\phi_s}\mathbf{d}_{\theta_s,\phi_s}^H + \phi_V \mathbf{\Gamma_v}\end{aligned} \tag{7.3}$$

and, in the case of the spherically isotropic (diffuse) noise field, (7.3) becomes

$$\mathbf{\Phi_y} = \phi_X \mathbf{d}_{\theta_s,\phi_s}\mathbf{d}_{\theta_s,\phi_s}^H + \phi_{\mathrm{d}}\mathbf{\Gamma}_{\mathrm{d}}, \tag{7.4}$$

where ϕ_{d} is the variance of the diffuse noise and $\mathbf{\Gamma}_{\mathrm{d}}$ is defined in Chap. 3. We recall that the input SNR or the SNR corresponding to $\mathbf{y}$ is

$$\mathrm{iSNR} = \frac{\phi_X}{\phi_V}. \tag{7.5}$$

7.2 First-Order Differential Beamforming

From the observed signal vector in (7.1), we can form the following $M-1$ first-order linear spatial difference equations [1]:

$$\begin{aligned}Y_{(1),i} &= A_{0,i}Y_i + A_{1,i}Y_{i+1} \qquad (7.6)\\ &= \left(A_{0,i}D_{\theta_s,\phi_s,i} + A_{1,i}D_{\theta_s,\phi_s,i+1}\right)X + A_{0,i}V_i + A_{1,i}V_{i+1},\end{aligned}$$

for $i = 1, 2, \ldots, M-1$, where the subscript $_{(1)}$ stands for first order and $A_{0,i}$, $A_{1,i} \neq 0$, $i = 1, 2, \ldots, M-1$ are some real-valued parameters, which should be chosen appropriately and carefully so that $Y_{(1),i}$ is a first-order spatial derivative of the observations. The values of these parameters depend mostly on the array geometry; for example, for a uniform linear array (ULA), we have $A_{0,i} = -1$ and $A_{1,i} = 1$ [2], and for a nonuniform linear array (NULA), we have $A_{0,i} = -\left(\delta_{i+1} - \delta_i\right)^{-1}$ and $A_{1,i} = \left(\delta_{i+1} - \delta_i\right)^{-1}$ [3], where δ_i is the spacing between the ith and first microphones with $\delta_1 = 0$. In a vector form, (7.6) becomes

$$\begin{aligned}\mathbf{y}_{(1)} &= \mathbf{\Delta}_{(1)}\mathbf{y} \qquad (7.7)\\ &= \mathbf{x}_{(1)} + \mathbf{v}_{(1)}\\ &= \mathbf{\Delta}_{(1)}\mathbf{d}_{\theta_s,\phi_s}X + \mathbf{\Delta}_{(1)}\mathbf{v},\end{aligned}$$

where

$$\mathbf{\Delta}_{(1)} = \begin{bmatrix} A_{0,1} & A_{1,1} & 0 & \cdots & 0 \\ 0 & A_{0,2} & A_{1,2} & \cdots & 0 \\ \vdots & \vdots & \ddots & \ddots & \vdots \\ 0 & 0 & \cdots & A_{0,M-1} & A_{1,M-1} \end{bmatrix} \tag{7.8}$$

is the first-order difference operator (upper bidiagonal matrix) of size $(M-1) \times M$, which can be seen as the first-order derivative operator. Compared to the conventional approach (as exposed in Chap. 3), where direct pressure on sensors of different quantities, $\mathbf{y}$, $\mathbf{x}$, and $\mathbf{v}$, is handled, here we will process first-order pressure difference through the quantities $\mathbf{y}_{(1)}$, $\mathbf{x}_{(1)}$, and $\mathbf{v}_{(1)}$. We deduce that the $(M-1) \times (M-1)$ covariance matrix of $\mathbf{y}_{(1)}$ is

$$\begin{aligned} \mathbf{\Phi}_{\mathbf{y}_{(1)}} &= \mathbf{\Delta}_{(1)} \mathbf{\Phi}_{\mathbf{y}} \mathbf{\Delta}_{(1)}^H \\ &= \phi_X \mathbf{\Delta}_{(1)} \mathbf{d}_{\theta_s,\phi_s} \mathbf{d}_{\theta_s,\phi_s}^H \mathbf{\Delta}_{(1)}^H + \mathbf{\Delta}_{(1)} \mathbf{\Phi}_{\mathbf{v}} \mathbf{\Delta}_{(1)}^H \\ &= \mathbf{\Phi}_{\mathbf{x}_{(1)}} + \mathbf{\Phi}_{\mathbf{v}_{(1)}}. \end{aligned} \tag{7.9}$$

Then, from (7.9), we find that the SNR corresponding to $\mathbf{y}_{(1)}$ is

$$\begin{aligned} \mathrm{SNR}_{\mathbf{y}_{(1)}} &= \frac{\mathrm{tr}\left(\mathbf{\Phi}_{\mathbf{x}_{(1)}}\right)}{\mathrm{tr}\left(\mathbf{\Phi}_{\mathbf{v}_{(1)}}\right)} \\ &= \mathrm{iSNR} \times \frac{\mathbf{d}_{\theta_s,\phi_s}^H \mathbf{\Delta}_{(1)}^H \mathbf{\Delta}_{(1)} \mathbf{d}_{\theta_s,\phi_s}}{\mathrm{tr}\left(\mathbf{\Delta}_{(1)} \mathbf{\Gamma}_{\mathbf{v}} \mathbf{\Delta}_{(1)}^H\right)}. \end{aligned} \tag{7.10}$$

As a result, the SNR gain (between $\mathbf{y}_{(1)}$ and $\mathbf{y}$) is

$$\begin{aligned} \mathcal{G}_{(1)} &= \frac{\mathrm{SNR}_{\mathbf{y}_{(1)}}}{\mathrm{iSNR}} \\ &= \frac{\mathbf{d}_{\theta_s,\phi_s}^H \mathbf{\Delta}_{(1)}^H \mathbf{\Delta}_{(1)} \mathbf{d}_{\theta_s,\phi_s}}{\mathrm{tr}\left(\mathbf{\Delta}_{(1)} \mathbf{\Gamma}_{\mathbf{v}} \mathbf{\Delta}_{(1)}^H\right)}. \end{aligned} \tag{7.11}$$

For white and diffuse noises, $\mathcal{G}_{(1)}$ becomes

$$\mathcal{W}_{(1)} = \frac{\mathbf{d}_{\theta_s,\phi_s}^H \mathbf{\Delta}_{(1)}^H \mathbf{\Delta}_{(1)} \mathbf{d}_{\theta_s,\phi_s}}{\mathrm{tr}\left(\mathbf{\Delta}_{(1)} \mathbf{\Delta}_{(1)}^H\right)} \tag{7.12}$$

and

$$\mathcal{D}_{(1)} = \frac{\mathbf{d}_{\theta_s,\phi_s}^H \boldsymbol{\Delta}_{(1)}^H \boldsymbol{\Delta}_{(1)} \mathbf{d}_{\theta_s,\phi_s}}{\text{tr}\left(\boldsymbol{\Delta}_{(1)} \boldsymbol{\Gamma}_{\text{d}} \boldsymbol{\Delta}_{(1)}^H\right)}, \tag{7.13}$$

respectively. For an appropriate choice of the parameters $A_{0,i}, A_{1,i}, \ i = 1, 2, \ldots, M-1$, we should expect $\mathcal{D}_{(1)}$ to be much greater than 1, at the expense of $\mathcal{W}_{(1)}$ to be smaller than 1, i.e., white noise amplification. We can also define the power pattern corresponding to $\mathbf{y}_{(1)}$ as

$$\left|\mathcal{B}_{\theta_s,\phi_s,(1)}\right|^2 = \mathbf{d}_{\theta,\phi}^H \boldsymbol{\Delta}_{(1)}^H \boldsymbol{\Delta}_{(1)} \mathbf{d}_{\theta,\phi}. \tag{7.14}$$

In general, the parameters $A_{0,i}, A_{1,i}, \ i = 1, 2, \ldots, M-1$ should be found in such a way that the power pattern forms a first-order dipole in the direction of the desired signal.

Given the new observed signal vector, $\mathbf{y}_{(1)}$ (of length $M-1$), first-order differential beamforming can be performed by applying a complex-valued linear filter, $\mathbf{h}_{(1)}$, to $\mathbf{y}_{(1)}$, i.e.,

$$\begin{aligned} Z_{(1)} &= \sum_{i=1}^{M-1} H_{(1),i}^* Y_{(1),i} \\ &= \mathbf{h}_{(1)}^H \mathbf{y}_{(1)} \\ &= X_{(1),\text{fd}} + V_{(1),\text{rn}}, \end{aligned} \tag{7.15}$$

where $Z_{(1)}$ is the first-order differential beamformer output signal ,

$$\mathbf{h}_{(1)} = \begin{bmatrix} H_{(1),1} & H_{(1),2} & \cdots & H_{(1),M-1} \end{bmatrix}^T \tag{7.16}$$

is the first-order differential filter of length $M-1$,

$$X_{(1),\text{fd}} = X \mathbf{h}_{(1)}^H \boldsymbol{\Delta}_{(1)} \mathbf{d}_{\theta_s,\phi_s} \tag{7.17}$$

is the filtered desired signal, and

$$V_{(1),\text{rn}} = \mathbf{h}_{(1)}^H \boldsymbol{\Delta}_{(1)} \mathbf{v} \tag{7.18}$$

is the residual noise. We immediately see from (7.17) that the distortionless constraint is

$$\mathbf{h}_{(1)}^H \boldsymbol{\Delta}_{(1)} \mathbf{d}_{\theta_s,\phi_s} = 1. \tag{7.19}$$

From (7.15), we can compute the variance of $Z_{(1)}$ as

$$\begin{aligned}\phi_{Z_{(1)}} &= \mathbf{h}_{(1)}^H \mathbf{\Phi}_{\mathbf{y}_{(1)}} \mathbf{h}_{(1)} \\ &= \mathbf{h}_{(1)}^H \mathbf{\Delta}_{(1)} \mathbf{\Phi}_{\mathbf{y}} \mathbf{\Delta}_{(1)}^H \mathbf{h}_{(1)} \\ &= \phi_{X_{(1),\mathrm{fd}}} + \phi_{V_{(1),\mathrm{rn}}},\end{aligned} \tag{7.20}$$

where

$$\phi_{X_{(1),\mathrm{fd}}} = \phi_X \left| \mathbf{h}_{(1)}^H \mathbf{\Delta}_{(1)} \mathbf{d}_{\theta_s,\phi_s} \right|^2, \tag{7.21}$$

$$\phi_{V_{(1),\mathrm{rn}}} = \mathbf{h}_{(1)}^H \mathbf{\Delta}_{(1)} \mathbf{\Phi}_{\mathbf{v}} \mathbf{\Delta}_{(1)}^H \mathbf{h}_{(1)}. \tag{7.22}$$

Then, from the above, it is not hard to see that the gain in SNR is

$$\begin{aligned}\mathcal{G}\left(\mathbf{h}_{(1)}\right) &= \frac{\mathrm{oSNR}\left(\mathbf{h}_{(1)}\right)}{\mathrm{iSNR}} \\ &= \frac{\left| \mathbf{h}_{(1)}^H \mathbf{\Delta}_{(1)} \mathbf{d}_{\theta_s,\phi_s} \right|^2}{\mathbf{h}_{(1)}^H \mathbf{\Delta}_{(1)} \mathbf{\Gamma}_{\mathbf{v}} \mathbf{\Delta}_{(1)}^H \mathbf{h}_{(1)}},\end{aligned} \tag{7.23}$$

where

$$\mathrm{oSNR}\left(\mathbf{h}_{(1)}\right) = \mathrm{iSNR} \times \frac{\left| \mathbf{h}_{(1)}^H \mathbf{\Delta}_{(1)} \mathbf{d}_{\theta_s,\phi_s} \right|^2}{\mathbf{h}_{(1)}^H \mathbf{\Delta}_{(1)} \mathbf{\Gamma}_{\mathbf{v}} \mathbf{\Delta}_{(1)}^H \mathbf{h}_{(1)}} \tag{7.24}$$

is the output SNR. As a result, the WNG and DF are, respectively,

$$\mathcal{W}\left(\mathbf{h}_{(1)}\right) = \frac{\left| \mathbf{h}_{(1)}^H \mathbf{\Delta}_{(1)} \mathbf{d}_{\theta_s,\phi_s} \right|^2}{\mathbf{h}_{(1)}^H \mathbf{\Delta}_{(1)} \mathbf{\Delta}_{(1)}^H \mathbf{h}_{(1)}} \tag{7.25}$$

and

$$\mathcal{D}\left(\mathbf{h}_{(1)}\right) = \frac{\left| \mathbf{h}_{(1)}^H \mathbf{\Delta}_{(1)} \mathbf{d}_{\theta_s,\phi_s} \right|^2}{\mathbf{h}_{(1)}^H \mathbf{\Delta}_{(1)} \mathbf{\Gamma}_{\mathrm{d}} \mathbf{\Delta}_{(1)}^H \mathbf{h}_{(1)}}. \tag{7.26}$$

Finally, the last measure of interest is the beampattern:

$$\mathcal{B}_{\theta,\phi}\left(\mathbf{h}_{(1)}\right) = \mathbf{d}_{\theta,\phi}^H \mathbf{\Delta}_{(1)}^H \mathbf{h}_{(1)}. \tag{7.27}$$

Now, we can derive some useful and obvious optimal first-order differential fixed and adaptive beamformers.

The first fixed beamformer is deduced from the maximization of the WNG in (7.25), i.e.,

$$\min_{\mathbf{h}_{(1)}} \mathbf{h}_{(1)}^H \mathbf{\Delta}_{(1)} \mathbf{\Delta}_{(1)}^H \mathbf{h}_{(1)} \quad \text{subject to} \quad \mathbf{h}_{(1)}^H \mathbf{\Delta}_{(1)} \mathbf{d}_{\theta_s,\phi_s} = 1. \tag{7.28}$$

From the previous minimization, we get the first-order differential maximum WNG (MWNG) beamformer:

$$\mathbf{h}_{(1),\text{MWNG}} = \frac{\left(\mathbf{\Delta}_{(1)} \mathbf{\Delta}_{(1)}^H\right)^{-1} \mathbf{\Delta}_{(1)} \mathbf{d}_{\theta_s,\phi_s}}{\mathbf{d}_{\theta_s,\phi_s}^H \mathbf{\Delta}_{(1)}^H \left(\mathbf{\Delta}_{(1)} \mathbf{\Delta}_{(1)}^H\right)^{-1} \mathbf{\Delta}_{(1)} \mathbf{d}_{\theta_s,\phi_s}}. \tag{7.29}$$

It is clear that the beampattern, WNG, and DF of $\mathbf{h}_{(1),\text{MWNG}}$ are, respectively,

$$\mathcal{B}_{\theta,\phi}\left(\mathbf{h}_{(1),\text{MWNG}}\right) = \frac{\mathbf{d}_{\theta,\phi}^H \mathbf{\Delta}_{(1)}^H \left(\mathbf{\Delta}_{(1)} \mathbf{\Delta}_{(1)}^H\right)^{-1} \mathbf{\Delta}_{(1)} \mathbf{d}_{\theta_s,\phi_s}}{\mathbf{d}_{\theta_s,\phi_s}^H \mathbf{\Delta}_{(1)}^H \left(\mathbf{\Delta}_{(1)} \mathbf{\Delta}_{(1)}^H\right)^{-1} \mathbf{\Delta}_{(1)} \mathbf{d}_{\theta_s,\phi_s}}, \tag{7.30}$$

$$\mathcal{W}\left(\mathbf{h}_{(1),\text{MWNG}}\right) = \mathbf{d}_{\theta_s,\phi_s}^H \mathbf{\Delta}_{(1)}^H \left(\mathbf{\Delta}_{(1)} \mathbf{\Delta}_{(1)}^H\right)^{-1} \mathbf{\Delta}_{(1)} \mathbf{d}_{\theta_s,\phi_s}, \tag{7.31}$$

and

$$\mathcal{D}\left(\mathbf{h}_{(1),\text{MWNG}}\right) = \frac{\left[\mathbf{d}_{\theta_s,\phi_s}^H \mathbf{\Delta}_{(1)}^H \left(\mathbf{\Delta}_{(1)} \mathbf{\Delta}_{(1)}^H\right)^{-1} \mathbf{\Delta}_{(1)} \mathbf{d}_{\theta_s,\phi_s}\right]^2}{\mathbf{d}_{\theta_s,\phi_s}^H \mathbf{\Delta}_{(1)}^H \left(\mathbf{\Delta}_{(1)} \mathbf{\Delta}_{(1)}^H\right)^{-1} \mathbf{\Delta}_{(1)} \mathbf{\Gamma}_{\text{d}} \mathbf{\Delta}_{(1)}^H \left(\mathbf{\Delta}_{(1)} \mathbf{\Delta}_{(1)}^H\right)^{-1} \mathbf{\Delta}_{(1)} \mathbf{d}_{\theta_s,\phi_s}}. \tag{7.32}$$

The second fixed beamformer of interest is the one derived from the maximization of the DF in (7.26). This maximization is equivalent to

$$\min_{\mathbf{h}_{(1)}} \mathbf{h}_{(1)}^H \mathbf{\Delta}_{(1)} \mathbf{\Gamma}_{\text{d}} \mathbf{\Delta}_{(1)}^H \mathbf{h}_{(1)} \quad \text{subject to} \quad \mathbf{h}_{(1)}^H \mathbf{\Delta}_{(1)} \mathbf{d}_{\theta_s,\phi_s} = 1, \tag{7.33}$$

from which we deduce the first-order differential maximum DF (MDF) beamformer:

$$\mathbf{h}_{(1),\text{MDF}} = \frac{\left(\mathbf{\Delta}_{(1)} \mathbf{\Gamma}_{\text{d}} \mathbf{\Delta}_{(1)}^H\right)^{-1} \mathbf{\Delta}_{(1)} \mathbf{d}_{\theta_s,\phi_s}}{\mathbf{d}_{\theta_s,\phi_s}^H \mathbf{\Delta}_{(1)}^H \left(\mathbf{\Delta}_{(1)} \mathbf{\Gamma}_{\text{d}} \mathbf{\Delta}_{(1)}^H\right)^{-1} \mathbf{\Delta}_{(1)} \mathbf{d}_{\theta_s,\phi_s}}. \tag{7.34}$$

As a result, the beampattern, WNG, and DF of $\mathbf{h}_{(1),\text{MDF}}$ are, respectively,

$$\mathcal{B}_{\theta,\phi}\left(\mathbf{h}_{(1),\text{MDF}}\right) = \frac{\mathbf{d}_{\theta,\phi}^H \boldsymbol{\Delta}_{(1)}^H \left(\boldsymbol{\Delta}_{(1)} \boldsymbol{\Gamma}_{\text{d}} \boldsymbol{\Delta}_{(1)}^H\right)^{-1} \boldsymbol{\Delta}_{(1)} \mathbf{d}_{\theta_s,\phi_s}}{\mathbf{d}_{\theta_s,\phi_s}^H \boldsymbol{\Delta}_{(1)}^H \left(\boldsymbol{\Delta}_{(1)} \boldsymbol{\Gamma}_{\text{d}} \boldsymbol{\Delta}_{(1)}^H\right)^{-1} \boldsymbol{\Delta}_{(1)} \mathbf{d}_{\theta_s,\phi_s}}, \tag{7.35}$$

$$\mathcal{W}\left(\mathbf{h}_{(1),\text{MDF}}\right) = \frac{\left[\mathbf{d}_{\theta_s,\phi_s}^H \boldsymbol{\Delta}_{(1)}^H \left(\boldsymbol{\Delta}_{(1)} \boldsymbol{\Gamma}_{\text{d}} \boldsymbol{\Delta}_{(1)}^H\right)^{-1} \boldsymbol{\Delta}_{(1)} \mathbf{d}_{\theta_s,\phi_s}\right]^2}{\mathbf{d}_{\theta_s,\phi_s}^H \boldsymbol{\Delta}_{(1)}^H \left(\boldsymbol{\Delta}_{(1)} \boldsymbol{\Gamma}_{\text{d}} \boldsymbol{\Delta}_{(1)}^H\right)^{-2} \boldsymbol{\Delta}_{(1)} \mathbf{d}_{\theta_s,\phi_s}}, \tag{7.36}$$

and

$$\mathcal{D}\left(\mathbf{h}_{(1),\text{MDF}}\right) = \mathbf{d}_{\theta_s,\phi_s}^H \boldsymbol{\Delta}_{(1)}^H \left(\boldsymbol{\Delta}_{(1)} \boldsymbol{\Gamma}_{\text{d}} \boldsymbol{\Delta}_{(1)}^H\right)^{-1} \boldsymbol{\Delta}_{(1)} \mathbf{d}_{\theta_s,\phi_s}. \tag{7.37}$$

The MDF beamformer will lead to a high value of the DF but at the expense of some white noise amplification, while the MWNG beamformer will lead to a large value of the WNG but at the expense of some low value of the DF. Therefore, in practice, it is important to be able to compromise between WNG and DF. To do so, we propose to exploit the joint diagonalization technique of two Hermitian matrices. The two Hermitian matrices $\boldsymbol{\Delta}_{(1)} \mathbf{d}_{\theta_s,\phi_s} \mathbf{d}_{\theta_s,\phi_s}^H \boldsymbol{\Delta}_{(1)}^H$ and $\boldsymbol{\Delta}_{(1)} \boldsymbol{\Gamma}_{\text{d}} \boldsymbol{\Delta}_{(1)}^H$, which appear in the definition of the DF, can be jointly diagonalized as follows [4]:

$$\mathbf{T}_{(1)}^H \boldsymbol{\Delta}_{(1)} \mathbf{d}_{\theta_s,\phi_s} \mathbf{d}_{\theta_s,\phi_s}^H \boldsymbol{\Delta}_{(1)}^H \mathbf{T}_{(1)} = \boldsymbol{\Lambda}_{(1)}, \tag{7.38}$$

$$\mathbf{T}_{(1)}^H \boldsymbol{\Delta}_{(1)} \boldsymbol{\Gamma}_{\text{d}} \boldsymbol{\Delta}_{(1)}^H \mathbf{T}_{(1)} = \mathbf{I}_{M-1}, \tag{7.39}$$

where

$$\mathbf{T}_{(1)} = \begin{bmatrix} \mathbf{t}_{(1),1} & \mathbf{t}_{(1),2} & \cdots & \mathbf{t}_{(1),M-1} \end{bmatrix} \tag{7.40}$$

is a full-rank square matrix of size $(M-1) \times (M-1)$;

$$\mathbf{t}_{(1),1} = \frac{\left(\boldsymbol{\Delta}_{(1)} \boldsymbol{\Gamma}_{\text{d}} \boldsymbol{\Delta}_{(1)}^H\right)^{-1} \boldsymbol{\Delta}_{(1)} \mathbf{d}_{\theta_s,\phi_s}}{\sqrt{\mathbf{d}_{\theta_s,\phi_s}^H \boldsymbol{\Delta}_{(1)}^H \left(\boldsymbol{\Delta}_{(1)} \boldsymbol{\Gamma}_{\text{d}} \boldsymbol{\Delta}_{(1)}^H\right)^{-1} \boldsymbol{\Delta}_{(1)} \mathbf{d}_{\theta_s,\phi_s}}} \tag{7.41}$$

is the first eigenvector of the matrix $\left(\boldsymbol{\Delta}_{(1)} \boldsymbol{\Gamma}_{\text{d}} \boldsymbol{\Delta}_{(1)}^H\right)^{-1} \boldsymbol{\Delta}_{(1)} \mathbf{d}_{\theta_s,\phi_s} \mathbf{d}_{\theta_s,\phi_s}^H \boldsymbol{\Delta}_{(1)}^H$;

$$\boldsymbol{\Lambda}_{(1)} = \text{diag}\left(\lambda_{(1),1}, 0, \ldots, 0\right) \tag{7.42}$$

is a diagonal matrix of size $(M-1) \times (M-1)$;

$$\lambda_{(1),1} = \mathbf{d}_{\theta_s,\phi_s}^H \mathbf{\Delta}_{(1)}^H \left(\mathbf{\Delta}_{(1)} \mathbf{\Gamma}_d \mathbf{\Delta}_{(1)}^H \right)^{-1} \mathbf{\Delta}_{(1)} \mathbf{d}_{\theta_s,\phi_s} \quad (7.43)$$

is the only nonnull eigenvalue of $\left(\mathbf{\Delta}_{(1)} \mathbf{\Gamma}_d \mathbf{\Delta}_{(1)}^H \right)^{-1} \mathbf{\Delta}_{(1)} \mathbf{d}_{\theta_s,\phi_s} \mathbf{d}_{\theta_s,\phi_s}^H \mathbf{\Delta}_{(1)}^H$, whose corresponding eigenvector is $\mathbf{t}_{(1),1}$; and $\mathbf{I}_{M-1}$ is the $(M-1) \times (M-1)$ identity matrix. It can be checked from (7.38) that

$$\mathbf{t}_{(1),i}^H \mathbf{\Delta}_{(1)} \mathbf{d}_{\theta_s,\phi_s} = 0, \; i = 2, 3, \ldots, M-1. \quad (7.44)$$

Let us define the matrix of size $(M-1) \times P$:

$$\mathbf{T}_{(1),1:P} = \left[\mathbf{t}_{(1),1} \;\; \mathbf{t}_{(1),2} \;\; \cdots \;\; \mathbf{t}_{(1),P} \right], \quad (7.45)$$

with $1 \leq P \leq M-1$. The third and last fixed beamformer that we consider in this section has the following form:

$$\mathbf{h}_{(1),1:P} = \mathbf{T}_{(1),1:P} \boldsymbol{\alpha}, \quad (7.46)$$

where

$$\boldsymbol{\alpha} = \left[\alpha_1 \;\; \alpha_2 \;\; \cdots \;\; \alpha_P \right]^T \neq \mathbf{0} \quad (7.47)$$

is a vector of length P. Substituting (7.46) into (7.15), we find that

$$\begin{aligned} Z_{(1)} &= \boldsymbol{\alpha}^H \mathbf{T}_{(1),1:P}^H \mathbf{\Delta}_{(1)} \mathbf{d}_{\theta_s,\phi_s} X + \boldsymbol{\alpha}^H \mathbf{T}_{(1),1:P}^H \mathbf{\Delta}_{(1)} \mathbf{v} \\ &= \alpha_1^* \sqrt{\lambda_{(1),1}} X + \boldsymbol{\alpha}^H \mathbf{T}_{(1),1:P}^H \mathbf{\Delta}_{(1)} \mathbf{v}. \end{aligned} \quad (7.48)$$

Since the distortionless constraint is desired, it is clear from the previous expression that we always choose

$$\alpha_1 = \frac{1}{\sqrt{\lambda_{(1),1}}}. \quad (7.49)$$

Now, we need to determine the other elements of $\boldsymbol{\alpha}$. Thanks to the nature of the beamformer given in (7.46), we can express the WNG and DF, respectively, as

$$\begin{aligned} \mathcal{W}\left(\mathbf{h}_{(1),1:P}\right) &= \frac{\left| \mathbf{h}_{(1),1:P}^H \mathbf{\Delta}_{(1)} \mathbf{d}_{\theta_s,\phi_s} \right|^2}{\mathbf{h}_{(1),1:P}^H \mathbf{\Delta}_{(1)} \mathbf{\Delta}_{(1)}^H \mathbf{h}_{(1),1:P}} \\ &= \frac{\left| \boldsymbol{\alpha}^H \mathbf{T}_{(1),1:P}^H \mathbf{\Delta}_{(1)} \mathbf{d}_{\theta_s,\phi_s} \right|^2}{\boldsymbol{\alpha}^H \mathbf{T}_{(1),1:P}^H \mathbf{\Delta}_{(1)} \mathbf{\Delta}_{(1)}^H \mathbf{T}_{(1),1:P} \boldsymbol{\alpha}} \end{aligned} \quad (7.50)$$

and

$$\mathcal{D}\left(\mathbf{h}_{(1),1:P}\right) = \frac{\left|\mathbf{h}_{(1),1:P}^H \mathbf{\Delta}_{(1)} \mathbf{d}_{\theta_s,\phi_s}\right|^2}{\mathbf{h}_{(1),1:P}^H \mathbf{\Delta}_{(1)} \mathbf{\Gamma}_{\mathrm{d}} \mathbf{\Delta}_{(1)}^H \mathbf{h}_{(1),1:P}} \tag{7.51}$$

$$= \frac{\left|\boldsymbol{\alpha}^H \mathbf{T}_{(1),1:P}^H \mathbf{\Delta}_{(1)} \mathbf{d}_{\theta_s,\phi_s}\right|^2}{\boldsymbol{\alpha}^H \mathbf{T}_{(1),1:P}^H \mathbf{\Delta}_{(1)} \mathbf{\Gamma}_{\mathrm{d}} \mathbf{\Delta}_{(1)}^H \mathbf{T}_{(1),1:P} \boldsymbol{\alpha}}$$

$$= \frac{\left|\boldsymbol{\alpha}^H \mathbf{T}_{(1),1:P}^H \mathbf{\Delta}_{(1)} \mathbf{d}_{\theta_s,\phi_s}\right|^2}{\boldsymbol{\alpha}^H \boldsymbol{\alpha}}.$$

Now, if we want to compromise between WNG and DF, we need to maximize the WNG given above, i.e.,

$$\min_{\boldsymbol{\alpha}} \boldsymbol{\alpha}^H \mathbf{T}_{(1),1:P}^H \mathbf{\Delta}_{(1)} \mathbf{\Delta}_{(1)}^H \mathbf{T}_{(1),1:P} \boldsymbol{\alpha} \quad \text{subject to} \quad \boldsymbol{\alpha}^H \mathbf{T}_{(1),1:P}^H \mathbf{\Delta}_{(1)} \mathbf{d}_{\theta_s,\phi_s} = 1. \tag{7.52}$$

Substituting the obtained $\boldsymbol{\alpha}$ from (7.52) into (7.46), we easily deduce that the compromising beamformer is

$$\mathbf{h}_{(1),1:P} = \frac{\left(\mathbf{\Delta}_{(1)} \mathbf{\Delta}_{(1)}^H\right)^{-1} \mathbf{\Delta}_{(1)} \mathbf{P}_{\mathbf{T}_{(1),1:P}} \mathbf{d}_{\theta_s,\phi_s}}{\mathbf{d}_{\theta_s,\phi_s}^H \mathbf{P}_{\mathbf{T}_{(1),1:P}} \mathbf{d}_{\theta_s,\phi_s}}, \tag{7.53}$$

where

$$\mathbf{P}_{\mathbf{T}_{(1),1:P}} = \mathbf{\Delta}_{(1)}^H \mathbf{T}_{(1),1:P} \left(\mathbf{T}_{(1),1:P}^H \mathbf{\Delta}_{(1)} \mathbf{\Delta}_{(1)}^H \mathbf{T}_{(1),1:P}\right)^{-1} \mathbf{T}_{(1),1:P}^H \mathbf{\Delta}_{(1)}. \tag{7.54}$$

For $P = 1$, we get

$$\mathbf{h}_{(1),1:1} = \frac{\mathbf{t}_{(1),1}}{\mathbf{d}_{\theta_s,\phi_s}^H \mathbf{\Delta}_{(1)}^H \mathbf{t}_{(1),1}} \tag{7.55}$$

$$= \mathbf{h}_{(1),\mathrm{MDF}},$$

which is the maximum DF beamformer, and for $P = M - 1$, we obtain

$$\mathbf{h}_{(1),1:M-1} = \frac{\left(\mathbf{\Delta}_{(1)} \mathbf{\Delta}_{(1)}^H\right)^{-1} \mathbf{\Delta}_{(1)} \mathbf{d}_{\theta_s,\phi_s}}{\mathbf{d}_{\theta_s,\phi_s}^H \mathbf{\Delta}_{(1)}^H \left(\mathbf{\Delta}_{(1)} \mathbf{\Delta}_{(1)}^H\right)^{-1} \mathbf{\Delta}_{(1)} \mathbf{d}_{\theta_s,\phi_s}} \tag{7.56}$$

$$= \mathbf{h}_{(1),\mathrm{MWNG}},$$

which is the maximum WNG beamformer. Therefore, by adjusting the positive integer P, we can obtain different beamformers whose performances are in between the performances of $\mathbf{h}_{(1),\mathrm{MDF}}$ and $\mathbf{h}_{(1),\mathrm{MWNG}}$. With the beamformer in (7.53), the beampattern, WNG, and DF, are, respectively,

$$\begin{aligned}\mathcal{B}_{\theta,\phi}\left(\mathbf{h}_{(1),1:P}\right) &= \frac{\mathbf{d}_{\theta,\phi}^H \mathbf{\Delta}_{(1)}^H \left(\mathbf{\Delta}_{(1)}\mathbf{\Delta}_{(1)}^H\right)^{-1} \mathbf{\Delta}_{(1)} \mathbf{P}_{\mathbf{T}_{(1),1:P}} \mathbf{d}_{\theta_s,\phi_s}}{\mathbf{d}_{\theta_s,\phi_s}^H \mathbf{P}_{\mathbf{T}_{(1),1:P}} \mathbf{d}_{\theta_s,\phi_s}} \\ &= \frac{\mathbf{d}_{\theta,\phi}^H \mathbf{P}_{\mathbf{T}_{(1),1:P}} \mathbf{d}_{\theta_s,\phi_s}}{\mathbf{d}_{\theta_s,\phi_s}^H \mathbf{P}_{\mathbf{T}_{(1),1:P}} \mathbf{d}_{\theta_s,\phi_s}},\end{aligned} \tag{7.57}$$

$$\begin{aligned}\mathcal{W}\left(\mathbf{h}_{(1),1:P}\right) &= \frac{\left(\mathbf{d}_{\theta_s,\phi_s}^H \mathbf{P}_{\mathbf{T}_{(1),1:P}} \mathbf{d}_{\theta_s,\phi_s}\right)^2}{\mathbf{d}_{\theta_s,\phi_s}^H \mathbf{P}_{\mathbf{T}_{(1),1:P}} \mathbf{\Delta}_{(1)}^H \left(\mathbf{\Delta}_{(1)}\mathbf{\Delta}_{(1)}^H\right)^{-1} \mathbf{\Delta}_{(1)} \mathbf{P}_{\mathbf{T}_{(1),1:P}} \mathbf{d}_{\theta_s,\phi_s}} \\ &= \mathbf{d}_{\theta_s,\phi_s}^H \mathbf{P}_{\mathbf{T}_{(1),1:P}} \mathbf{d}_{\theta_s,\phi_s} \\ &= \lambda_{(1),1} \mathbf{i}^T \left(\mathbf{T}_{(1),1:P}^H \mathbf{\Delta}_{(1)} \mathbf{\Delta}_{(1)}^H \mathbf{T}_{(1),1:P}\right)^{-1} \mathbf{i},\end{aligned} \tag{7.58}$$

and

$$\mathcal{D}\left(\mathbf{h}_{(1),1:P}\right) = \lambda_{(1),1} \frac{\left[\mathbf{i}^T \left(\mathbf{T}_{(1),1:P}^H \mathbf{\Delta}_{(1)} \mathbf{\Delta}_{(1)}^H \mathbf{T}_{(1),1:P}\right)^{-1} \mathbf{i}\right]^2}{\mathbf{i}^T \left(\mathbf{T}_{(1),1:P}^H \mathbf{\Delta}_{(1)} \mathbf{\Delta}_{(1)}^H \mathbf{T}_{(1),1:P}\right)^{-2} \mathbf{i}}, \tag{7.59}$$

where $\mathbf{i}$ is the first column of the $P \times P$ identity matrix $\mathbf{I}_P$.

We can also derive first-order differential adaptive beamformers. The first one is the MVDR beamformer, which can be obtained in our context by maximizing the output SNR in (7.24) or the gain in SNR in (7.23), i.e.,

$$\min_{\mathbf{h}_{(1)}} \mathbf{h}_{(1)}^H \mathbf{\Delta}_{(1)} \mathbf{\Gamma}_{\mathbf{v}} \mathbf{\Delta}_{(1)}^H \mathbf{h}_{(1)} \quad \text{subject to} \quad \mathbf{h}_{(1)}^H \mathbf{\Delta}_{(1)} \mathbf{d}_{\theta_s,\phi_s} = 1. \tag{7.60}$$

Then, we get the first-order differential MVDR beamformer:

$$\mathbf{h}_{(1),\mathrm{MVDR}} = \frac{\left(\mathbf{\Delta}_{(1)} \mathbf{\Gamma}_{\mathbf{v}} \mathbf{\Delta}_{(1)}^H\right)^{-1} \mathbf{\Delta}_{(1)} \mathbf{d}_{\theta_s,\phi_s}}{\mathbf{d}_{\theta_s,\phi_s}^H \mathbf{\Delta}_{(1)}^H \left(\mathbf{\Delta}_{(1)} \mathbf{\Gamma}_{\mathbf{v}} \mathbf{\Delta}_{(1)}^H\right)^{-1} \mathbf{\Delta}_{(1)} \mathbf{d}_{\theta_s,\phi_s}} \tag{7.61}$$

$$= \frac{\left(\mathbf{\Delta}_{(1)}\mathbf{\Phi}_{\mathbf{y}}\mathbf{\Delta}_{(1)}^H\right)^{-1}\mathbf{\Delta}_{(1)}\mathbf{d}_{\theta_s,\phi_s}}{\mathbf{d}_{\theta_s,\phi_s}^H\mathbf{\Delta}_{(1)}^H\left(\mathbf{\Delta}_{(1)}\mathbf{\Phi}_{\mathbf{y}}\mathbf{\Delta}_{(1)}^H\right)^{-1}\mathbf{\Delta}_{(1)}\mathbf{d}_{\theta_s,\phi_s}}.$$

Another interesting adaptive beamformer can be derived from the MSE criterion:

$$\begin{aligned}\mathcal{J}\left(\mathbf{h}_{(1)}\right) &= E\left(\left|Z_{(1)} - X\right|^2\right) \qquad (7.62)\\ &= \phi_X + \mathbf{h}_{(1)}^H\mathbf{\Delta}_{(1)}\mathbf{\Phi}_{\mathbf{y}}\mathbf{\Delta}_{(1)}^H\mathbf{h}_{(1)} - \phi_X\mathbf{h}_{(1)}^H\mathbf{\Delta}_{(1)}\mathbf{d}_{\theta_s,\phi_s} - \phi_X\mathbf{d}_{\theta_s,\phi_s}^H\mathbf{\Delta}_{(1)}^H\mathbf{h}_{(1)}.\end{aligned}$$

The minimization of the previous expression leads to the first-order differential Wiener beamformer:

$$\begin{aligned}\mathbf{h}_{(1),\mathrm{W}} &= \phi_X\left(\mathbf{\Delta}_{(1)}\mathbf{\Phi}_{\mathbf{y}}\mathbf{\Delta}_{(1)}^H\right)^{-1}\mathbf{\Delta}_{(1)}\mathbf{d}_{\theta_s,\phi_s}\\ &= \frac{\phi_X\left(\mathbf{\Delta}_{(1)}\mathbf{\Phi}_{\mathbf{v}}\mathbf{\Delta}_{(1)}^H\right)^{-1}\mathbf{\Delta}_{(1)}\mathbf{d}_{\theta_s,\phi_s}}{1 + \phi_X\mathbf{d}_{\theta_s,\phi_s}^H\mathbf{\Delta}_{(1)}^H\left(\mathbf{\Delta}_{(1)}\mathbf{\Phi}_{\mathbf{v}}\mathbf{\Delta}_{(1)}^H\right)^{-1}\mathbf{\Delta}_{(1)}\mathbf{d}_{\theta_s,\phi_s}}. \qquad (7.63)\end{aligned}$$

A useful way to write the Wiener beamformer is

$$\mathbf{h}_{(1),\mathrm{W}} = H_{\mathrm{W}}\left(\mathbf{\Delta}_{(1)}\mathbf{\Gamma}_{\mathbf{y}}\mathbf{\Delta}_{(1)}^H\right)^{-1}\mathbf{\Delta}_{(1)}\mathbf{d}_{\theta_s,\phi_s}, \qquad (7.64)$$

where

$$H_{\mathrm{W}} = \frac{\mathrm{iSNR}}{1 + \mathrm{iSNR}} \qquad (7.65)$$

is the single-channel Wiener gain (see Chap. 2) and

$$\mathbf{\Gamma}_{\mathbf{y}} = \frac{\mathbf{\Phi}_{\mathbf{y}}}{\phi_Y} \qquad (7.66)$$

is the coherence matrix of $\mathbf{y}$ (see Chap. 3). In practice, it is easy to estimate $\mathbf{\Gamma}_{\mathbf{y}}$ while there are some robust approaches to approximately calculate H_{W} (see again Chap. 3).

An adaptive beamformer that links the MVDR and Wiener beamformers is the first-order differential tradeoff beamformer:

$$\mathbf{h}_{(1),\mathrm{T},\mu} = \frac{\phi_X\left(\mathbf{\Delta}_{(1)}\mathbf{\Phi}_{\mathbf{v}}\mathbf{\Delta}_{(1)}^H\right)^{-1}\mathbf{\Delta}_{(1)}\mathbf{d}_{\theta_s,\phi_s}}{\mu + \phi_X\mathbf{d}_{\theta_s,\phi_s}^H\mathbf{\Delta}_{(1)}^H\left(\mathbf{\Delta}_{(1)}\mathbf{\Phi}_{\mathbf{v}}\mathbf{\Delta}_{(1)}^H\right)^{-1}\mathbf{\Delta}_{(1)}\mathbf{d}_{\theta_s,\phi_s}}, \qquad (7.67)$$

where $\mu \geq 0$ is a tuning parameter. It can be shown that $\mathbf{h}_{(1),\mathrm{T},\mu}$ can be rewritten as

$$\mathbf{h}_{(1),\mathrm{T},\mu} = \frac{H_{\mathrm{W}} \left(\mathbf{\Delta}_{(1)} \mathbf{\Gamma}_{\mathbf{y}} \mathbf{\Delta}_{(1)}^H \right)^{-1} \mathbf{\Delta}_{(1)} \mathbf{d}_{\theta_{\mathrm{s}},\phi_{\mathrm{s}}}}{\mu + (1-\mu) H_{\mathrm{W}} \mathbf{d}_{\theta_{\mathrm{s}},\phi_{\mathrm{s}}}^H \mathbf{\Delta}_{(1)}^H \left(\mathbf{\Delta}_{(1)} \mathbf{\Gamma}_{\mathbf{y}} \mathbf{\Delta}_{(1)}^H \right)^{-1} \mathbf{\Delta}_{(1)} \mathbf{d}_{\theta_{\mathrm{s}},\phi_{\mathrm{s}}}}. \quad (7.68)$$

We observe that for

- $\mu = 1$, $\mathbf{h}_{(1),\mathrm{T},1} = \mathbf{h}_{(1),\mathrm{W}}$, which is the Wiener beamformer;
- $\mu = 0$, $\mathbf{h}_{(1),\mathrm{T},0} = \mathbf{h}_{(1),\mathrm{MVDR}}$, which is the MVDR beamformer;
- $\mu > 1$, results in a beamformer with low residual noise (from a broadband perspective) at the expense of high desired signal distortion (as compared to Wiener); and
- $\mu < 1$, results in a beamformer with high residual noise and low desired signal distortion (as compared to Wiener).

With the matrix $\mathbf{\Delta}_{(1)}$, we can place a null in any desired direction $(\theta_0, \phi_0) \neq (\theta_{\mathrm{s}}, \phi_{\mathrm{s}})$. Now, let us assume that we want to place another null in the direction $(\theta_{\mathrm{i}}, \phi_{\mathrm{i}}) \neq (\theta_{\mathrm{s}}, \phi_{\mathrm{s}})$. Considering also the distortionless constraint, we can build the constraint equation as

$$\mathbf{C}_{\mathrm{s,i}}^H \mathbf{\Delta}_{(1)}^H \mathbf{h}_{(1)} = \begin{bmatrix} 1 \\ 0 \end{bmatrix}, \quad (7.69)$$

where

$$\mathbf{C}_{\mathrm{s,i}} = \begin{bmatrix} \mathbf{d}_{\theta_{\mathrm{s}},\phi_{\mathrm{s}}} & \mathbf{d}_{\theta_{\mathrm{i}},\phi_{\mathrm{i}}} \end{bmatrix} \quad (7.70)$$

is the constraint matrix of size $M \times 2$ whose columns are linearly independent. The most convenient way to solve this problem is by minimizing the power of the residual noise subject to (7.69), i.e.,

$$\min_{\mathbf{h}_{(1)}} \mathbf{h}_{(1)}^H \mathbf{\Delta}_{(1)} \mathbf{\Gamma}_{\mathbf{v}} \mathbf{\Delta}_{(1)}^H \mathbf{h}_{(1)} \quad \text{subject to} \quad \mathbf{C}_{\mathrm{s,i}}^H \mathbf{\Delta}_{(1)}^H \mathbf{h}_{(1)} = \begin{bmatrix} 1 \\ 0 \end{bmatrix}. \quad (7.71)$$

Then, we find the first-order differential LCMV beamformer:

$$\mathbf{h}_{(1),\mathrm{LCMV}} = \left(\mathbf{\Delta}_{(1)} \mathbf{\Gamma}_{\mathbf{v}} \mathbf{\Delta}_{(1)}^H \right)^{-1} \mathbf{\Delta}_{(1)} \mathbf{C}_{\mathrm{s,i}} \times \left[\mathbf{C}_{\mathrm{s,i}}^H \mathbf{\Delta}_{(1)}^H \left(\mathbf{\Delta}_{(1)} \mathbf{\Gamma}_{\mathbf{v}} \mathbf{\Delta}_{(1)}^H \right)^{-1} \mathbf{\Delta}_{(1)} \mathbf{C}_{\mathrm{s,i}} \right]^{-1} \begin{bmatrix} 1 \\ 0 \end{bmatrix}, \quad (7.72)$$

which depends on the statistics of the noise only.

7.3 Higher-Order Differential Beamforming

Let $N \geq 1$ be the order of the difference equations, with $M \geq N + 1$. The case $N = 1$ corresponds to the first order studied in the previous section.

From the M entries of the observed signal vector, $\mathbf{y}$, we can arrange the following $(M - N)$ Nth-order linear spatial difference equations [1]:

$$\begin{aligned} Y_{(N),i} &= A_{0,i} Y_i + A_{1,i} Y_{i+1} + \cdots + A_{N,i} Y_{i+N} \\ &= \left(A_{0,i} D_{\theta_s,\phi_s,i} + A_{1,i} D_{\theta_s,\phi_s,i+1} + \cdots + A_{N,i} D_{\theta_s,\phi_s,i+N} \right) X \\ &\quad + A_{0,i} V_i + A_{1,i} V_{i+1} + \cdots + A_{N,i} V_{i+N}, \end{aligned} \tag{7.73}$$

for $i = 1, 2, \ldots, M - N$, where the subscript $_{(N)}$ stands for Nth order and $A_{n,i} \neq 0$, $n = 0, 1, \ldots, N$, $i = 1, 2, \ldots, M - N$ are some real-valued parameters, which should be chosen appropriately and carefully so that $Y_{(N),i}$ is an Nth-order spatial derivative of the observations. The values of these parameters depend mostly on the array geometry [2, 3]. The $M - N$ previous equations can be written in a vector form as

$$\begin{aligned} \mathbf{y}_{(N)} &= \mathbf{\Delta}_{(N)} \mathbf{y} \\ &= \mathbf{x}_{(N)} + \mathbf{v}_{(N)} \\ &= \mathbf{\Delta}_{(N)} \mathbf{d}_{\theta_s,\phi_s} X + \mathbf{\Delta}_{(N)} \mathbf{v}, \end{aligned} \tag{7.74}$$

where

$$\mathbf{\Delta}_{(N)} = \begin{bmatrix} A_{0,1} & A_{1,1} & \cdots & A_{N,1} & 0 & \cdots & 0 \\ 0 & A_{0,2} & A_{1,2} & \cdots & A_{N,2} & \cdots & 0 \\ \vdots & \vdots & \ddots & \ddots & \cdots & \ddots & \vdots \\ 0 & 0 & \cdots & A_{0,M-N} & A_{1,M-N} & \cdots & A_{N,M-N} \end{bmatrix} \tag{7.75}$$

is the Nth-order difference operator [upper $(N + 1)$-diagonal matrix] of size $(M - N) \times M$, which can be seen as the Nth-order derivative operator. From (7.74), we find that the $(M - N) \times (M - N)$ covariance matrix of $\mathbf{y}_{(N)}$ is

$$\begin{aligned} \mathbf{\Phi}_{\mathbf{y}_{(N)}} &= \mathbf{\Delta}_{(N)} \mathbf{\Phi}_{\mathbf{y}} \mathbf{\Delta}_{(N)}^H \\ &= \phi_X \mathbf{\Delta}_{(N)} \mathbf{d}_{\theta_s,\phi_s} \mathbf{d}_{\theta_s,\phi_s}^H \mathbf{\Delta}_{(N)}^H + \phi_V \mathbf{\Delta}_{(N)} \mathbf{\Gamma}_{\mathbf{v}} \mathbf{\Delta}_{(N)}^H \\ &= \mathbf{\Phi}_{\mathbf{x}_{(N)}} + \mathbf{\Phi}_{\mathbf{v}_{(N)}}, \end{aligned} \tag{7.76}$$

where $\mathbf{\Phi}_{\mathbf{x}_{(N)}}$ and $\mathbf{\Phi}_{\mathbf{v}_{(N)}}$ are the covariance matrices of $\mathbf{x}_{(N)}$ and $\mathbf{v}_{(N)}$, respectively.

Thanks to the newly constructed observed signal vector, $\mathbf{y}_{(N)}$ of length $M - N$, we can perform Nth-order differential beamforming by applying a complex-valued linear filter to this vector, i.e.,

$$
\begin{aligned}
Z_{(N)} &= \sum_{i=1}^{M-N} H_{(N),i}^{*} Y_{(N),i} \\
&= \mathbf{h}_{(N)}^{H} \mathbf{y}_{(N)} \\
&= X_{(N),\mathrm{fd}} + V_{(N),\mathrm{rn}},
\end{aligned}
\tag{7.77}
$$

where $Z_{(N)}$ is the Nth-order differential beamformer output signal ,

$$
\mathbf{h}_{(N)} = \begin{bmatrix} H_{(N),1} & H_{(N),2} & \cdots & H_{(N),M-N} \end{bmatrix}^{T} \tag{7.78}
$$

is the Nth-order differential filter of length $M - N$,

$$
X_{(N),\mathrm{fd}} = X \mathbf{h}_{(N)}^{H} \boldsymbol{\Delta}_{(N)} \mathbf{d}_{\theta_{\mathrm{s}},\phi_{\mathrm{s}}} \tag{7.79}
$$

is the filtered desired signal, and

$$
V_{(N),\mathrm{rn}} = \mathbf{h}_{(N)}^{H} \boldsymbol{\Delta}_{(N)} \mathbf{v} \tag{7.80}
$$

is the residual noise. We see from (7.79) that the distortionless constraint is

$$
\mathbf{h}_{(N)}^{H} \boldsymbol{\Delta}_{(N)} \mathbf{d}_{\theta_{\mathrm{s}},\phi_{\mathrm{s}}} = 1. \tag{7.81}
$$

From (7.77), we can compute the variance of $Z_{(N)}$ as

$$
\begin{aligned}
\phi_{Z_{(N)}} &= \mathbf{h}_{(N)}^{H} \boldsymbol{\Phi}_{\mathbf{y}_{(N)}} \mathbf{h}_{(N)} \\
&= \mathbf{h}_{(N)}^{H} \boldsymbol{\Delta}_{(N)} \boldsymbol{\Phi}_{\mathbf{y}} \boldsymbol{\Delta}_{(N)}^{H} \mathbf{h}_{(N)} \\
&= \phi_{X_{(N),\mathrm{fd}}} + \phi_{V_{(N),\mathrm{rn}}},
\end{aligned}
\tag{7.82}
$$

where

$$
\phi_{X_{(N),\mathrm{fd}}} = \phi_X \left| \mathbf{h}_{(N)}^{H} \boldsymbol{\Delta}_{(N)} \mathbf{d}_{\theta_{\mathrm{s}},\phi_{\mathrm{s}}} \right|^2, \tag{7.83}
$$

$$
\phi_{V_{(N),\mathrm{rn}}} = \mathbf{h}_{(N)}^{H} \boldsymbol{\Delta}_{(N)} \boldsymbol{\Phi}_{\mathbf{v}} \boldsymbol{\Delta}_{(N)}^{H} \mathbf{h}_{(N)}. \tag{7.84}
$$

Then, the most important performance measures in this context are

- output SNR:

$$\text{oSNR}\left(\mathbf{h}_{(N)}\right) = \text{iSNR} \times \frac{\left|\mathbf{h}_{(N)}^H \mathbf{\Delta}_{(N)} \mathbf{d}_{\theta_s,\phi_s}\right|^2}{\mathbf{h}_{(N)}^H \mathbf{\Delta}_{(N)} \mathbf{\Gamma}_{\mathbf{v}} \mathbf{\Delta}_{(N)}^H \mathbf{h}_{(N)}}, \tag{7.85}$$

- gain in SNR:

$$\mathcal{G}\left(\mathbf{h}_{(N)}\right) = \frac{\left|\mathbf{h}_{(N)}^H \mathbf{\Delta}_{(N)} \mathbf{d}_{\theta_s,\phi_s}\right|^2}{\mathbf{h}_{(N)}^H \mathbf{\Delta}_{(N)} \mathbf{\Gamma}_{\mathbf{v}} \mathbf{\Delta}_{(N)}^H \mathbf{h}_{(N)}}, \tag{7.86}$$

- WNG:

$$\mathcal{W}\left(\mathbf{h}_{(N)}\right) = \frac{\left|\mathbf{h}_{(N)}^H \mathbf{\Delta}_{(N)} \mathbf{d}_{\theta_s,\phi_s}\right|^2}{\mathbf{h}_{(N)}^H \mathbf{\Delta}_{(N)} \mathbf{\Delta}_{(N)}^H \mathbf{h}_{(N)}}, \tag{7.87}$$

- DF:

$$\mathcal{D}\left(\mathbf{h}_{(N)}\right) = \frac{\left|\mathbf{h}_{(N)}^H \mathbf{\Delta}_{(N)} \mathbf{d}_{\theta_s,\phi_s}\right|^2}{\mathbf{h}_{(N)}^H \mathbf{\Delta}_{(N)} \mathbf{\Gamma}_{\mathrm{d}} \mathbf{\Delta}_{(N)}^H \mathbf{h}_{(N)}}, \tag{7.88}$$

- and beampattern:

$$\mathcal{B}_{\theta,\phi}\left(\mathbf{h}_{(N)}\right) = \mathbf{d}_{\theta,\phi}^H \mathbf{\Delta}_{(N)}^H \mathbf{h}_{(N)}. \tag{7.89}$$

The optimal differential beamformers derived in the previous section can all be obtained here as well; we only need to replace $\mathbf{\Delta}_{(1)}$ by $\mathbf{\Delta}_{(N)}$, so they are not re-derived in this section.

7.4 Illustrative Examples

In this section, we study some examples of differential beamformers. The arrays that we choose to use are NULAs with M omnidirectional microphones, where we consider four NULAs with different values of M. The first NULA consists of $M = 4$ microphones (denoted as NULA-I), where the spacing between the first microphone (reference sensor) and the other *three* microphones are, respectively, 0.4 cm, 1.7 cm, and 3.0 cm (these positions are arbitrarily chosen but with the constraint that the spacing between neighboring microphones is not too large). The second NULA consists of $M = 5$ microphones (denoted as NULA-II), where the spacing between the first microphone and the other *four* microphones are, respectively, 0.4 cm,

2.7 cm, 3.5 cm, and 4.0 cm. The third NULA consists of $M = 6$ microphones (denoted as NULA-III), where the spacing between the first microphone and the other *five* microphones are, respectively, 0.4 cm, 1.7 cm, 2.7 cm, 3.5 cm, and 5.0 cm. Finally, the fourth NULA consists of $M = 8$ microphones (denoted as NULA-IV), where the spacing between the first microphone and the other *seven* microphones are, respectively, 0.4 cm, 1.7 cm, 2.7 cm, 3.5 cm, 4.5 cm, 5.5 cm, and 7.0 cm. In the rest, when we choose $M = 4$, 5, 6, and 8, we mean to use NULA-I, NULA-II, NULA-III, and NULA-IV, respectively. We assume that the desired source signal propagates from the endfire direction of the NULAs, i.e., $(\theta_s, \phi_s) = (90°, 0°)$.

We first demonstrate the performance of the first-order differential MWNG beamformer, $\mathbf{h}_{(1),\text{MWNG}}$, given in (7.29). Figure 7.1 shows plots of the power patterns, $\left|\mathcal{B}_{\theta,\phi}\left(\mathbf{h}_{(1)}\right)\right|^2$, of this beamformer versus the azimuth angle for $M = 4$ and different frequencies. As seen, the power patterns are almost frequency invariant and they all have a null at $\phi = 90°$, which is caused by the essence of the first-order differential operator. Figure 7.2 shows plots of the DF, $\mathcal{D}\left(\mathbf{h}_{(1)}\right)$, and WNG, $\mathcal{W}\left(\mathbf{h}_{(1)}\right)$, of this beamformer as a function of frequency for different values of M. It is observed that the WNG increases, while the DF stays almost the same as the value of M increases.

We then demonstrate the performance of the first-order differential MDF beamformer, $\mathbf{h}_{(1),\text{MDF}}$, given in (7.34). Figure 7.3 shows plots of the power patterns, $\left|\mathcal{B}_{\theta,\phi}\left(\mathbf{h}_{(1)}\right)\right|^2$, of this beamformer versus the azimuth angle for $M = 4$ and different frequencies. It can be observed that the power patterns are almost frequency invariant and their main beam points in the desired direction. The differential MDF beamformer leads to a narrower main beam and more nulls in the beampattern than the differential MWNG beamformer, which obviously means higher directivity. Figure 7.4 shows plots of the DF, $\mathcal{D}\left(\mathbf{h}_{(1)}\right)$, and WNG, $\mathcal{W}\left(\mathbf{h}_{(1)}\right)$, of this beamformer as a function of frequency for different values of M. As seen, the DF is almost frequency invariant over the studied frequency band and is much higher than the differential MWNG beamformer. It is also observed that the DF increases, while the WNG decreases when M increases. Generally, the WNG is considerably less than 0 dB, which indicates that this beamformer suffers from significant white noise amplification.

To demonstrate the performance of the compromising beamformer, $\mathbf{h}_{(1),1:P}$, given in (7.53), we choose $M = 5$. Figure 7.5 shows plots of the power patterns, $\left|\mathcal{B}_{\theta,\phi}\left(\mathbf{h}_{(1)}\right)\right|^2$, of this beamformer versus the azimuth angle for $P = 2$ and different frequencies. As seen, the power patterns are almost frequency invariant and their main beam points in the desired direction. Figure 7.6 shows plots of the power patterns, $\left|\mathcal{B}_{\theta,\phi}\left(\mathbf{h}_{(1)}\right)\right|^2$, of this beamformer versus the azimuth angle for $f = 2000$ Hz and different values of P. As seen, the smaller is the value of P, the narrower are the main beams and more nulls are in the beampatterns. Figure 7.7 shows plots of the DF, $\mathcal{D}\left(\mathbf{h}_{(1)}\right)$, and WNG, $\mathcal{W}\left(\mathbf{h}_{(1)}\right)$, of this beamformer as a function of frequency for different values of P. It can be clearly observed that this beamformer is able to achieve a tradeoff between the DF and WNG, i.e., as the value

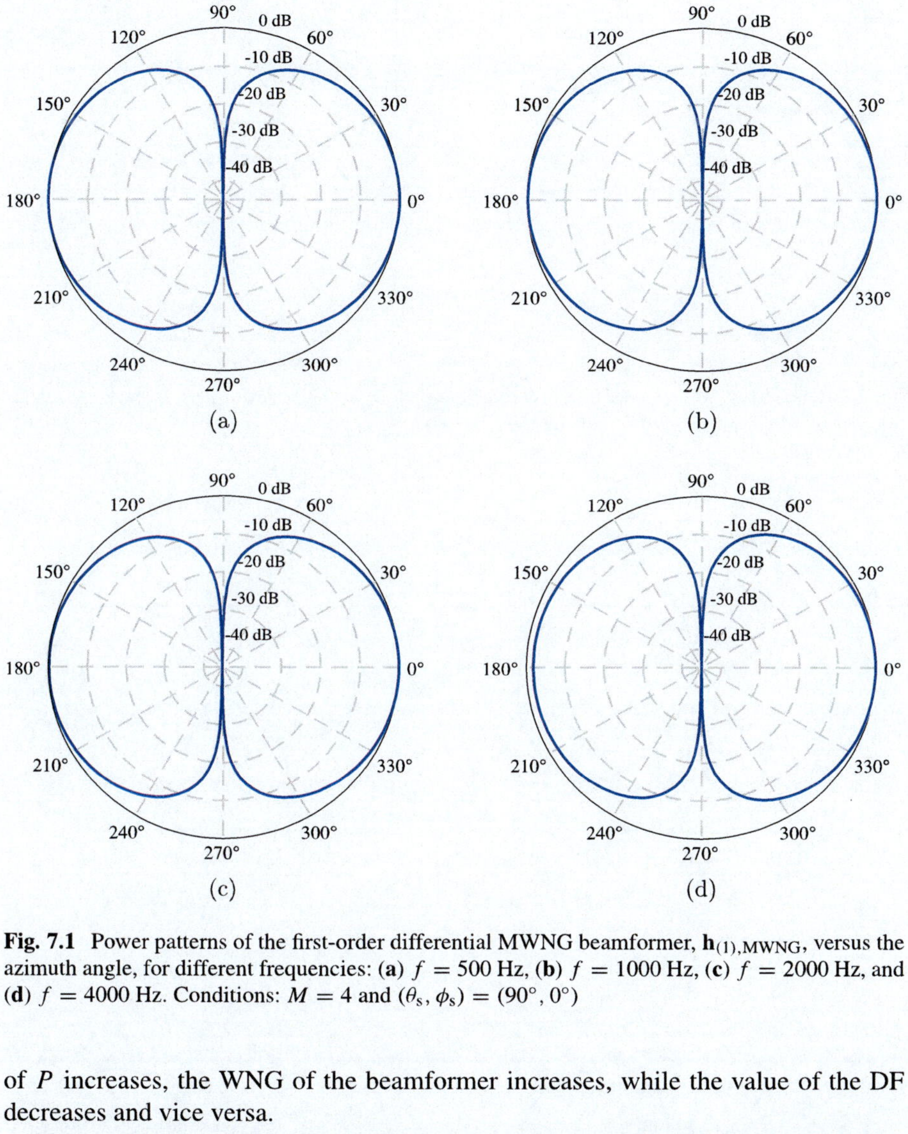

Fig. 7.1 Power patterns of the first-order differential MWNG beamformer, $\mathbf{h}_{(1),\mathrm{MWNG}}$, versus the azimuth angle, for different frequencies: (**a**) $f = 500$ Hz, (**b**) $f = 1000$ Hz, (**c**) $f = 2000$ Hz, and (**d**) $f = 4000$ Hz. Conditions: $M = 4$ and $(\theta_s, \phi_s) = (90°, 0°)$

of P increases, the WNG of the beamformer increases, while the value of the DF decreases and vice versa.

Now, we study the performance of the first-order differential adaptive beamformers. We consider an environment with point source noises and white noise, where the two statistically independent interferences impinge on the array from two different directions $(\theta_{i,1}, \phi_{i,1})$ and $(\theta_{i,2}, \phi_{i,2})$. The two interferences are incoherent with the desired signal. We choose $(\theta_s, \phi_s) = (90°, 0°)$, $(\theta_{i,1}, \phi_{i,1}) = (90°, 150°)$, and $(\theta_{i,2}, \phi_{i,2}) = (90°, 210°)$. The covariance matrix of the noise signal, $\mathbf{\Phi}_{\mathbf{v}}$, is given in (3.75), with $\alpha = 10^{-4}$ ($\alpha > 0$ is related to the ratio between the powers of the white noise and point source noise). The variance of the desired signal is set to 1

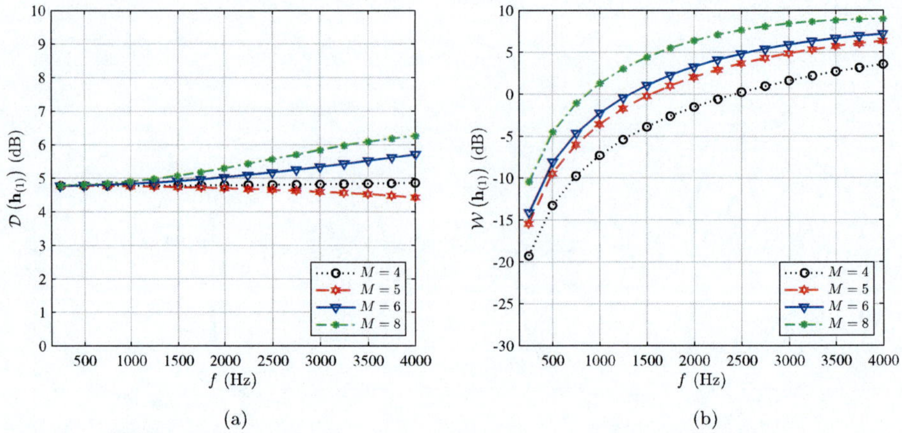

Fig. 7.2 DF and WNG of the first-order differential MWNG beamformer, $\mathbf{h}_{(1),\mathrm{MWNG}}$, as a function of frequency, for different numbers of sensors, M: (**a**) DF and (**b**) WNG. Condition: $(\theta_s, \phi_s) = (90^\circ, 0^\circ)$

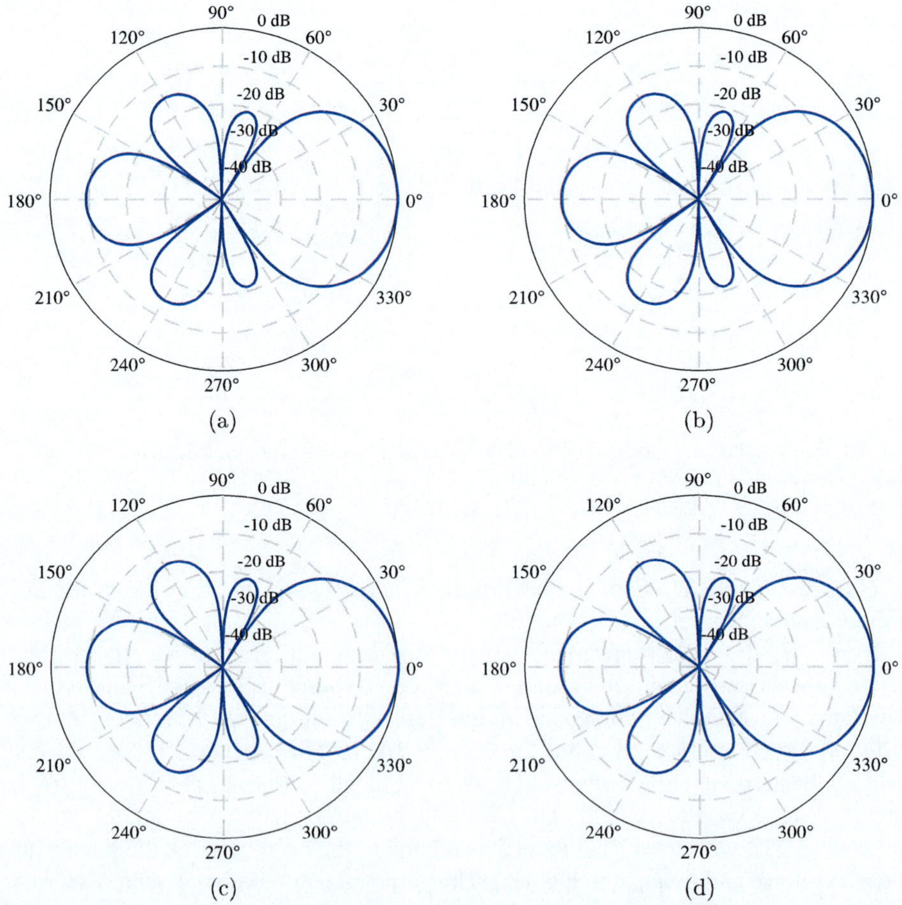

Fig. 7.3 Power patterns of the first-order differential MDF beamformer, $\mathbf{h}_{(1),\mathrm{MDF}}$, versus the azimuth angle, for different frequencies: (**a**) $f = 500$ Hz, (**b**) $f = 1000$ Hz, (**c**) $f = 2000$ Hz, and (**d**) $f = 4000$ Hz. Conditions: $M = 4$ and $(\theta_s, \phi_s) = (90^\circ, 0^\circ)$

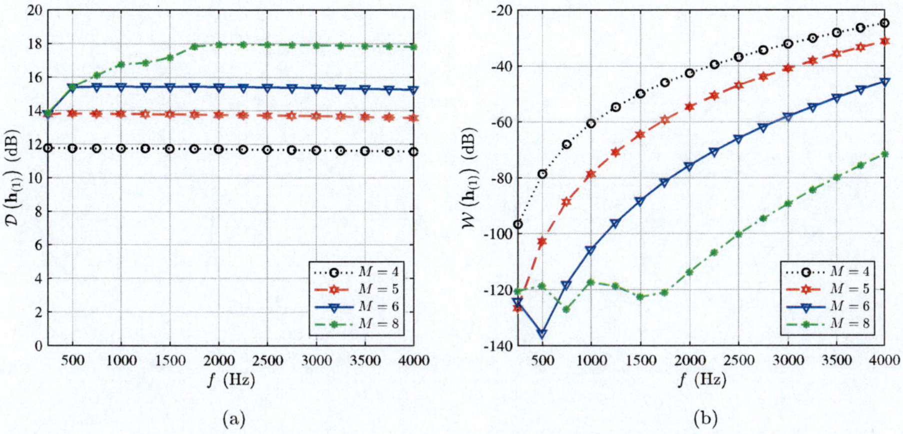

Fig. 7.4 DF and WNG of the first-order differential MDF beamformer, $\mathbf{h}_{(1),\mathrm{MDF}}$, as a function of frequency, for different numbers of sensors, M: (**a**) DF and (**b**) WNG. Condition: $(\theta_\mathrm{s}, \phi_\mathrm{s}) = (90°, 0°)$

and the variance of the noise is computed according to the specified input SNR and value of α.

To demonstrate the performance of the first-order differential MVDR beamformer, $\mathbf{h}_{(1),\mathrm{MVDR}}$, given in (7.61), we choose iSNR = 10 dB. Figure 7.8 shows plots of the power patterns, $\left|\mathcal{B}_{\theta,\phi}\left(\mathbf{h}_{(1)}\right)\right|^2$, of this beamformer versus the azimuth angle for $M = 4$ and different frequencies. As seen, the main beam of the power patterns points in the direction of the desired signal and there are nulls in the direction of the interferences. Figure 7.9 shows plots of the SNR gain, $\mathcal{G}\left(\mathbf{h}_{(1)}\right)$, and WNG, $\mathcal{W}\left(\mathbf{h}_{(1)}\right)$, of this beamformer as a function of frequency for different values of M. As seen, both the SNR gain and WNG increase as the value of M increases.

To show the performance of the first-order differential Wiener beamformer, $\mathbf{h}_{(1),\mathrm{W}}$, given in (7.63), we choose iSNR = 10 dB. Figure 7.10 shows plots of the SNR gain, $\mathcal{G}\left(\mathbf{h}_{(1)}\right)$, and WNG, $\mathcal{W}\left(\mathbf{h}_{(1)}\right)$, of this beamformer as a function of frequency for different values of M. It is observed that both the SNR gain and WNG increase when M increases. It is also observed that the first-order differential MVDR and Wiener beamformers have the same SNR gain and WNG. This is because these two beamformers are identical up to a scaling factor and are special cases of the tradeoff beamformer, i.e., $\mathbf{h}_{(1),\mathrm{MVDR}} = \mathbf{h}_{(1),\mathrm{T},0}$ and $\mathbf{h}_{(1),\mathrm{W}} = \mathbf{h}_{(1),\mathrm{T},1}$. As explained in Chap. 3, the SNR gain of the tradeoff beamformer is independent of the parameter μ from a narrowband perspective.

To demonstrate the performance of the first-order differential tradeoff beamformer, $\mathbf{h}_{(1),\mathrm{T},\mu}$, given in (7.67), we choose $M = 4$. We use broadband performance metrics to better evaluate the performance of this beamformer, where the broadband SNR gain and desired signal distortion index are, respectively, defined in a similar way to (3.78) and (3.80). Figure 7.11 shows plots of the broadband

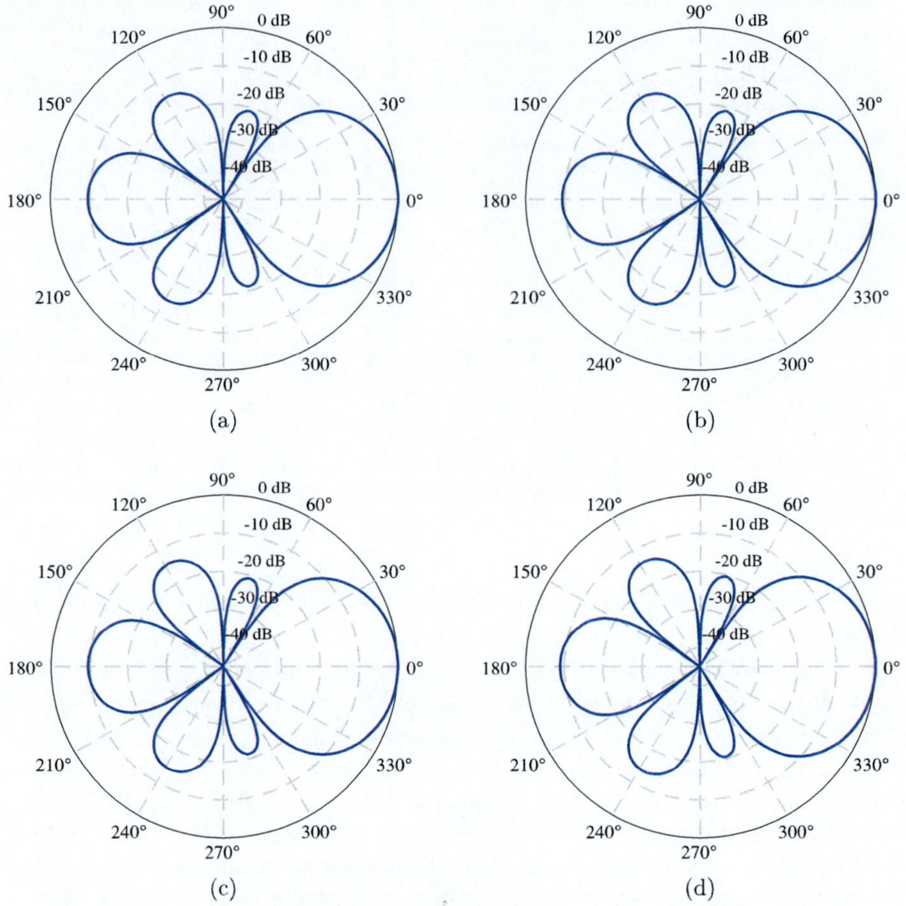

Fig. 7.5 Power patterns of the compromising beamformer, $\mathbf{h}_{(1),1:P}$, versus the azimuth angle, for different frequencies: (**a**) $f = 500$ Hz, (**b**) $f = 1000$ Hz, (**c**) $f = 2000$ Hz, and (**d**) $f = 4000$ Hz. Conditions: $M = 5$, $P = 2$, and $(\theta_s, \phi_s) = (90^\circ, 0^\circ)$

SNR gain, $\overline{\mathcal{G}\left(\mathbf{h}_{(1)}\right)}$, and broadband desired signal distortion index, $\overline{\upsilon\left(\mathbf{h}_{(1)}\right)}$, of this beamformer, as a function of the input SNR, for different values of μ. It can be seen that both the SNR gain and desired signal distortion index increase as the value of μ increases, which indicates that the tradeoff beamformer achieves a good compromise between noise reduction and desired signal distortion from a broadband perspective.

To show the performance of the first-order differential LCMV beamformer, $\mathbf{h}_{(1),\mathrm{LCMV}}$, given in (7.72), we choose iSNR = 10 dB and set a null constraint in the direction $(\theta, \phi) = (90^\circ, 150^\circ)$. Figure 7.12 shows plots of the power patterns, $\left|\mathcal{B}_{\theta,\phi}\left(\mathbf{h}_{(1)}\right)\right|^2$, of this beamformer versus the azimuth angle for $M = 4$ and different frequencies. As seen, the main beam of the power patterns points in the direction

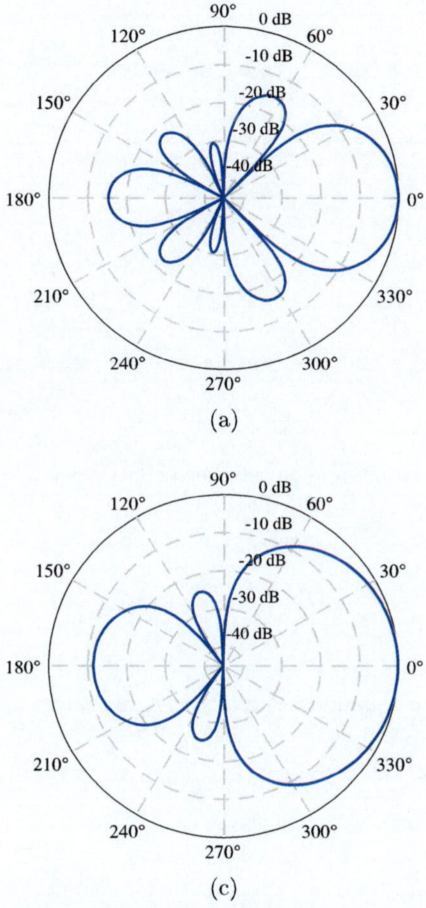

Fig. 7.6 Power patterns of the compromising beamformer, $\mathbf{h}_{(1),1:P}$, versus the azimuth angle, for different values of P: **(a)** $P = 1$, **(b)** $P = 2$, **(c)** $P = 3$, and **(d)** $P = 4$. Conditions: $M = 5$, $f = 2000$ Hz, and $(\theta_s, \phi_s) = (90°, 0°)$

of the desired signal, and there are nulls in the interference direction and $\phi = 90°$. Figure 7.13 shows plots of the SNR gain, $\mathcal{G}\left(\mathbf{h}_{(1)}\right)$, and WNG, $\mathcal{W}\left(\mathbf{h}_{(1)}\right)$, of this beamformer as a function of frequency for different values of M. It is observed that both the SNR gain and WNG increase as the value of M increases.

Finally, we study some examples of second- and third-order differential beamformers. For the second-order linear spatial difference equations with a NULA, we have

$$A_{0,i} = \left[(\delta_i - \delta_{i+1})(\delta_i - \delta_{i+2})\right]^{-1},$$

$$A_{1,i} = \left[(\delta_{i+1} - \delta_i)(\delta_{i+1} - \delta_{i+2})\right]^{-1},$$

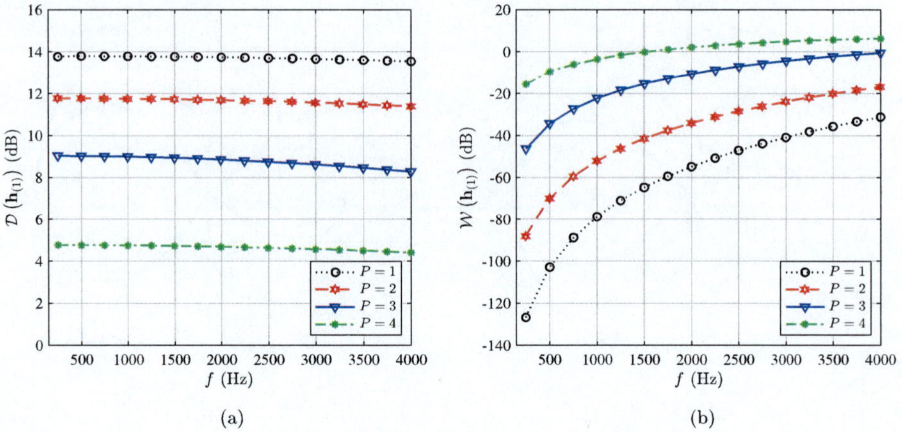

Fig. 7.7 DF and WNG of the compromising beamformer, $\mathbf{h}_{(1),1:P}$, as a function of frequency, for different values of P: (**a**) DF and (**b**) WNG. Conditions: $M = 5$ and $(\theta_s, \phi_s) = (90°, 0°)$

$$A_{2,i} = [(\delta_{i+2} - \delta_i)(\delta_{i+2} - \delta_{i+1})]^{-1},$$
$$i = 1, 2, \ldots, M-2$$

and for the third-order linear spatial difference equations with a NULA, we have

$$\begin{aligned} A_{0,i} &= [(\delta_i - \delta_{i+1})(\delta_i - \delta_{i+2})(\delta_i - \delta_{i+3})]^{-1}, \\ A_{1,i} &= [(\delta_{i+1} - \delta_i)(\delta_{i+1} - \delta_{i+2})(\delta_{i+1} - \delta_{i+3})]^{-1}, \\ &\vdots \\ A_{3,i} &= [(\delta_{i+3} - \delta_i)(\delta_{i+3} - \delta_{i+1})(\delta_{i+3} - \delta_{i+2})]^{-1}, \\ i &= 1, 2, \ldots, M-3. \end{aligned}$$

Then, the second-order difference operator, $\mathbf{\Delta}_{(2)}$ of size $(M-2) \times M$, and the third-order difference operator, $\mathbf{\Delta}_{(3)}$ of size $(M-3) \times M$, can be constructed according to (7.75). As explained in Sect. 7.3, the optimal Nth-order differential beamformers can be easily obtained by replacing $\mathbf{\Delta}_{(1)}$ by $\mathbf{\Delta}_{(N)}$. Consequently, we get the second-order differential MWNG beamformer:

$$\mathbf{h}_{(2),\mathrm{MWNG}} = \frac{\left(\mathbf{\Delta}_{(2)}\mathbf{\Delta}_{(2)}^H\right)^{-1}\mathbf{\Delta}_{(2)}\mathbf{d}_{\theta_s,\phi_s}}{\mathbf{d}_{\theta_s,\phi_s}^H\mathbf{\Delta}_{(2)}^H\left(\mathbf{\Delta}_{(2)}\mathbf{\Delta}_{(2)}^H\right)^{-1}\mathbf{\Delta}_{(2)}\mathbf{d}_{\theta_s,\phi_s}} \tag{7.90}$$

and the third-order differential MWNG beamformer:

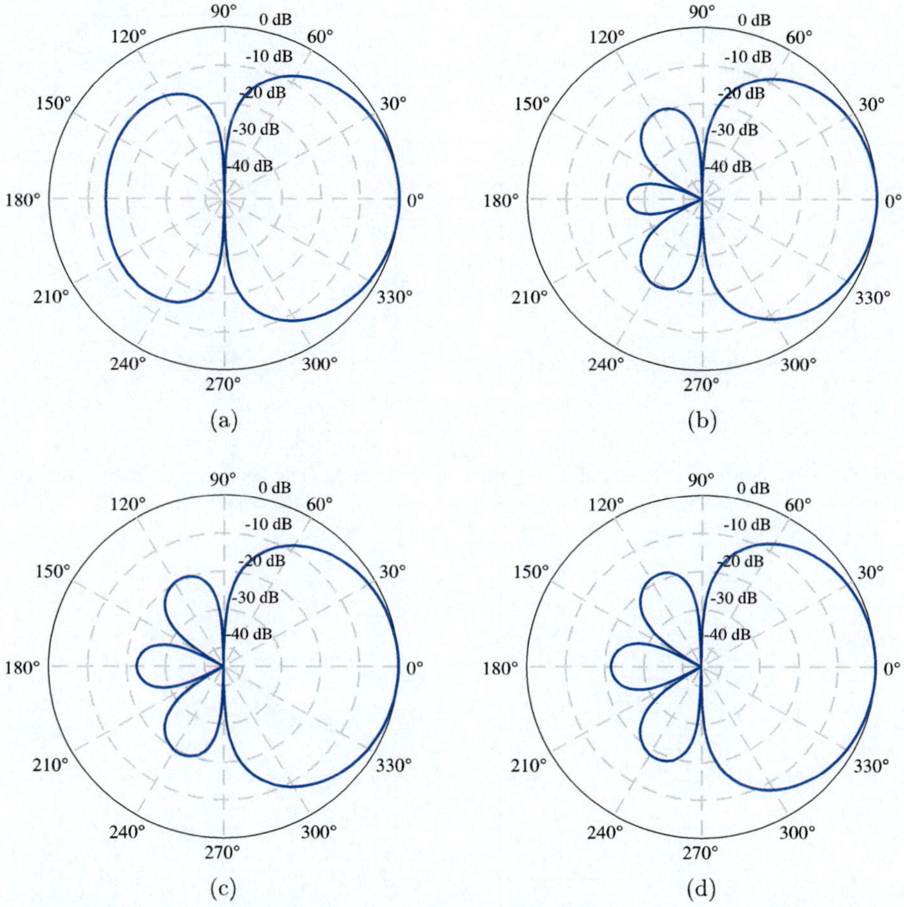

Fig. 7.8 Power patterns of the first-order differential MVDR beamformer, $\mathbf{h}_{(1),\mathrm{MVDR}}$, versus the azimuth angle, for different frequencies: (**a**) $f = 500$ Hz, (**b**) $f = 1000$ Hz, (**c**) $f = 2000$ Hz, and (**d**) $f = 4000$ Hz. Conditions: $M = 4$, iSNR = 10 dB, $\alpha = 10^{-4}$, and $(\theta_s, \phi_s) = (90°, 0°)$

$$\mathbf{h}_{(3),\mathrm{MWNG}} = \frac{\left(\mathbf{\Delta}_{(3)}\mathbf{\Delta}_{(3)}^H\right)^{-1}\mathbf{\Delta}_{(3)}\mathbf{d}_{\theta_s,\phi_s}}{\mathbf{d}_{\theta_s,\phi_s}^H\mathbf{\Delta}_{(3)}^H\left(\mathbf{\Delta}_{(3)}\mathbf{\Delta}_{(3)}^H\right)^{-1}\mathbf{\Delta}_{(3)}\mathbf{d}_{\theta_s,\phi_s}}. \tag{7.91}$$

Figure 7.14 shows plots of the power patterns, $\left|\mathcal{B}_{\theta,\phi}\left(\mathbf{h}_{(2)}\right)\right|^2$, of the second-order differential MWNG beamformer, $\mathbf{h}_{(2),\mathrm{MWNG}}$, given in (7.90), versus the azimuth angle for $M = 4$ and different frequencies. As seen, the power patterns are almost frequency invariant and have a null at $\phi = 90°$. In comparison, the main beam of the power patterns is narrower than the first-order one in Fig. 7.1. Figure 7.15 shows plots of the DF, $\mathcal{D}\left(\mathbf{h}_{(2)}\right)$, and WNG, $\mathcal{W}\left(\mathbf{h}_{(2)}\right)$, of this beamformer as a function of

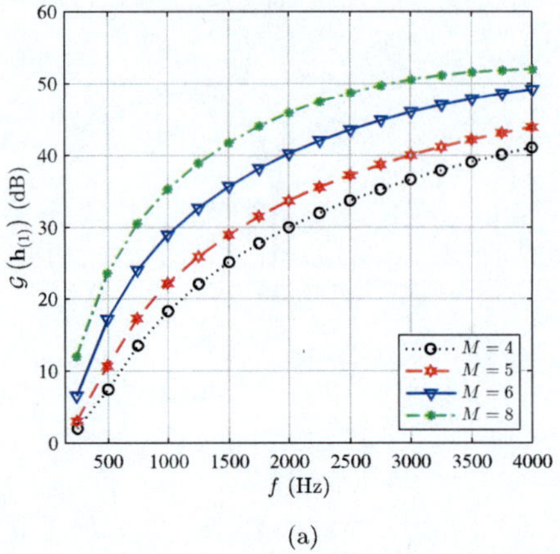

Fig. 7.9 SNR gain and WNG of the first-order differential MVDR beamformer, $\mathbf{h}_{(1),\mathrm{MVDR}}$, as a function of frequency, for different numbers of sensors, M: (**a**) SNR gain and (**b**) WNG. Conditions: iSNR = 10 dB, $\alpha = 10^{-4}$, and $(\theta_s, \phi_s) = (90°, 0°)$

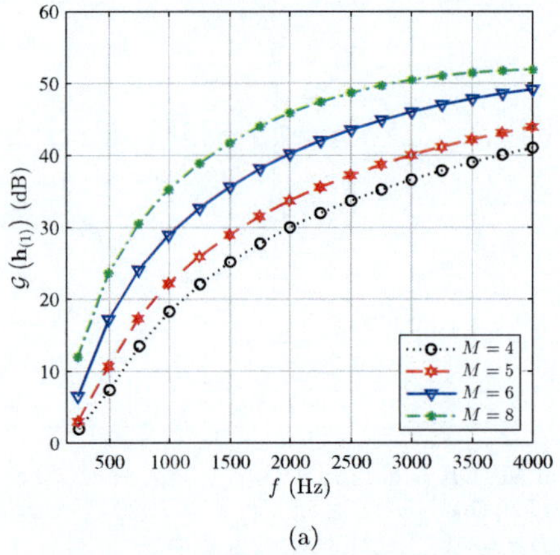

Fig. 7.10 SNR gain and WNG of the first-order differential Wiener beamformer, $\mathbf{h}_{(1),\mathrm{W}}$, as a function of frequency, for different numbers of sensors, M: (**a**) SNR gain and (**b**) WNG. Conditions: iSNR = 10 dB, $\alpha = 10^{-4}$, and $(\theta_s, \phi_s) = (90°, 0°)$

frequency for different values of M. It is observed that the WNG increases, while the DF stays almost the same as the value of M increases. In comparison, the DF is higher, while the WNG is lower than the first-order ones in Fig. 7.2.

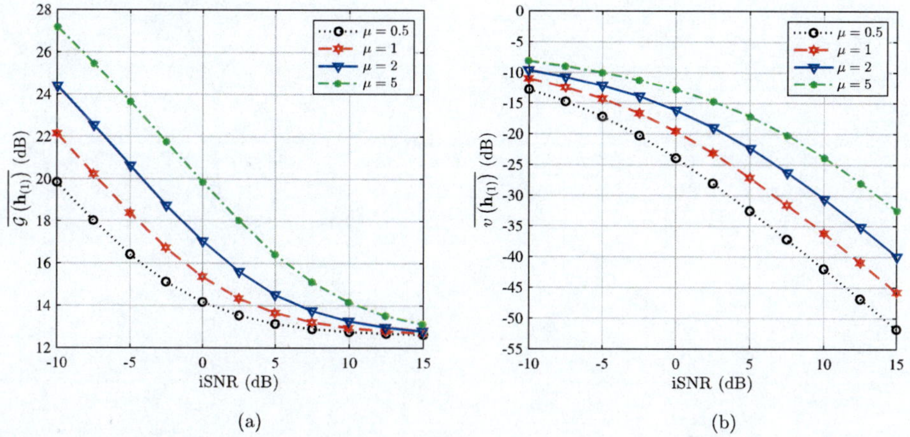

Fig. 7.11 Broadband SNR gain and desired signal distortion index of the first-order differential tradeoff beamformer, $\mathbf{h}_{(1),\mathrm{T},\mu}$, as a function of the input SNR, for different values of μ: (**a**) broadband SNR gain and (**b**) broadband desired signal distortion index. Conditions: $\alpha = 10^{-4}$ and $(\theta_s, \phi_s) = (90°, 0°)$

Figure 7.16 shows plots of the power patterns, $\left|\mathcal{B}_{\theta,\phi}\left(\mathbf{h}_{(3)}\right)\right|^2$, of the third-order differential MWNG beamformer, $\mathbf{h}_{(3),\mathrm{MWNG}}$, given in (7.91), versus the azimuth angle for $M = 4$ and different frequencies. Figure 7.17 shows plots of the DF, $\mathcal{D}\left(\mathbf{h}_{(3)}\right)$, and WNG, $\mathcal{W}\left(\mathbf{h}_{(3)}\right)$, of this beamformer as a function of frequency for different values of M. As seen, the power patterns are almost frequency invariant and have narrower main beams than the second-order ones in Fig. 7.14. The DF is higher, while the WNG is lower than the second-order ones in Fig. 7.15.

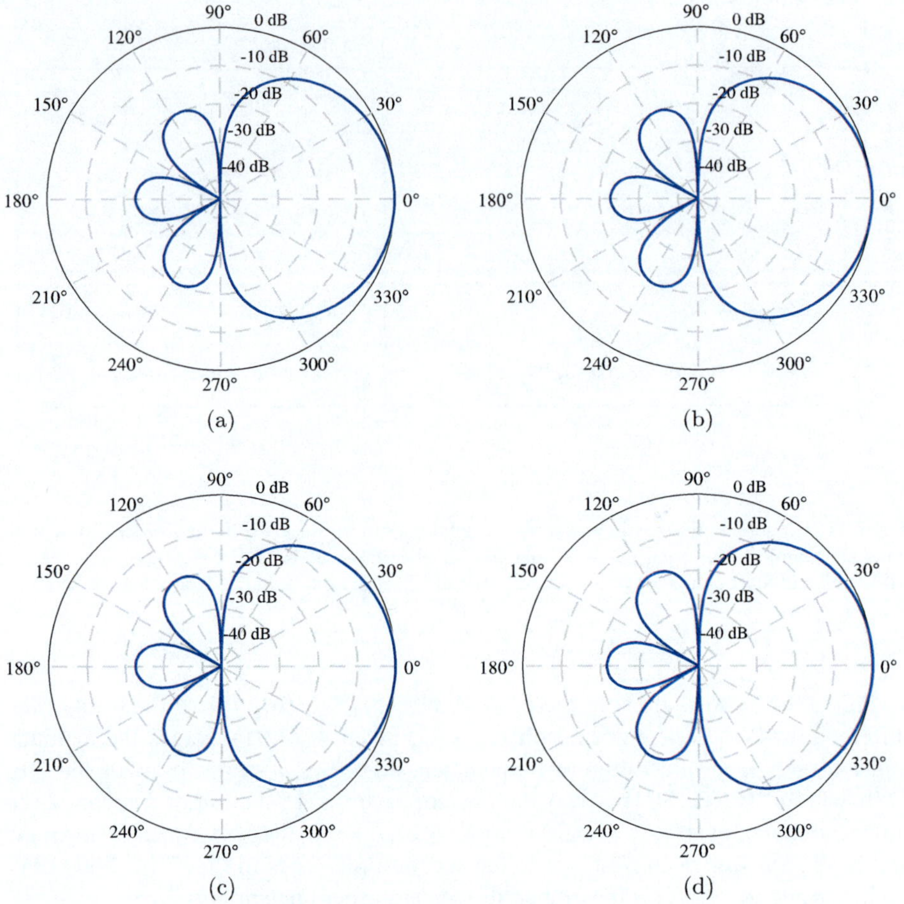

Fig. 7.12 Power patterns of the first-order differential LCMV beamformer, $\mathbf{h}_{(1),\mathrm{LCMV}}$, versus the azimuth angle, for different frequencies: (**a**) $f = 500$ Hz, (**b**) $f = 1000$ Hz, (**c**) $f = 2000$ Hz, and (**d**) $f = 4000$ Hz. Conditions: $M = 4$, iSNR = 10 dB, $\alpha = 10^{-4}$, and $(\theta_\mathrm{s}, \phi_\mathrm{s}) = (90°, 0°)$

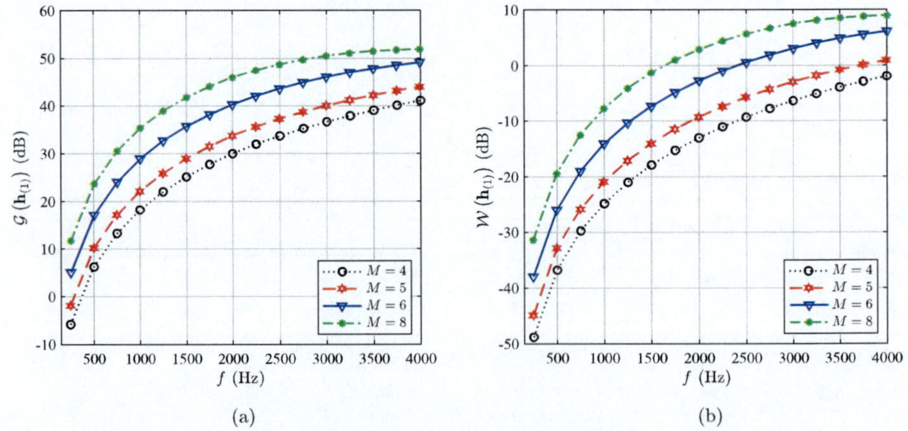

Fig. 7.13 SNR gain and WNG of the first-order differential LCMV beamformer, $\mathbf{h}_{(1),\mathrm{LCMV}}$, as a function of frequency, for different numbers of sensors, M: (**a**) SNR gain and (**b**) WNG. Conditions: iSNR = 10 dB, $\alpha = 10^{-4}$, and $(\theta_\mathrm{s}, \phi_\mathrm{s}) = (90°, 0°)$

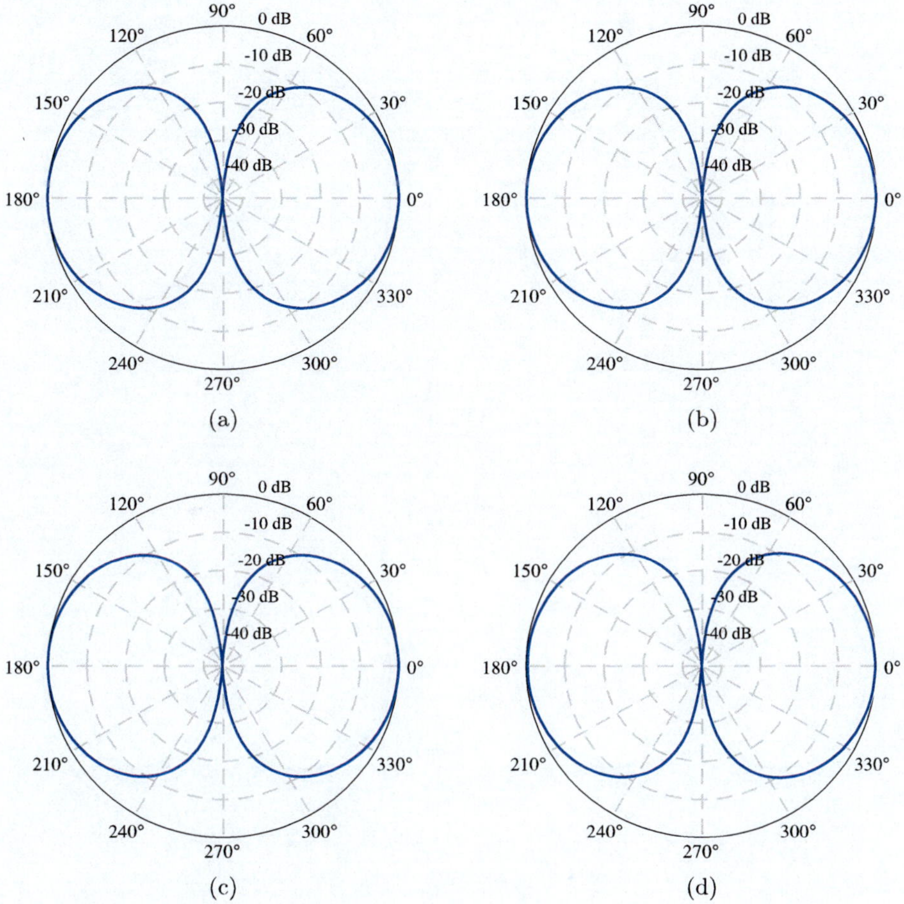

Fig. 7.14 Power patterns of the second-order differential MWNG beamformer, $\mathbf{h}_{(2),\text{MWNG}}$, versus the azimuth angle, for different frequencies: (a) $f = 500$ Hz, (b) $f = 1000$ Hz, (c) $f = 2000$ Hz, and (d) $f = 4000$ Hz. Conditions: $M = 4$ and $(\theta_\text{s}, \phi_\text{s}) = (90°, 0°)$

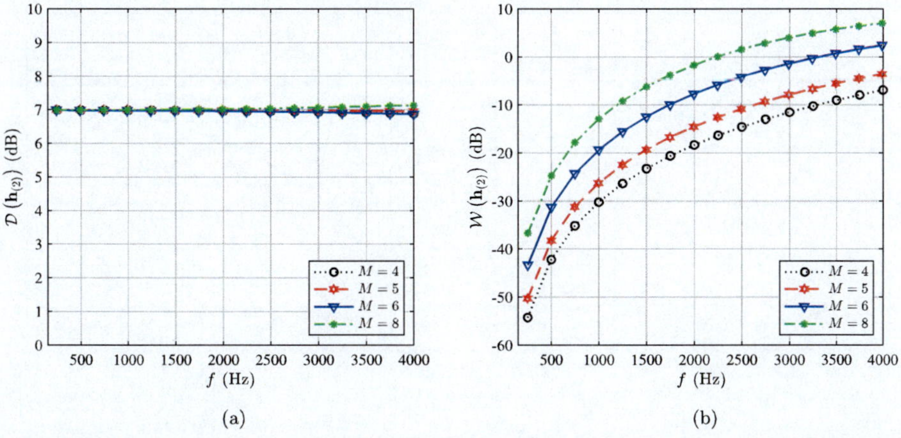

Fig. 7.15 DF and WNG of the second-order differential MWNG beamformer, $\mathbf{h}_{(2),\text{MWNG}}$, as a function of frequency, for different numbers of sensors, M: (a) DF and (b) WNG. Condition: $(\theta_\text{s}, \phi_\text{s}) = (90°, 0°)$

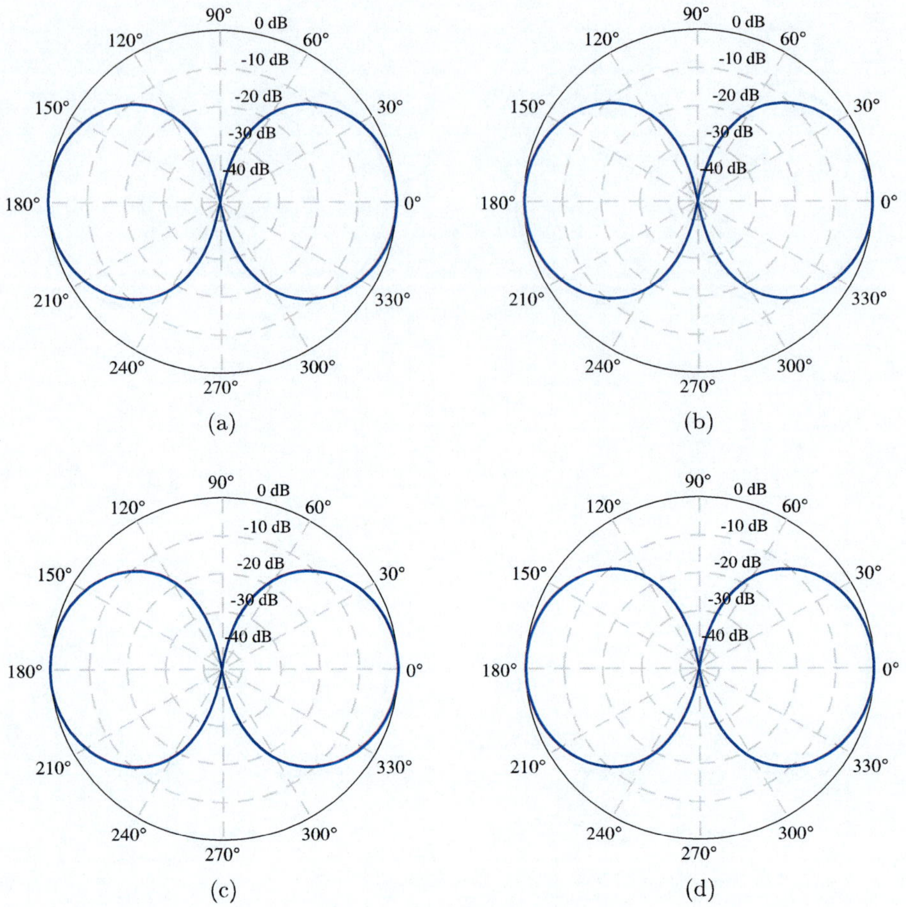

Fig. 7.16 Power patterns of the third-order differential MWNG beamformer, $\mathbf{h}_{(3),\mathrm{MWNG}}$, versus the azimuth angle, for different frequencies: (**a**) $f = 500$ Hz, (**b**) $f = 1000$ Hz, (**c**) $f = 2000$ Hz, and (**d**) $f = 4000$ Hz. Conditions: $M = 4$ and $(\theta_s, \phi_s) = (90°, 0°)$

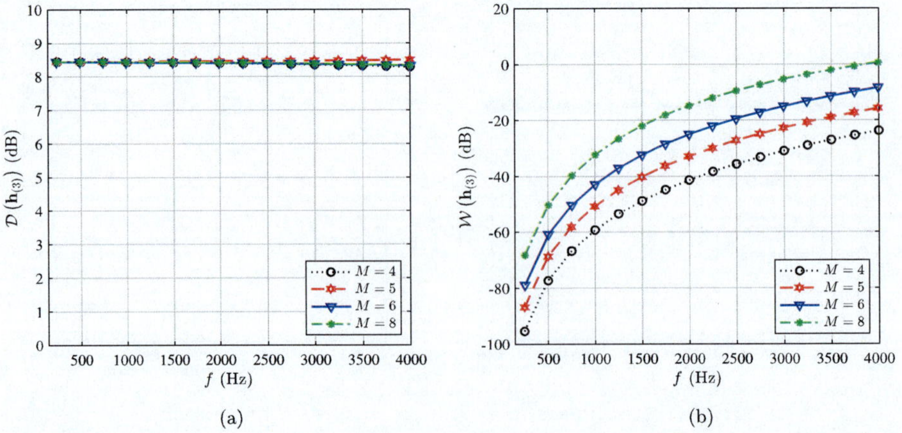

Fig. 7.17 DF and WNG of the third-order differential MWNG beamformer, $\mathbf{h}_{(3),\mathrm{MWNG}}$, as a function of frequency, for different numbers of sensors, M: (**a**) DF and (**b**) WNG. Condition: $(\theta_s, \phi_s) = (90°, 0°)$

References

1. J. Benesty, I. Cohen, J. Chen, *Array Beamforming with Linear Difference Equations* (Springer Nature, Cham, Switzerland, 2021)
2. G. Huang, J. Benesty, I. Cohen, J. Chen, A simple theory and new method of differential beamforming with uniform linear microphone arrays. IEEE/ACM Trans. Audio Speech Lang. Process. **28**(1), 1079–1093 (2020)
3. J. Jin, J. Benesty, G. Huang, J. Chen, On differential beamforming with nonuniform linear microphone arrays. IEEE/ACM Trans. Audio Speech Lang. Process. **30**, 1840–1852 (2022)
4. J.N. Franklin, *Matrix Theory* (Prentice-Hall, Englewood Cliffs, NJ, 1968)

Chapter 8
Adaptive Noise Cancellation

In some applications, it may be more convenient to find a noise reference, which can then be used to adaptively cancel the additive noise at microphones. This chapter is concerned with this problem. Furthermore, as it will be explained, adaptive noise cancellation gives another insightful perspective of distortionless adaptive beamforming.

8.1 Signal Model

Again, in this part, we consider the signal model of Chap. 3 (with a 3-D array of M microphones), i.e.,

$$\begin{aligned}\mathbf{y} &= \begin{bmatrix} Y_1 & Y_2 & \cdots & Y_M \end{bmatrix}^T \\ &= \mathbf{x} + \mathbf{v} \\ &= \mathbf{d}_{\theta_s,\phi_s} X + \mathbf{v},\end{aligned} \tag{8.1}$$

where

$$\mathbf{d}_{\theta_s,\phi_s} = \begin{bmatrix} e^{\jmath 2\pi f \mathbf{a}_{\theta_s,\phi_s}^T \mathbf{r}_1/c} & e^{\jmath 2\pi f \mathbf{a}_{\theta_s,\phi_s}^T \mathbf{r}_2/c} & \cdots & e^{\jmath 2\pi f \mathbf{a}_{\theta_s,\phi_s}^T \mathbf{r}_M/c} \end{bmatrix}^T \tag{8.2}$$

is the steering vector of length M. The covariance matrix of $\mathbf{y}$ is

$$\begin{aligned}\mathbf{\Phi}_{\mathbf{y}} &= E\left(\mathbf{y}\mathbf{y}^H\right) \\ &= \mathbf{\Phi}_{\mathbf{x}} + \mathbf{\Phi}_{\mathbf{v}} \\ &= \phi_X \mathbf{d}_{\theta_s,\phi_s} \mathbf{d}_{\theta_s,\phi_s}^H + \mathbf{\Phi}_{\mathbf{v}}\end{aligned}$$

J. Benesty et al., *Microphone Arrays*, Springer Topics in Signal Processing 22,
https://doi.org/10.1007/978-3-031-36974-2_8

$$= \phi_X \mathbf{d}_{\theta_s,\phi_s}\mathbf{d}_{\theta_s,\phi_s}^H + \phi_V \mathbf{\Gamma}_\mathbf{v}, \tag{8.3}$$

where $\mathbf{\Phi_x}$ and $\mathbf{\Phi_v}$ are the covariance matrices of $\mathbf{x}$ and $\mathbf{v}$, respectively, and $\mathbf{\Gamma_v}$ is the coherence matrix of the noise.

Our objective in the rest of this chapter is to find a noise reference given the observed signal vector, $\mathbf{y}$, and the steering vector, $\mathbf{d}_{\theta_s,\phi_s}$. And then find a way to adaptively cancel the additive noise at microphones, thanks to this noise reference.

8.2 Arrangement of a Noise Reference

The most fundamental constituent of a noise cancellation system is the arrangement of a noise reference [1], i.e., a signal that contains only some linear combinations of the microphone noise terms, which can then be used to cancel the noise from one of the sensor signals. Of course, this noise reference and the noise at the microphones should be somewhat coherent in order to get some noise cancellation.

Let

$$\begin{aligned}\mathbf{D}_{\theta_s,\phi_s} &= \text{diag}\left(\mathbf{d}^*_{\theta_s,\phi_s}\right) &(8.4)\\ &= \text{diag}\left(e^{-\jmath 2\pi f \mathbf{a}^T_{\theta_s,\phi_s}\mathbf{r}_1/c}, e^{-\jmath 2\pi f \mathbf{a}^T_{\theta_s,\phi_s}\mathbf{r}_2/c}, \ldots, e^{-\jmath 2\pi f \mathbf{a}^T_{\theta_s,\phi_s}\mathbf{r}_M/c}\right)\end{aligned}$$

be an $M \times M$ diagonal matrix. Left multiplying both sides of (8.1) by $\mathbf{D}_{\theta_s,\phi_s}$, we get

$$\begin{aligned}\widetilde{\mathbf{y}} &= \begin{bmatrix}\widetilde{Y}_1 & \widetilde{Y}_2 & \cdots & \widetilde{Y}_M\end{bmatrix}^T\\ &= \mathbf{D}_{\theta_s,\phi_s}\mathbf{y}\\ &= \mathbf{1}X + \widetilde{\mathbf{v}}, &(8.5)\end{aligned}$$

where $\mathbf{1}$ is a vector containing only ones and $\widetilde{\mathbf{v}} = \mathbf{D}_{\theta_s,\phi_s}\mathbf{v}$ is defined similarly to $\widetilde{\mathbf{y}}$.

Now, let us define the matrix of size $(M-1) \times M$:

$$\mathbf{\Delta} = \begin{bmatrix} -1 & 1 & 0 & \cdots & 0\\ 0 & -1 & 1 & \cdots & 0\\ \vdots & \vdots & \ddots & \ddots & \vdots\\ 0 & 0 & \cdots & -1 & 1\end{bmatrix}. \tag{8.6}$$

Left multiplying both sides of (8.5) by $\mathbf{\Delta}$, we obtain the noise reference:

$$\begin{aligned}\mathbf{v}_{\text{NR}} &= \mathbf{\Delta}\widetilde{\mathbf{y}}\\ &= \mathbf{\Delta}\mathbf{1}X + \mathbf{\Delta}\widetilde{\mathbf{v}}\end{aligned}$$

$$= \mathbf{\Delta}\widetilde{\mathbf{v}}, \tag{8.7}$$

which is a vector of length $M - 1$ and, as expected, does not depend on X since $\mathbf{\Delta 1} = \mathbf{0}$. Eventually, a good way to know how adaptive noise cancellation performs is from the coherence matrix of size $(M - 1) \times M$:

$$\mathbf{C}_{\mathbf{v}_{\mathrm{NR}}\widetilde{\mathbf{v}}} = \mathbf{\Phi}_{\mathbf{v}_{\mathrm{NR}}}^{-1/2}\mathbf{\Phi}_{\mathbf{v}_{\mathrm{NR}}\widetilde{\mathbf{v}}}\mathbf{\Phi}_{\widetilde{\mathbf{v}}}^{-1/2}, \tag{8.8}$$

where $\mathbf{\Phi}_{\mathbf{v}_{\mathrm{NR}}} = E\left(\mathbf{v}_{\mathrm{NR}}\mathbf{v}_{\mathrm{NR}}^H\right)$, $\mathbf{\Phi}_{\mathbf{v}_{\mathrm{NR}}\widetilde{\mathbf{v}}} = E\left(\mathbf{v}_{\mathrm{NR}}\widetilde{\mathbf{v}}^H\right)$, and $\mathbf{\Phi}_{\widetilde{\mathbf{v}}} = E\left(\widetilde{\mathbf{v}}\widetilde{\mathbf{v}}^H\right)$.

Assuming that Microphone 1 is the reference, we can also find a single noise reference as follows:

$$\begin{aligned} V_{\mathrm{NR}} &= \widetilde{Y}_1 - \frac{1}{M}\mathbf{1}^T\widetilde{\mathbf{y}} \\ &= X + \widetilde{V}_1 - \frac{1}{M}\left(MX + \mathbf{1}^T\widetilde{\mathbf{v}}\right) \\ &= \widetilde{V}_1 - \frac{1}{M}\mathbf{1}^T\widetilde{\mathbf{v}}, \end{aligned} \tag{8.9}$$

which, as expected, depends only on the noise terms, and the coherence function between V_{NR} and $\widetilde{V}_1$, i.e.,

$$\gamma_{V_{\mathrm{NR}}\widetilde{V}_1} = \frac{E\left(V_{\mathrm{NR}}\widetilde{V}_1^*\right)}{\sqrt{E\left(|V_{\mathrm{NR}}|^2\right)E\left(|\widetilde{V}_1|^2\right)}}, \tag{8.10}$$

is a good indicator of the behavior of an adaptive noise canceller.

8.3 Noise Cancellation

We start this section with the single noise reference; with this approach, noise cancellation is limited, in general, since only a gain is used, but it is simple to study and understand and gives a good idea on the functioning of adaptive noise cancellation.

An estimate of the desired signal, X, is

$$\begin{aligned} Z &= \widetilde{Y}_1 - G^* V_{\mathrm{NR}} \\ &= X + \left(\widetilde{V}_1 - G^* V_{\mathrm{NR}}\right), \end{aligned} \tag{8.11}$$

where G is a complex-valued gain and V_{NR} depends only on the observed signals. We see from the previous equation that this approach is distortionless. We deduce that the variance of Z is

$$\phi_Z = \phi_X + \phi_V + |G|^2 \phi_{V_{\mathrm{NR}}} - G^* \phi_{V_{\mathrm{NR}} \widetilde{V}_1} - G \phi^*_{V_{\mathrm{NR}} \widetilde{V}_1}, \tag{8.12}$$

where ϕ_V and $\phi_{V_{\mathrm{NR}}}$ are the variances of $\widetilde{V}_1$ and V_{NR}, respectively, and $\phi_{V_{\mathrm{NR}} \widetilde{V}_1} = E\left(V_{\mathrm{NR}} \widetilde{V}_1^*\right)$. As a result, the output SNR and SNR gain are, respectively,

$$\mathrm{oSNR}\,(G) = \frac{\phi_X}{\phi_V + |G|^2 \phi_{V_{\mathrm{NR}}} - G^* \phi_{V_{\mathrm{NR}} \widetilde{V}_1} - G \phi^*_{V_{\mathrm{NR}} \widetilde{V}_1}} \tag{8.13}$$

and

$$\mathcal{G}\,(G) = \frac{1}{1 + |G|^2 \dfrac{\phi_{V_{\mathrm{NR}}}}{\phi_V} - G^* \dfrac{\phi_{V_{\mathrm{NR}} \widetilde{V}_1}}{\phi_V} - G \dfrac{\phi^*_{V_{\mathrm{NR}} \widetilde{V}_1}}{\phi_V}}. \tag{8.14}$$

Now, to find the optimal gain, we need to maximize $\mathcal{G}\,(G)$ or, equivalently, minimize its denominator. We easily get

$$G_{\mathrm{O}} = \frac{\phi_{V_{\mathrm{NR}} \widetilde{V}_1}}{\phi_{V_{\mathrm{NR}}}}. \tag{8.15}$$

Substituting (8.15) into (8.14) results in

$$\mathcal{G}\,(G_{\mathrm{O}}) = \frac{1}{1 - \left|\gamma_{V_{\mathrm{NR}} \widetilde{V}_1}\right|^2}, \tag{8.16}$$

where we observe the importance of the MSCF between V_{NR} and $\widetilde{V}_1$, i.e., the performance of G_{O}, as far as the SNR gain is concerned, depends only on $\left|\gamma_{V_{\mathrm{NR}} \widetilde{V}_1}\right|^2$.

The study of adaptive noise cancellation with $\mathbf{v}_{\mathrm{NR}}$ is certainly more interesting since we have more degrees of freedom. Let $\mathbf{g}$ be a complex-valued filter of length $M - 1$. An estimate of X is

$$\begin{aligned} Z &= \overline{\mathbf{1}}^T \widetilde{\mathbf{y}} - \mathbf{g}^H \mathbf{v}_{\mathrm{NR}} \\ &= X + \left(\overline{\mathbf{1}}^T \widetilde{\mathbf{v}} - \mathbf{g}^H \mathbf{v}_{\mathrm{NR}}\right), \end{aligned} \tag{8.17}$$

where $\overline{\mathbf{1}} = \frac{1}{M}\mathbf{1}$. Again, as in the single noise reference case, this technique is distortionless. The variance of Z is

$$\phi_Z = \phi_X + \overline{\mathbf{1}}^T \mathbf{\Phi}_{\widetilde{\mathbf{v}}} \overline{\mathbf{1}} + \mathbf{g}^H \mathbf{\Phi}_{\mathbf{v}_{\mathrm{NR}}} \mathbf{g} - \mathbf{g}^H \mathbf{\Phi}_{\mathbf{v}_{\mathrm{NR}} \widetilde{\mathbf{v}}} \overline{\mathbf{1}} - \overline{\mathbf{1}}^T \mathbf{\Phi}^H_{\mathbf{v}_{\mathrm{NR}} \widetilde{\mathbf{v}}} \mathbf{g}. \tag{8.18}$$

Then, it is clear that the output SNR is

$$\text{oSNR}(\mathbf{g}) = \frac{\phi_X}{\overline{\mathbf{1}}^T \mathbf{\Phi}_{\widetilde{\mathbf{v}}} \overline{\mathbf{1}} + \mathbf{g}^H \mathbf{\Phi}_{\mathbf{v}_{\text{NR}}} \mathbf{g} - \mathbf{g}^H \mathbf{\Phi}_{\mathbf{v}_{\text{NR}} \widetilde{\mathbf{v}}} \overline{\mathbf{1}} - \overline{\mathbf{1}}^T \mathbf{\Phi}_{\mathbf{v}_{\text{NR}} \widetilde{\mathbf{v}}}^H \mathbf{g}}, \tag{8.19}$$

whose maximization leads to the optimal adaptive noise canceller:

$$\mathbf{g}_{\text{O}} = \mathbf{\Phi}_{\mathbf{v}_{\text{NR}}}^{-1} \mathbf{\Phi}_{\mathbf{v}_{\text{NR}} \widetilde{\mathbf{v}}} \overline{\mathbf{1}}. \tag{8.20}$$

Substituting (8.20) into (8.19), we get

$$\begin{aligned} \text{oSNR}(\mathbf{g}_{\text{O}}) &= \frac{\phi_X}{\overline{\mathbf{1}}^T \mathbf{\Phi}_{\widetilde{\mathbf{v}}} \overline{\mathbf{1}} - \overline{\mathbf{1}}^T \mathbf{\Phi}_{\mathbf{v}_{\text{NR}} \widetilde{\mathbf{v}}}^H \mathbf{\Phi}_{\mathbf{v}_{\text{NR}}}^{-1} \mathbf{\Phi}_{\mathbf{v}_{\text{NR}} \widetilde{\mathbf{v}}} \overline{\mathbf{1}}} \\ &= \frac{\phi_X}{\overline{\mathbf{1}}^T \mathbf{\Phi}_{\widetilde{\mathbf{v}}} \overline{\mathbf{1}} - \overline{\mathbf{1}}^T \mathbf{\Phi}_{\widetilde{\mathbf{v}}}^{1/2} \mathbf{C}_{\mathbf{v}_{\text{NR}} \widetilde{\mathbf{v}}}^H \mathbf{C}_{\mathbf{v}_{\text{NR}} \widetilde{\mathbf{v}}} \mathbf{\Phi}_{\widetilde{\mathbf{v}}}^{1/2} \overline{\mathbf{1}}} \\ &= \frac{\phi_X}{\overline{\mathbf{1}}^T \mathbf{\Phi}_{\widetilde{\mathbf{v}}}^{1/2} \left(\mathbf{I}_M - \mathbf{C}_{\mathbf{v}_{\text{NR}} \widetilde{\mathbf{v}}}^H \mathbf{C}_{\mathbf{v}_{\text{NR}} \widetilde{\mathbf{v}}} \right) \mathbf{\Phi}_{\widetilde{\mathbf{v}}}^{1/2} \overline{\mathbf{1}}}, \end{aligned} \tag{8.21}$$

where we see the importance of the coherence matrix, $\mathbf{C}_{\mathbf{v}_{\text{NR}} \widetilde{\mathbf{v}}}$, in the maximization of the output SNR.

The great advantage of a noise cancellation system over a multichannel noise reduction approach [2, 3] is that we do not need to wait for silences to get an estimate of the noise statistics. As a result, the former approach should work better than the latter one with nonstationary noises.

From a theoretical point of view, the desired signal, X, is not distorted since it is not filtered at all. Only the noise terms are filtered. However, because of reverberation and other imperfections, the noise reference may contain some speech. Therefore, we should expect some desired signal cancellation or distortion. This cancellation should be proportional to the amount of speech present in the noise reference.

Also, from a theoretical point of view, the above optimal adaptive noise cancellation filter is equivalent to the MVDR beamformer (see Chap. 3); however, in practice, the two may behave differently.

8.4 Low-Rank Noise Cancellation

In this section, we briefly discuss low-rank noise cancellation (see also Chap. 5).

Assume that the filter, $\mathbf{g}$ of length $M - 1 = M_1 M_2$, is low rank (see Chap. 5), i.e.,

$$\mathbf{g} = \sum_{p=1}^{P} \mathbf{g}_{2,p}^{*} \otimes \mathbf{g}_{1,p}, \tag{8.22}$$

where $P \ll M - 1$, and the filters $\mathbf{g}_{1,p}$ and $\mathbf{g}_{2,p}$ have lengths M_1 and M_2, respectively. Then, (8.22) can also be expressed as

$$\begin{aligned} \mathbf{g} &= \sum_{p=1}^{P} \mathbf{G}_{2,p}^{*} \mathbf{g}_{1,p} \\ &= \underline{\mathbf{G}}_2 \underline{\mathbf{g}}_1 \qquad (8.23) \\ &= \sum_{p=1}^{P} \mathbf{G}_{1,p} \mathbf{g}_{2,p}^{*} \\ &= \underline{\mathbf{G}}_1 \underline{\mathbf{g}}_2, \qquad (8.24) \end{aligned}$$

where

$$\mathbf{G}_{2,p}^{*} = \mathbf{g}_{2,p}^{*} \otimes \mathbf{I}_{M_1},$$
$$\mathbf{G}_{1,p} = \mathbf{I}_{M_2} \otimes \mathbf{g}_{1,p}$$

are matrices of sizes $(M - 1) \times M_1$ and $(M - 1) \times M_2$, respectively;

$$\underline{\mathbf{G}}_2 = \begin{bmatrix} \mathbf{G}_{2,1}^{*} & \mathbf{G}_{2,2}^{*} & \cdots & \mathbf{G}_{2,P}^{*} \end{bmatrix},$$
$$\underline{\mathbf{G}}_1 = \begin{bmatrix} \mathbf{G}_{1,1} & \mathbf{G}_{1,2} & \cdots & \mathbf{G}_{1,P} \end{bmatrix}$$

are matrices of sizes $(M - 1) \times PM_1$ and $(M - 1) \times PM_2$, respectively; and

$$\underline{\mathbf{g}}_1 = \begin{bmatrix} \mathbf{g}_{1,1}^{T} & \mathbf{g}_{1,2}^{T} & \cdots & \mathbf{g}_{1,P}^{T} \end{bmatrix}^{T},$$
$$\underline{\mathbf{g}}_2 = \begin{bmatrix} \mathbf{g}_{2,1}^{H} & \mathbf{g}_{2,2}^{H} & \cdots & \mathbf{g}_{2,P}^{H} \end{bmatrix}^{T}$$

are filters of lengths PM_1 and PM_2, respectively.

Now, instead of carrying out noise cancellation with $\mathbf{g}$ as in the previous section, we propose to perform this task with $\underline{\mathbf{g}}_1$ and $\underline{\mathbf{g}}_2$. Therefore, an estimate of X is

$$Z = \overline{\mathbf{1}}^{T} \widetilde{\mathbf{y}} - \underline{\mathbf{g}}_1^{H} \underline{\mathbf{G}}_2^{H} \mathbf{v}_{\mathrm{NR}} \tag{8.25}$$
$$= \overline{\mathbf{1}}^{T} \widetilde{\mathbf{y}} - \underline{\mathbf{g}}_2^{H} \underline{\mathbf{G}}_1^{H} \mathbf{v}_{\mathrm{NR}}. \tag{8.26}$$

Assuming that $\underline{\mathbf{g}}_2$ is fixed, we can express the output SNR as

$$\text{oSNR}\left(\underline{\mathbf{g}}_1|\underline{\mathbf{g}}_2\right) = \frac{\phi_X}{\overline{\mathbf{1}}^T\boldsymbol{\Phi}_{\widetilde{\mathbf{v}}}\overline{\mathbf{1}} + \underline{\mathbf{g}}_1^H\underline{\mathbf{G}}_2^H\boldsymbol{\Phi}_{\mathbf{v}_{\text{NR}}}\underline{\mathbf{G}}_2\underline{\mathbf{g}}_1 - \underline{\mathbf{g}}_1^H\underline{\mathbf{G}}_2^H\boldsymbol{\Phi}_{\mathbf{v}_{\text{NR}}\widetilde{\mathbf{v}}}\overline{\mathbf{1}} - \overline{\mathbf{1}}^T\boldsymbol{\Phi}_{\mathbf{v}_{\text{NR}}\widetilde{\mathbf{v}}}^H\underline{\mathbf{G}}_2\underline{\mathbf{g}}_1}. \tag{8.27}$$

In the same way, by assuming that $\underline{\mathbf{g}}_1$ is fixed, we can define the output SNR as

$$\text{oSNR}\left(\underline{\mathbf{g}}_2|\underline{\mathbf{g}}_1\right) = \frac{\phi_X}{\overline{\mathbf{1}}^T\boldsymbol{\Phi}_{\widetilde{\mathbf{v}}}\overline{\mathbf{1}} + \underline{\mathbf{g}}_2^H\underline{\mathbf{G}}_1^H\boldsymbol{\Phi}_{\mathbf{v}_{\text{NR}}}\underline{\mathbf{G}}_1\underline{\mathbf{g}}_2 - \underline{\mathbf{g}}_2^H\underline{\mathbf{G}}_1^H\boldsymbol{\Phi}_{\mathbf{v}_{\text{NR}}\widetilde{\mathbf{v}}}\overline{\mathbf{1}} - \overline{\mathbf{1}}^T\boldsymbol{\Phi}_{\mathbf{v}_{\text{NR}}\widetilde{\mathbf{v}}}^H\underline{\mathbf{G}}_1\underline{\mathbf{g}}_2}. \tag{8.28}$$

The optimal filters are obtained from the maximization of the output SNRs. We get

$$\underline{\mathbf{g}}_{1,\text{O}} = \left(\underline{\mathbf{G}}_2^H\boldsymbol{\Phi}_{\mathbf{v}_{\text{NR}}}\underline{\mathbf{G}}_2\right)^{-1}\underline{\mathbf{G}}_2^H\boldsymbol{\Phi}_{\mathbf{v}_{\text{NR}}\widetilde{\mathbf{v}}}\overline{\mathbf{1}}, \tag{8.29}$$

$$\underline{\mathbf{g}}_{2,\text{O}} = \left(\underline{\mathbf{G}}_1^H\boldsymbol{\Phi}_{\mathbf{v}_{\text{NR}}}\underline{\mathbf{G}}_1\right)^{-1}\underline{\mathbf{G}}_1^H\boldsymbol{\Phi}_{\mathbf{v}_{\text{NR}}\widetilde{\mathbf{v}}}\overline{\mathbf{1}}. \tag{8.30}$$

Using the ALS algorithm, these two optimal filters will converge to the low-rank optimal adaptive noise canceller just after a couple of iterations:

$$\mathbf{g}_{\text{LRO}} = \sum_{p=1}^{P}\mathbf{g}_{2,p,\text{O}}^{*}\otimes\mathbf{g}_{1,p,\text{O}}, \tag{8.31}$$

where

$$\underline{\mathbf{g}}_{1,\text{O}} = \left[\mathbf{g}_{1,1,\text{O}}^T\ \mathbf{g}_{1,2,\text{O}}^T\ \cdots\ \mathbf{g}_{1,P,\text{O}}^T\right]^T,$$

$$\underline{\mathbf{g}}_{2,\text{O}} = \left[\mathbf{g}_{2,1,\text{O}}^H\ \mathbf{g}_{2,2,\text{O}}^H\ \cdots\ \mathbf{g}_{2,P,\text{O}}^H\right]^T.$$

8.5 Illustrative Examples

In this section, we study some examples of adaptive noise cancellers. Again, in this part, we consider a 3-D cube array consisting of $M = M_0^3$ microphones, as shown in Fig. 3.1, with $\delta = 1.0$ cm and $(\theta_\text{s}, \phi_\text{s}) = (80^\circ, 0^\circ)$. We consider an environment with point source noises and white noise, where the two statistically independent interferences impinge on the cube array from two different directions $\left(\theta_{\text{i},1}, \phi_{\text{i},1}\right) =$

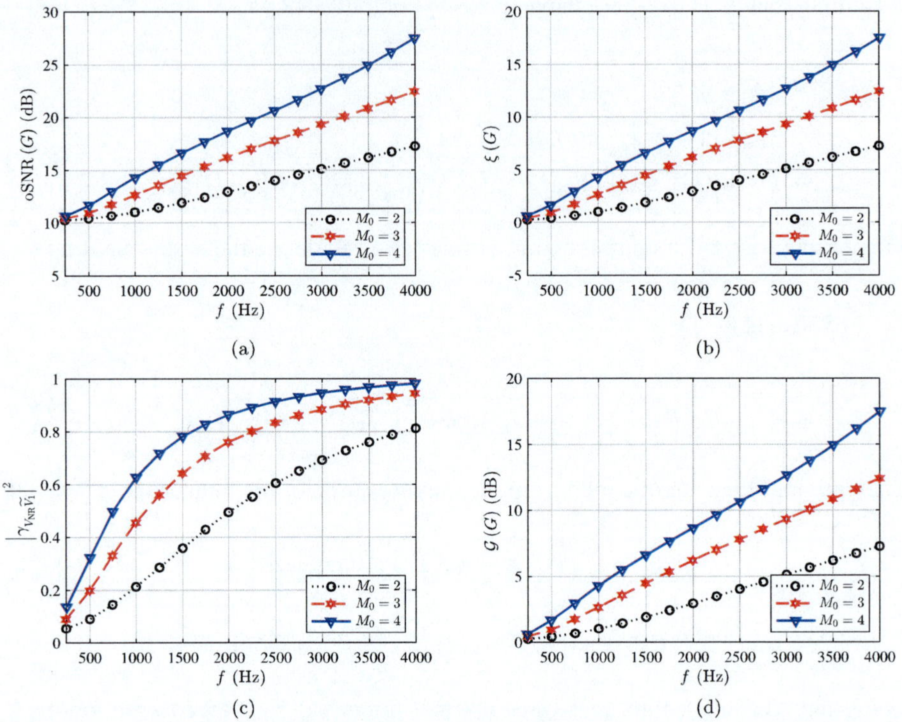

Fig. 8.1 Performance of the optimal noise cancellation gain, G_{O}, as a function of frequency, for different numbers of sensors, $M = M_0^3$: (**a**) output SNR, (**b**) noise reduction factor, (**c**) MSCF between V_{NR} and $\widetilde{V}_1$, and (**d**) SNR gain. Conditions: $\delta = 1.0$ cm, $\alpha = 0.1$, iSNR $= 10$ dB, and $(\theta_{\mathrm{s}}, \phi_{\mathrm{s}}) = (80^\circ, 0^\circ)$

$(80^\circ, 120^\circ)$ and $\left(\theta_{\mathrm{i},2}, \phi_{\mathrm{i},2}\right) = (80^\circ, 150^\circ)$. The two interferences are incoherent with the desired signal. The covariance matrix of the noise signal, $\mathbf{\Phi}_{\mathbf{v}}$, is given in (3.75). The variance of the desired signal is set to 1 and the variance of the noise is computed according to the specified input SNR and value of α ($\alpha > 0$ is related to the ratio between the powers of the white noise and point source noises).

The first simulation is to demonstrate the performance of the optimal noise cancellation gain, G_{O}, given in (8.15), where the value of α is set to 0.1. Figure 8.1 shows plots of the output SNR, oSNR (G), noise reduction factor, $\xi(G)$, MSCF between V_{NR} and $\widetilde{V}_1$, $\left|\gamma_{V_{\mathrm{NR}}\widetilde{V}_1}\right|^2$, and SNR gain, $\mathcal{G}(G)$, as a function of frequency, for iSNR $= 10$ dB and different numbers of sensors $M = M_0^3$. It can be observed that the noise cancellation gain improves the output SNR and reduces the noise. For a given frequency, the output SNR, noise reduction factor, MSCF, and SNR gain all increase as the value of M increases. Figure 8.2 shows plots of the output SNR, oSNR (G), noise reduction factor, $\xi(G)$, MSCF between V_{NR} and $\widetilde{V}_1$, $\left|\gamma_{V_{\mathrm{NR}}\widetilde{V}_1}\right|^2$,

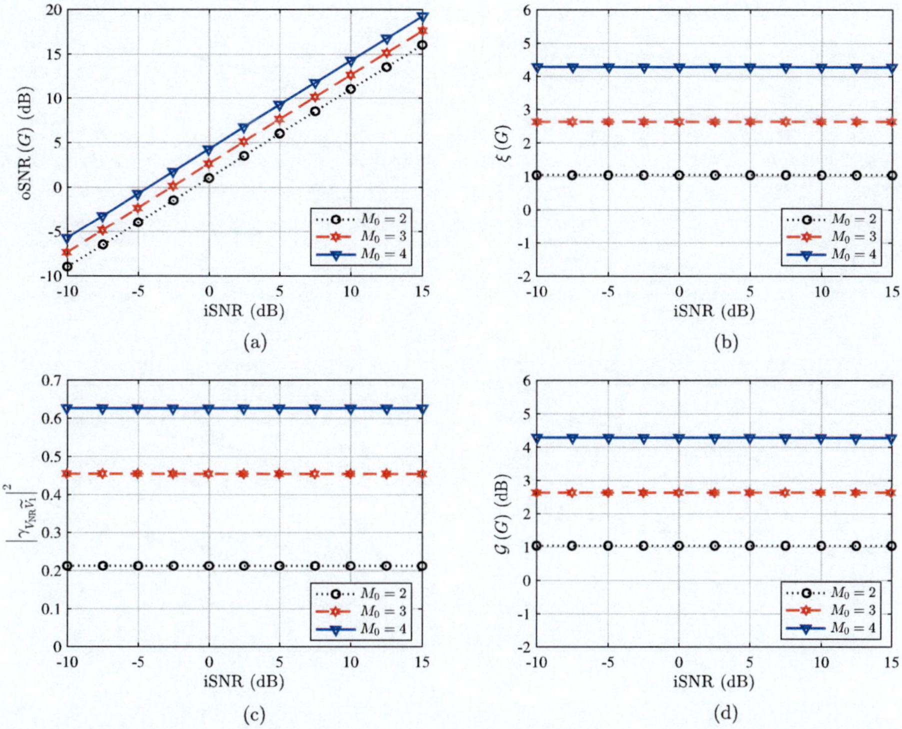

Fig. 8.2 Performance of the optimal noise cancellation gain, G_O, as a function of the input SNR, for different numbers of sensors, $M = M_0^3$: **(a)** output SNR, **(b)** noise reduction factor, **(c)** MSCF between V_{NR} and $\widetilde{V}_1$, and **(d)** SNR gain. Conditions: $\delta = 1.0$ cm, $\alpha = 0.1$, $f = 1000$ Hz, and $(\theta_s, \phi_s) = (80°, 0°)$

and SNR gain, $\mathcal{G}(G)$, as a function of the input SNR, for $f = 1000$ Hz and different numbers of sensors $M = M_0^3$. As seen, for a specified input SNR, the output SNR, noise reduction factor, MSCF, and SNR gain all increase as the value of M increases. It is also observed that the higher is the value of the MSCF between V_{NR} and $\widetilde{V}_1$, the higher is the SNR gain.

We then study the performance of the optimal adaptive noise canceller, $\mathbf{g}_O$, given in (8.20). Figure 8.3 shows plots of the output SNR, oSNR $(\mathbf{g})$, as a function of frequency, for iSNR $= 10$ dB, different numbers of sensors, $M = M_0^3$, and different values of α. It can be clearly observed that this canceller significantly improves the output SNR, and, for a given frequency, the output SNR increases as the value of M increases. Figure 8.4 shows plots of the output SNR, oSNR $(\mathbf{g})$, of this noise canceller, as a function of the input SNR, for $f = 1000$ Hz, different numbers of sensors, $M = M_0^3$, and different values of α. Similarly, the output SNR is significantly improved and, for a given input SNR, the output SNR increases as the value of M increases. It is also observed that, for a given frequency or input SNR, the output SNR decreases when α increases. This is because a larger value

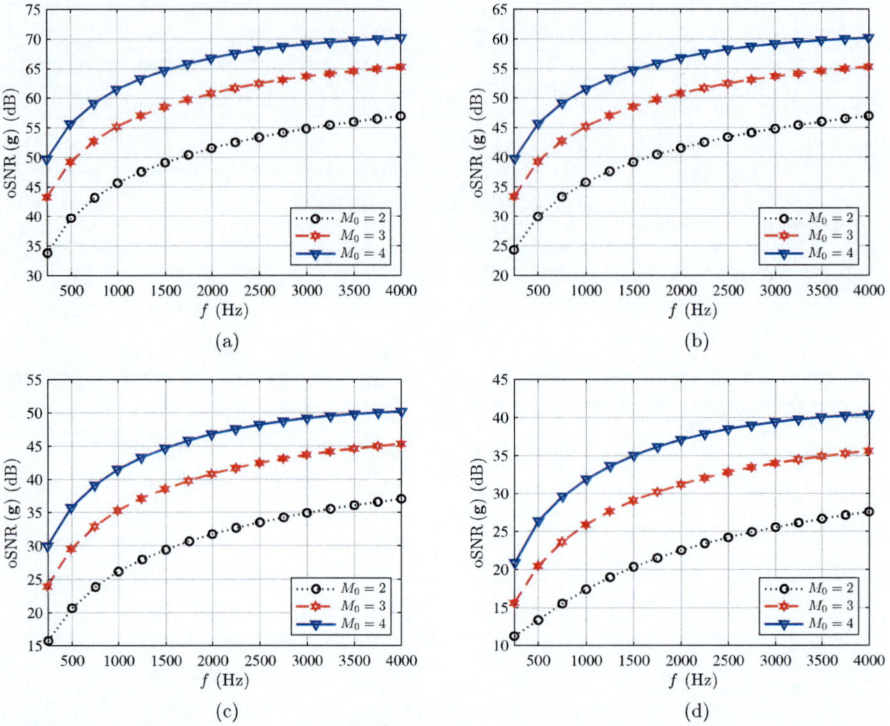

Fig. 8.3 Output SNR of the optimal adaptive noise canceller, $\mathbf{g}_\mathrm{O}$, as a function of frequency, for different numbers of sensors, $M = M_0^3$, and different values of α: (**a**) $\alpha = 10^{-4}$, (**b**) $\alpha = 10^{-3}$, (**c**) $\alpha = 10^{-2}$, and (**d**) $\alpha = 10^{-1}$. Conditions: $\delta = 1.0$ cm, iSNR = 10 dB, and $(\theta_\mathrm{s}, \phi_\mathrm{s}) = (80°, 0°)$

of α means that noise signals at the two sensors are more white and less coherent; as a result, the adaptive noise canceller is less effective in suppressing noise. In comparison to the optimal noise cancellation gain, G_O, the optimal adaptive noise canceller, $\mathbf{g}_\mathrm{O}$, achieves a higher output SNR in the same conditions. This shows that noise cancellation performance is limited when using a single noise cancellation gain, but can be significantly improved by using a noise cancellation filter.

Finally, we study the performance of the low-rank optimal adaptive noise canceller, $\mathbf{g}_\mathrm{LRO}$, given in (8.31). We choose $M_0 = 4$, i.e., $M = 64$, so that the length of the low-rank filter, $\mathbf{g}$, is $M - 1 = 63$. The lengths of the two series of short filters, i.e., $\mathbf{h}_{1,p}$ and $\mathbf{h}_{2,p}$, are chosen as $M_1 = 9$ and $M_2 = 7$, respectively. The two optimal filters are optimized by using the ALS algorithm, where the filter $\underline{\mathbf{g}}_{1,\mathrm{O}}$ is initialized as $\mathbf{g}_{1,p,\mathrm{O}} = \eta \mathbf{i}_{M_2,p}$, $p = 1, 2, \ldots, P$, with η being an arbitrary constant (randomly generated), and $\mathbf{i}_{M_2,p}$ being a vector of length M_2 whose pth element is $1/M_2$ and all other elements are 0. In the implementation of a matrix inversion, a small regularization parameter, 10^{-6}, is added to the diagonal elements of the matrix. The number of iterations is chosen to be 5. Figure 8.5 shows plots of

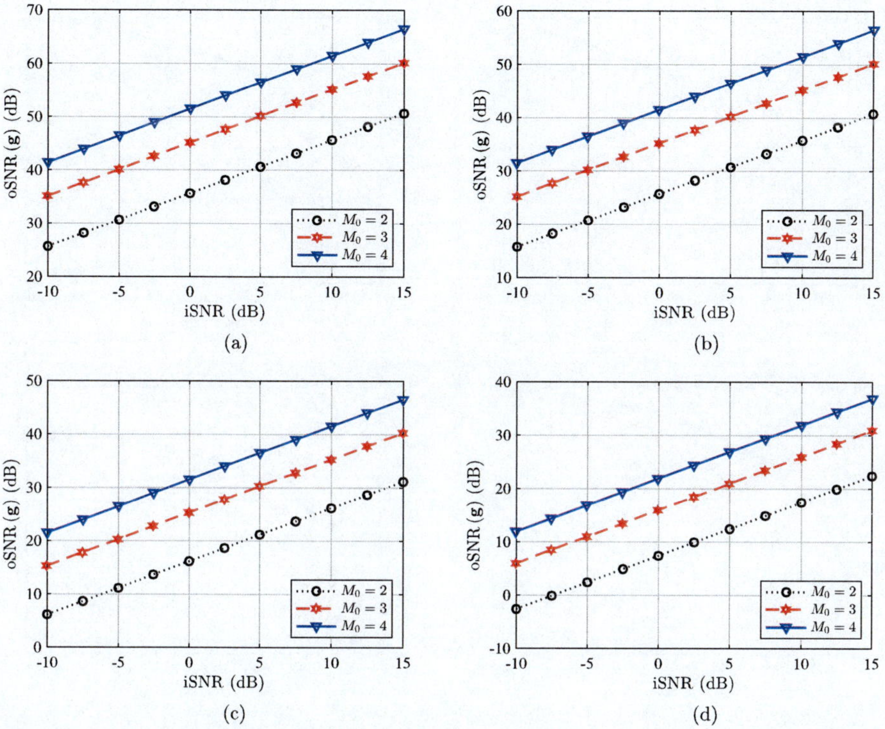

Fig. 8.4 Output SNR of the optimal adaptive noise canceller, $\mathbf{g}_\mathrm{O}$, as a function of the input SNR, for different numbers of sensors, $M = M_0^3$, and different values of α: **(a)** $\alpha = 10^{-4}$, **(b)** $\alpha = 10^{-3}$, **(c)** $\alpha = 10^{-2}$, and **(d)** $\alpha = 10^{-1}$. Conditions: $\delta = 1.0$ cm, $f = 1000$ Hz, and $(\theta_\mathrm{s}, \phi_\mathrm{s}) = (80°, 0°)$

the output SNR, oSNR $(\mathbf{g})$, as a function of frequency, for iSNR $= 10$ dB, different values of P, and different values of α. Figure 8.6 shows plots of the output SNR, oSNR $(\mathbf{g})$, as a function of the input SNR, for $f = 1000$ Hz, different values of P, and different values of α. The results are based on the average of ten Monte Carlo experiments, i.e., randomly generating the constant η. It can be seen that the low-rank optimal adaptive noise canceller significantly improves the output SNR in all conditions. For a given frequency or input SNR, it is clearly observed that the output SNR increases with the value of P. This is easy to understand, the larger is the value of P, the more the number of parameters in the adaptive noise canceller, $\mathbf{g}_\mathrm{LRO}$, and, as a result, more degrees of freedom can be used for SNR maximization.

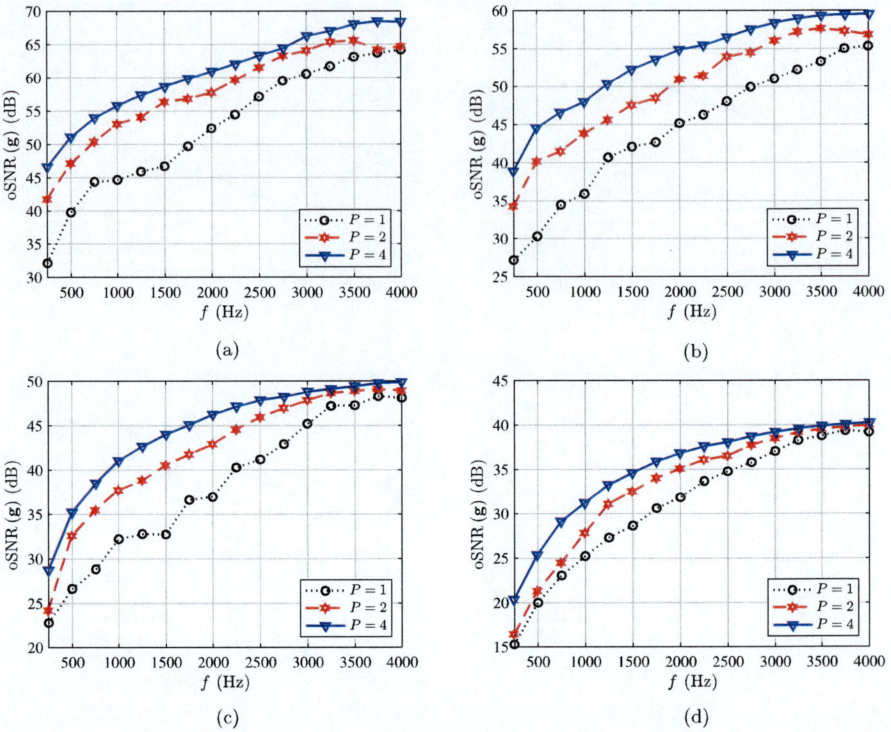

Fig. 8.5 Output SNR of the low-rank optimal adaptive noise canceller, $\mathbf{g}_{\mathrm{LRO}}$, as a function of frequency, for different values of P, and different values of α: (**a**) $\alpha = 10^{-4}$, (**b**) $\alpha = 10^{-3}$, (**c**) $\alpha = 10^{-2}$, and (**d**) $\alpha = 10^{-1}$. Conditions: $M_0 = 4$, $\delta = 1.0$ cm, iSNR = 10 dB, and $(\theta_{\mathrm{s}}, \phi_{\mathrm{s}}) = (80^\circ, 0^\circ)$

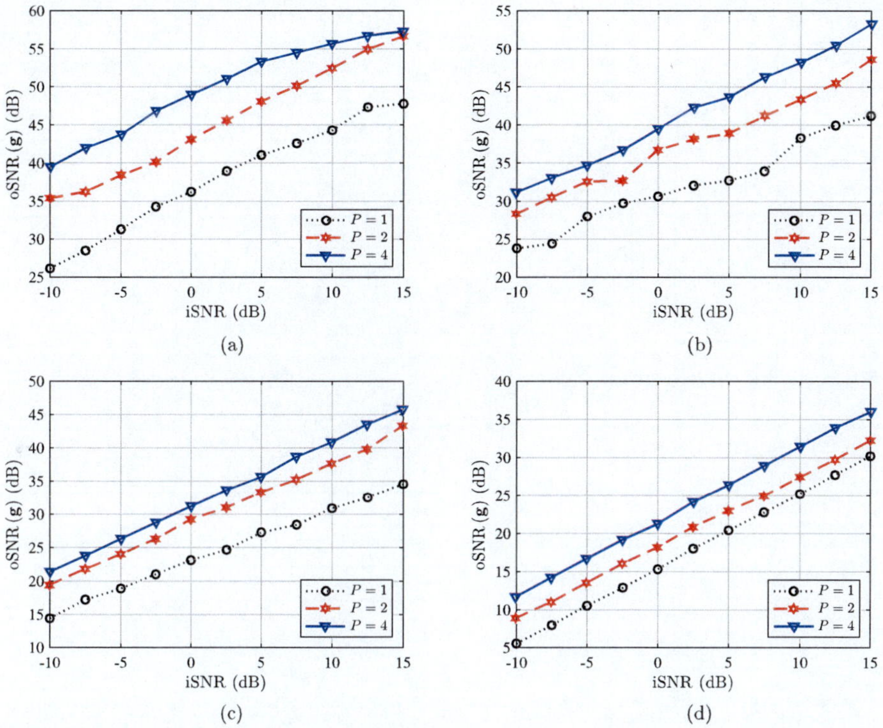

Fig. 8.6 Output SNR of the low-rank optimal adaptive noise canceller, $\mathbf{g}_{\mathrm{LRO}}$, as a function of the input SNR, for different values of P, and different values of α: (**a**) $\alpha = 10^{-4}$, (**b**) $\alpha = 10^{-3}$, (**c**) $\alpha = 10^{-2}$, and (**d**) $\alpha = 10^{-1}$. Conditions: $M_0 = 4$, $\delta = 1.0$ cm, $f = 1000$ Hz, and $(\theta_s, \phi_s) = (80^\circ, 0^\circ)$

References

1. B. Widrow et al., Adaptive noise cancelling: principles and applications. Proc. IEEE **63**, 1692–1716 (1975)
2. J. Benesty, J. Chen, *Optimal Time-Domain Noise Reduction Filters—A Theoretical Study*. Springer Briefs in Electrical and Computer Engineering (Springer Berlin, Heidelberg, 2011)
3. J. Benesty, J. Chen, E. Habets, *Speech Enhancement in the STFT Domain*. Springer Briefs in Electrical and Computer Engineering (Springer Berlin, Heidelberg, 2011)

Chapter 9
Binaural Beamforming

In this chapter, we give a fresh perspective on binaural beamforming, where within the same process, we also try to take advantage of our binaural hearing system. Basically, after beamforming, we wish to place the desired speech signal and noise in different positions in the perceptual space in order to possibly improve intelligibility as compared to the conventional (monaural) beamforming technique.

9.1 Signal Model

In this chapter, we consider the signal model of Chap. 3, where now the 3-D array is composed of $2M$ microphones instead of M. Then, the observed signal vector of length $2M$ is

$$\begin{aligned} \mathbf{y} &= \left[Y_1 \;\; Y_2 \;\; \cdots \;\; Y_{2M} \right]^T \\ &= \mathbf{x} + \mathbf{v} \\ &= \mathbf{d}_{\theta_s,\phi_s} X + \mathbf{v}, \end{aligned} \tag{9.1}$$

where

$$\mathbf{d}_{\theta_s,\phi_s} = \left[e^{\jmath 2\pi f \mathbf{a}_{\theta_s,\phi_s}^T \mathbf{r}_1/c} \;\; e^{\jmath 2\pi f \mathbf{a}_{\theta_s,\phi_s}^T \mathbf{r}_2/c} \;\; \cdots \;\; e^{\jmath 2\pi f \mathbf{a}_{\theta_s,\phi_s}^T \mathbf{r}_{2M}/c} \right]^T \tag{9.2}$$

is the steering vector of length $2M$. The $2M \times 2M$ covariance matrix of $\mathbf{y}$ is

$$\begin{aligned} \mathbf{\Phi}_{\mathbf{y}} &= E\left(\mathbf{y}\mathbf{y}^H \right) \\ &= \mathbf{\Phi}_{\mathbf{x}} + \mathbf{\Phi}_{\mathbf{v}} \end{aligned}$$

J. Benesty et al., *Microphone Arrays*, Springer Topics in Signal Processing 22,
https://doi.org/10.1007/978-3-031-36974-2_9

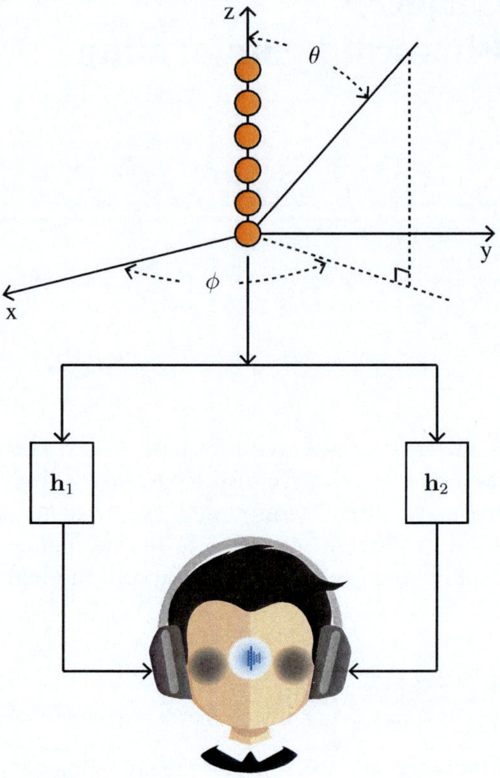

Fig. 9.1 Illustration of a binaural beamformer, which renders the desired signal (blue circle) and noise (gray circle) into different directions

$$= \phi_X \mathbf{d}_{\theta_s,\phi_s} \mathbf{d}_{\theta_s,\phi_s}^H + \mathbf{\Phi}_\mathbf{v}$$
$$= \phi_X \mathbf{d}_{\theta_s,\phi_s} \mathbf{d}_{\theta_s,\phi_s}^H + \phi_V \mathbf{\Gamma}_\mathbf{v}. \tag{9.3}$$

In the case of the spherically isotropic (diffuse) noise field, (9.3) becomes

$$\mathbf{\Phi}_\mathbf{y} = \phi_X \mathbf{d}_{\theta_s,\phi_s} \mathbf{d}_{\theta_s,\phi_s}^H + \phi_\mathrm{d} \mathbf{\Gamma}_\mathrm{d}, \tag{9.4}$$

where ϕ_d is the variance of the diffuse noise and the elements of the $2M \times 2M$ coherence matrix $\mathbf{\Gamma}_\mathrm{d}$ can be easily obtained (see Chap. 3).

Our objective in this chapter is to derive binaural beamformers, which can take advantage of the human binaural auditory system to separate the desired speech signal from noise as illustrated in Fig. 9.1, so that intelligibility from the beamformers' output signals will be perceived higher than the intelligibility of the outputs from the conventional (monaural) beamformers. To this end, we will construct two different and meaningful estimates of X, one for the left ear and the other for the right ear.

9.2 Motivation

Many studies in psychoacoustics have demonstrated the importance of the interaural coherence (IC) for speech perception and localization in the human auditory system [1–3]. When the IC of the binaural signals reaches its maximum value of 1 (i.e., the two channels are in phase), there is a precise region of the sound source, which is located in the middle of the head, when played back with headphones. In contrast, binaural signals with an IC equal to 0 (i.e., the two channels are in random phase) would be perceived in the respective side of the head [4, 5]. As a result, the binaural perception of a noisy speech signal, which is a mixture of a desired clean speech signal and background noise, generally falls into two categories: homophasic and heterophasic. The former refers to the case in which the desired speech and noise signal components in the binaural channels are both in phase, so both the speech and noise are perceived in the same zone. In contrast, the latter refers to the case in which the desired speech components in the binaural channels are in phase, while the noise signal components are in a random phase. In this scenario, the desired speech and noise would be perceived in different zones [6, 7]. Numerous experiments and evidences in psychoacoustics studies have shown that binaural signals with heterophasic representation have higher speech intelligibility than homophasic representation since the hearing system can better distinguish the desired signal from noise [8]. This inspired the design of a new type of binaural beamformers [9–11], which has demonstrated some potential in combining the noise reduction technique with perceptual properties for intelligibility enhancement. This is the focus of this chapter, where heterophasic beamformers are derived.

9.3 Binaural Beamforming and Performance Measures

The extension of the conventional (monaural) beamforming (see Chap. 3) to the binaural case can be accomplished by applying two complex-valued linear filters, $\mathbf{h}_1$ and $\mathbf{h}_2$ of length $2M$, to the observed signal vector, $\mathbf{y}$, i.e.,

$$\begin{aligned} Z_i &= \mathbf{h}_i^H \mathbf{y} \\ &= X \mathbf{h}_i^H \mathbf{d}_{\theta_s,\phi_s} + \mathbf{h}_i^H \mathbf{v} \\ &= X_{\mathrm{fd},i} + V_{\mathrm{rn},i}, \; i = 1, 2, \end{aligned} \tag{9.5}$$

where Z_1 and Z_2 are two different estimates of X. The variance of Z_i is then

$$\begin{aligned} \phi_{Z_i} &= \mathbf{h}_i^H \mathbf{\Phi}_{\mathbf{y}} \mathbf{h}_i \\ &= \phi_X \left| \mathbf{h}_i^H \mathbf{d}_{\theta_s,\phi_s} \right|^2 + \mathbf{h}_i^H \mathbf{\Phi}_{\mathbf{v}} \mathbf{h}_i \end{aligned} \tag{9.6}$$

$$= \phi_X \left| \mathbf{h}_i^H \mathbf{d}_{\theta_s,\phi_s} \right|^2 + \phi_V \mathbf{h}_i^H \mathbf{\Gamma}_{\mathbf{v}} \mathbf{h}_i .$$

We immediately see from (9.5) that the two distortionless constraints are

$$\mathbf{h}_i^H \mathbf{d}_{\theta_s,\phi_s} = 1, \ i = 1, 2. \tag{9.7}$$

We recall that the input SNR is

$$\text{iSNR} = \frac{\phi_X}{\phi_V}. \tag{9.8}$$

Using (9.6), we can define the binaural output SNR as

$$\text{oSNR}(\mathbf{h}_1, \mathbf{h}_2) = \frac{\phi_X}{\phi_V} \times \frac{\sum_{i=1}^{2} \left| \mathbf{h}_i^H \mathbf{d}_{\theta_s,\phi_s} \right|^2}{\sum_{i=1}^{2} \mathbf{h}_i^H \mathbf{\Gamma}_{\mathbf{v}} \mathbf{h}_i}. \tag{9.9}$$

In the particular case of $\mathbf{h}_1 = \mathbf{i}_i$ and $\mathbf{h}_2 = \mathbf{i}_j$, where $\mathbf{i}_i$ and $\mathbf{i}_j$ are, respectively, the ith and jth columns of the $2M \times 2M$ identity matrix, $\mathbf{I}_{2M}$, the binaural output SNR is equal to the input SNR, i.e., $\text{oSNR}\left(\mathbf{i}_i, \mathbf{i}_j\right) = \text{iSNR}$. From the two previous definitions, we get the binaural SNR gain:

$$\begin{aligned} \mathcal{G}(\mathbf{h}_1, \mathbf{h}_2) &= \frac{\text{oSNR}(\mathbf{h}_1, \mathbf{h}_2)}{\text{iSNR}} \\ &= \frac{\sum_{i=1}^{2} \left| \mathbf{h}_i^H \mathbf{d}_{\theta_s,\phi_s} \right|^2}{\sum_{i=1}^{2} \mathbf{h}_i^H \mathbf{\Gamma}_{\mathbf{v}} \mathbf{h}_i}. \end{aligned} \tag{9.10}$$

From the above definition, we can deduce two other important measures. They are

- the binaural WNG:

$$\mathcal{W}(\mathbf{h}_1, \mathbf{h}_2) = \frac{\sum_{i=1}^{2} \left| \mathbf{h}_i^H \mathbf{d}_{\theta_s,\phi_s} \right|^2}{\sum_{i=1}^{2} \mathbf{h}_i^H \mathbf{h}_i} \tag{9.11}$$

- and binaural DF:

$$\mathcal{D}(\mathbf{h}_1, \mathbf{h}_2) = \frac{\sum_{i=1}^{2} \left| \mathbf{h}_i^H \mathbf{d}_{\theta_s, \phi_s} \right|^2}{\sum_{i=1}^{2} \mathbf{h}_i^H \mathbf{\Gamma}_d \mathbf{h}_i}. \tag{9.12}$$

Another fundamental performance measure in beamforming is the beampattern. We define the binaural power beampattern as

$$\left| \mathcal{B}_{\theta_s, \phi_s}(\mathbf{h}_1, \mathbf{h}_2) \right|^2 = \frac{\sum_{i=1}^{2} \left| \mathbf{h}_i^H \mathbf{d}_{\theta, \phi} \right|^2}{2}. \tag{9.13}$$

Since we want to somehow separate the desired speech and noise in the perceptual space or, in other words, we want the desired speech and noise to be in different positions in space, the IC of the noise is going to be extremely useful in our context. It is well known that, in a multisource environment, the IC (or its modulus) is important for source localization since it is very strongly related to the two principal binaural cues, i.e., the interaural time difference (ITD) and interaural level difference (ILD), that the brain uses to localize sounds. Psychoacoustically, the localization performance decreases when the IC decreases [12].

Let A and B be two zero-mean complex-valued random variables. The coherence function between A and B is defined as

$$\gamma_{AB} = \frac{E(AB^*)}{\sqrt{E(|A|^2) E(|B|^2)}}. \tag{9.14}$$

It is clear that $0 \leq |\gamma_{AB}|^2 \leq 1$.

For any pair of different sensors (i, j), the input IC of the desired speech signal is simply the coherence function between X_i and X_j, i.e.,

$$\gamma_{X_i X_j} = \frac{E\left(X_i X_j^*\right)}{\sqrt{E(|X_i|^2) E\left(|X_j|^2\right)}}. \tag{9.15}$$

It is easy to check that $\left|\gamma_{X_i X_j}\right|^2 = 1$, meaning that the desired signal components at different microphones are fully coherent or in phase; thus, by listening to Y_i and Y_j with headphones, the desired speech signal will appear in the middle of the head [1–3].

For any pair of different sensors (i, j), the input IC of the noise is

$$\gamma_{V_i V_j} = \frac{E\left(V_i V_j^*\right)}{\sqrt{E\left(|V_i|^2\right) E\left(\left|V_j\right|^2\right)}} \tag{9.16}$$
$$= \frac{\mathbf{i}_i^T \mathbf{\Phi}_{\mathbf{v}} \mathbf{i}_j}{\sqrt{\mathbf{i}_i^T \mathbf{\Phi}_{\mathbf{v}} \mathbf{i}_i \times \mathbf{i}_j^T \mathbf{\Phi}_{\mathbf{v}} \mathbf{i}_j}}$$
$$= \frac{\mathbf{i}_i^T \mathbf{\Gamma}_{\mathbf{v}} \mathbf{i}_j}{\sqrt{\mathbf{i}_i^T \mathbf{\Gamma}_{\mathbf{v}} \mathbf{i}_i \times \mathbf{i}_j^T \mathbf{\Gamma}_{\mathbf{v}} \mathbf{i}_j}}$$
$$= \left(\mathbf{\Gamma}_{\mathbf{v}}\right)_{ij} .$$

Obviously, for white and diffuse noises, the input ICs are, respectively,

$$\gamma_{\mathrm{w},ij} = 0 \tag{9.17}$$

and

$$\gamma_{\mathrm{d},ij} = \left(\mathbf{\Gamma}_{\mathrm{d}}\right)_{ij} . \tag{9.18}$$

In the scenario in which $\left|\gamma_{V_i V_j}\right|^2 \ll 1$, the noise will appear on both sides of the head if we listen to Y_i and Y_j with headphones. In this case, the desired speech and noise signals are well separated in the perceptual space. We deduce that binaural listening can improve intelligibility as compared to monaural listening, without any processing, by just having two microphones in different spots.

In the rest of this chapter, it will be assumed that the two filters $\mathbf{h}_1$ and $\mathbf{h}_2$ are distortionless, i.e., $\mathbf{h}_1^H \mathbf{d}_{\theta_s,\phi_s} = \mathbf{h}_2^H \mathbf{d}_{\theta_s,\phi_s} = 1$.

Now, the output IC of the desired speech signal, i.e., the coherence function between $X_{\mathrm{fd},1}$ (in Z_1) and $X_{\mathrm{fd},2}$ (in Z_2), is

$$\gamma_{X_{\mathrm{fd},1} X_{\mathrm{fd},2}} = \frac{E\left(X_{\mathrm{fd},1} X_{\mathrm{fd},2}^*\right)}{\sqrt{E\left(\left|X_{\mathrm{fd},1}\right|^2\right) E\left(\left|X_{\mathrm{fd},2}\right|^2\right)}} \tag{9.19}$$
$$= \frac{\mathbf{h}_1^H \mathbf{d}_{\theta_s,\phi_s} \mathbf{d}_{\theta_s,\phi_s}^H \mathbf{h}_2}{\left|\mathbf{h}_1^H \mathbf{d}_{\theta_s,\phi_s}\right| \left|\mathbf{h}_2^H \mathbf{d}_{\theta_s,\phi_s}\right|} = 1.$$

Similarly, the output IC of the noise, i.e., the coherence function between $V_{\mathrm{rn},1}$ (in Z_1) and $V_{\mathrm{rn},2}$ (in Z_2), is

$$\gamma_{V_{\text{rn},1}V_{\text{rn},2}} = \frac{E\left(V_{\text{r},1}V_{\text{r},2}^*\right)}{\sqrt{E\left(|V_{\text{r},1}|^2\right)E\left(|V_{\text{r},2}|^2\right)}} \tag{9.20}$$

$$= \frac{\mathbf{h}_1^H \mathbf{\Gamma}_{\text{v}} \mathbf{h}_2}{\sqrt{\mathbf{h}_1^H \mathbf{\Gamma}_{\text{v}} \mathbf{h}_1 \times \mathbf{h}_2^H \mathbf{\Gamma}_{\text{v}} \mathbf{h}_2}}.$$

In case of diffuse noise, (9.20) becomes

$$\gamma_{\text{d}}(\mathbf{h}_1, \mathbf{h}_2) = \frac{\mathbf{h}_1^H \mathbf{\Gamma}_{\text{d}} \mathbf{h}_2}{\sqrt{\mathbf{h}_1^H \mathbf{\Gamma}_{\text{d}} \mathbf{h}_1 \times \mathbf{h}_2^H \mathbf{\Gamma}_{\text{d}} \mathbf{h}_2}} \tag{9.21}$$

and in case of white noise, (9.20) becomes

$$\gamma_{\text{w}}(\mathbf{h}_1, \mathbf{h}_2) = \frac{\mathbf{h}_1^H \mathbf{h}_2}{\sqrt{\mathbf{h}_1^H \mathbf{h}_1 \times \mathbf{h}_2^H \mathbf{h}_2}}. \tag{9.22}$$

In many approaches proposed in the literature, the two derived filters $\mathbf{h}_1$ and $\mathbf{h}_2$ are collinear, i.e.,

$$\mathbf{h}_1 = \varsigma \mathbf{h}_2, \tag{9.23}$$

where $\varsigma \neq 0$ is a complex-valued number. In this case, one can check that $\left|\gamma_{X_{\text{fd},1}X_{\text{fd},2}}\right|^2 = \left|\gamma_{V_{\text{rn},1}V_{\text{rn},2}}\right|^2 = 1$, meaning that all signals (desired speech source plus noises) will be perceived to be coming from the same direction. As a result, our brain will have difficulties to separate them with binaural presentation (from collinear filters); hence, intelligibility will certainly be affected. Needless to say that these techniques (based on collinearity) should be avoided in practice.

9.4 Examples of Binaural Beamformers

Since $\gamma_{X_{\text{fd},1}X_{\text{fd},2}} = 1$, we would like to find the two filters $\mathbf{h}_1$ and $\mathbf{h}_2$ in such a way that $\gamma_{V_{\text{rn},1}V_{\text{rn},2}} = 0$ or $\gamma_{\text{d}}(\mathbf{h}_1, \mathbf{h}_2) = 0$, implying that $\mathbf{h}_1^H \mathbf{\Gamma}_{\text{v}} \mathbf{h}_2 = 0$ or $\mathbf{h}_1^H \mathbf{\Gamma}_{\text{d}} \mathbf{h}_2 = 0$. With this approach, the desired speech and noise signals will be well separated in the perceptual space.

Let $\mathbf{\Gamma} \in \{\mathbf{\Gamma}_{\text{v}}, \mathbf{\Gamma}_{\text{d}}\}$. Thanks to the eigenvalue decomposition technique, the covariance matrix $\mathbf{\Gamma}$ can be factorized as

$$\mathbf{\Gamma} = \mathbf{U}\mathbf{\Lambda}\mathbf{U}^H, \tag{9.24}$$

where

$$\mathbf{U} = \begin{bmatrix} \mathbf{u}_1 & \mathbf{u}_2 & \cdots & \mathbf{u}_{2M} \end{bmatrix} \tag{9.25}$$

is a unitary (eigenvector) matrix, i.e., $\mathbf{U}^H\mathbf{U} = \mathbf{U}\mathbf{U}^H = \mathbf{I}_{2M}$, and

$$\mathbf{\Lambda} = \text{diag}\left(\lambda_1, \lambda_2, \ldots, \lambda_{2M}\right) \tag{9.26}$$

is a diagonal matrix consisting of the (real-valued and positive) eigenvalues, which are assumed to be arranged in a descending order, i.e., $\lambda_1 \geq \lambda_2 \geq \cdots \geq \lambda_{2M} > 0$. From $\mathbf{U}$, we consider the construction of two semi-unitary matrices of size $2M \times M$:

$$\mathbf{U}_+ = \begin{bmatrix} \mathbf{u}_{+,1} & \mathbf{u}_{+,2} & \cdots & \mathbf{u}_{+,M} \end{bmatrix}, \tag{9.27}$$

$$\mathbf{U}_- = \begin{bmatrix} \mathbf{u}_{-,1} & \mathbf{u}_{-,2} & \cdots & \mathbf{u}_{-,M} \end{bmatrix}, \tag{9.28}$$

where

$$\mathbf{u}_{+,m} = \frac{\mathbf{u}_m + \mathbf{u}_{2M-m+1}}{\sqrt{2}}, \tag{9.29}$$

$$\mathbf{u}_{-,m} = \frac{\mathbf{u}_m - \mathbf{u}_{2M-m+1}}{\sqrt{2}}. \tag{9.30}$$

One can verify that

$$\mathbf{U}_+^H\mathbf{U}_+ = \mathbf{U}_-^H\mathbf{U}_- = \mathbf{I}_M, \tag{9.31}$$

$$\mathbf{U}_+^H\mathbf{U}_- = \mathbf{U}_-^H\mathbf{U}_+ = \mathbf{0}_M. \tag{9.32}$$

By constructing the two following matrices:

$$\mathbf{Q}_+ = \mathbf{\Gamma}^{-1/2}\mathbf{U}_+, \tag{9.33}$$

$$\mathbf{Q}_- = \mathbf{\Gamma}^{-1/2}\mathbf{U}_-, \tag{9.34}$$

we can further deduce that

$$\mathbf{Q}_+^H\mathbf{\Gamma}\mathbf{Q}_+ = \mathbf{Q}_-^H\mathbf{\Gamma}\mathbf{Q}_- = \mathbf{I}_M, \tag{9.35}$$

$$\mathbf{Q}_+^H\mathbf{\Gamma}\mathbf{Q}_- = \mathbf{Q}_-^H\mathbf{\Gamma}\mathbf{Q}_+ = \mathbf{0}_M. \tag{9.36}$$

Next, we consider filters of the forms:

$$\mathbf{h}_1 = \mathbf{Q}_+\mathbf{h}_+ = \mathbf{\Gamma}^{-1/2}\mathbf{U}_+\mathbf{h}_+, \tag{9.37}$$

$$\mathbf{h}_2 = \mathbf{Q}_-\mathbf{h}_- = \mathbf{\Gamma}^{-1/2}\mathbf{U}_-\mathbf{h}_-, \tag{9.38}$$

where $\mathbf{h}_+$ and $\mathbf{h}_-$ are two filters of length M.

Let $\mathbf{Q}_{\mathrm{d},+} = \mathbf{\Gamma}_{\mathrm{d}}^{-1/2}\mathbf{U}_+$ and $\mathbf{Q}_{\mathrm{d},-} = \mathbf{\Gamma}_{\mathrm{d}}^{-1/2}\mathbf{U}_-$. Then, the binaural DF in (9.12) becomes

$$\mathcal{D}\left(\mathbf{h}_+, \mathbf{h}_-\right) = \frac{\left|\mathbf{h}_+^H\mathbf{Q}_{\mathrm{d},+}^H\mathbf{d}_{\theta_\mathrm{s},\phi_\mathrm{s}}\right|^2 + \left|\mathbf{h}_-^H\mathbf{Q}_{\mathrm{d},-}^H\mathbf{d}_{\theta_\mathrm{s},\phi_\mathrm{s}}\right|^2}{\mathbf{h}_+^H\mathbf{h}_+ + \mathbf{h}_-^H\mathbf{h}_-}. \tag{9.39}$$

From the maximization of the above expression, which is equivalent to solving the following optimization problems:

$$\min_{\mathbf{h}_+} \mathbf{h}_+^H\mathbf{h}_+ \quad \text{subject to} \quad \mathbf{h}_+^H\mathbf{Q}_{\mathrm{d},+}^H\mathbf{d}_{\theta_\mathrm{s},\phi_\mathrm{s}} = 1, \tag{9.40}$$

$$\min_{\mathbf{h}_-} \mathbf{h}_-^H\mathbf{h}_- \quad \text{subject to} \quad \mathbf{h}_-^H\mathbf{Q}_{\mathrm{d},-}^H\mathbf{d}_{\theta_\mathrm{s},\phi_\mathrm{s}} = 1, \tag{9.41}$$

we find that the solutions are

$$\mathbf{h}_+ = \frac{\mathbf{Q}_{\mathrm{d},+}^H\mathbf{d}_{\theta_\mathrm{s},\phi_\mathrm{s}}}{\mathbf{d}_{\theta_\mathrm{s},\phi_\mathrm{s}}^H\mathbf{Q}_{\mathrm{d},+}\mathbf{Q}_{\mathrm{d},+}^H\mathbf{d}_{\theta_\mathrm{s},\phi_\mathrm{s}}}, \tag{9.42}$$

$$\mathbf{h}_- = \frac{\mathbf{Q}_{\mathrm{d},-}^H\mathbf{d}_{\theta_\mathrm{s},\phi_\mathrm{s}}}{\mathbf{d}_{\theta_\mathrm{s},\phi_\mathrm{s}}^H\mathbf{Q}_{\mathrm{d},-}\mathbf{Q}_{\mathrm{d},-}^H\mathbf{d}_{\theta_\mathrm{s},\phi_\mathrm{s}}}. \tag{9.43}$$

As a result, the binaural MDF beamformer is

$$\mathbf{h}_{1,\mathrm{MDF}} = \frac{\mathbf{Q}_{\mathrm{d},+}\mathbf{Q}_{\mathrm{d},+}^H\mathbf{d}_{\theta_\mathrm{s},\phi_\mathrm{s}}}{\mathbf{d}_{\theta_\mathrm{s},\phi_\mathrm{s}}^H\mathbf{Q}_{\mathrm{d},+}\mathbf{Q}_{\mathrm{d},+}^H\mathbf{d}_{\theta_\mathrm{s},\phi_\mathrm{s}}}, \tag{9.44}$$

$$\mathbf{h}_{2,\mathrm{MDF}} = \frac{\mathbf{Q}_{\mathrm{d},-}\mathbf{Q}_{\mathrm{d},-}^H\mathbf{d}_{\theta_\mathrm{s},\phi_\mathrm{s}}}{\mathbf{d}_{\theta_\mathrm{s},\phi_\mathrm{s}}^H\mathbf{Q}_{\mathrm{d},-}\mathbf{Q}_{\mathrm{d},-}^H\mathbf{d}_{\theta_\mathrm{s},\phi_\mathrm{s}}}. \tag{9.45}$$

Let $\mathbf{Q}_{\mathbf{v},+} = \mathbf{\Gamma}_{\mathbf{v}}^{-1/2}\mathbf{U}_+$ and $\mathbf{Q}_{\mathbf{v},-} = \mathbf{\Gamma}_{\mathbf{v}}^{-1/2}\mathbf{U}_-$. Then, the binaural SNR gain in (9.10) becomes

$$\mathcal{G}\left(\mathbf{h}_+, \mathbf{h}_-\right) = \frac{\left|\mathbf{h}_+^H\mathbf{Q}_{\mathbf{v},+}^H\mathbf{d}_{\theta_\mathrm{s},\phi_\mathrm{s}}\right|^2 + \left|\mathbf{h}_-^H\mathbf{Q}_{\mathbf{v},-}^H\mathbf{d}_{\theta_\mathrm{s},\phi_\mathrm{s}}\right|^2}{\mathbf{h}_+^H\mathbf{h}_+ + \mathbf{h}_-^H\mathbf{h}_-}. \tag{9.46}$$

From the maximization of the above expression, which is equivalent to solving the following optimization problems:

$$\min_{\mathbf{h}_+} \mathbf{h}_+^H \mathbf{h}_+ \quad \text{subject to} \quad \mathbf{h}_+^H \mathbf{Q}_{\mathbf{v},+}^H \mathbf{d}_{\theta_s,\phi_s} = 1, \tag{9.47}$$

$$\min_{\mathbf{h}_-} \mathbf{h}_-^H \mathbf{h}_- \quad \text{subject to} \quad \mathbf{h}_-^H \mathbf{Q}_{\mathbf{v},-}^H \mathbf{d}_{\theta_s,\phi_s} = 1, \tag{9.48}$$

we find that the solutions are

$$\mathbf{h}_+ = \frac{\mathbf{Q}_{\mathbf{v},+}^H \mathbf{d}_{\theta_s,\phi_s}}{\mathbf{d}_{\theta_s,\phi_s}^H \mathbf{Q}_{\mathbf{v},+} \mathbf{Q}_{\mathbf{v},+}^H \mathbf{d}_{\theta_s,\phi_s}}, \tag{9.49}$$

$$\mathbf{h}_- = \frac{\mathbf{Q}_{\mathbf{v},-}^H \mathbf{d}_{\theta_s,\phi_s}}{\mathbf{d}_{\theta_s,\phi_s}^H \mathbf{Q}_{\mathbf{v},-} \mathbf{Q}_{\mathbf{v},-}^H \mathbf{d}_{\theta_s,\phi_s}}. \tag{9.50}$$

As a result, the binaural MVDR beamformer is

$$\mathbf{h}_{1,\mathrm{MVDR}} = \frac{\mathbf{Q}_{\mathbf{v},+} \mathbf{Q}_{\mathbf{v},+}^H \mathbf{d}_{\theta_s,\phi_s}}{\mathbf{d}_{\theta_s,\phi_s}^H \mathbf{Q}_{\mathbf{v},+} \mathbf{Q}_{\mathbf{v},+}^H \mathbf{d}_{\theta_s,\phi_s}}, \tag{9.51}$$

$$\mathbf{h}_{2,\mathrm{MVDR}} = \frac{\mathbf{Q}_{\mathbf{v},-} \mathbf{Q}_{\mathbf{v},-}^H \mathbf{d}_{\theta_s,\phi_s}}{\mathbf{d}_{\theta_s,\phi_s}^H \mathbf{Q}_{\mathbf{v},-} \mathbf{Q}_{\mathbf{v},-}^H \mathbf{d}_{\theta_s,\phi_s}}. \tag{9.52}$$

Let us define the matrix:

$$\mathbf{\Gamma}_\alpha = (1-\alpha)\mathbf{\Gamma}_\mathrm{d} + \alpha \mathbf{I}_M, \tag{9.53}$$

where $0 \leq \alpha \leq 1$. Now, $\mathbf{U}_+$ and $\mathbf{U}_-$ are constructed from the eigenvalue decomposition of $\mathbf{\Gamma}_\alpha$. Let $\mathbf{Q}_{\alpha,+} = \mathbf{\Gamma}_\alpha^{-1/2}\mathbf{U}_+$ and $\mathbf{Q}_{\alpha,-} = \mathbf{\Gamma}_\alpha^{-1/2}\mathbf{U}_-$. Then, the binaural WNG in (9.11) becomes

$$\mathcal{W}(\mathbf{h}_+, \mathbf{h}_-) = \frac{\left|\mathbf{h}_+^H \mathbf{Q}_{\alpha,+}^H \mathbf{d}_{\theta_s,\phi_s}\right|^2 + \left|\mathbf{h}_-^H \mathbf{Q}_{\alpha,-}^H \mathbf{d}_{\theta_s,\phi_s}\right|^2}{\mathbf{h}_+^H \mathbf{Q}_{\alpha,+}^H \mathbf{Q}_{\alpha,+} \mathbf{h}_+ + \mathbf{h}_-^H \mathbf{Q}_{\alpha,-}^H \mathbf{Q}_{\alpha,-} \mathbf{h}_-}. \tag{9.54}$$

The maximization of the binaural WNG is equivalent to solving the following optimization problems:

$$\min_{\mathbf{h}_+} \mathbf{h}_+^H \mathbf{Q}_{\alpha,+}^H \mathbf{Q}_{\alpha,+} \mathbf{h}_+ \quad \text{subject to} \quad \mathbf{h}_+^H \mathbf{Q}_{\alpha,+}^H \mathbf{d}_{\theta_s,\phi_s} = 1, \tag{9.55}$$

$$\min_{\mathbf{h}_-} \mathbf{h}_-^H \mathbf{Q}_{\alpha,-}^H \mathbf{Q}_{\alpha,-} \mathbf{h}_- \quad \text{subject to} \quad \mathbf{h}_-^H \mathbf{Q}_{\alpha,-}^H \mathbf{d}_{\theta_s,\phi_s} = 1, \tag{9.56}$$

for which the solutions are

$$\mathbf{h}_+ = \frac{\left(\mathbf{Q}_{\alpha,+}^H \mathbf{Q}_{\alpha,+}\right)^{-1} \mathbf{Q}_{\alpha,+}^H \mathbf{d}_{\theta_s,\phi_s}}{\mathbf{d}_{\theta_s,\phi_s}^H \mathbf{Q}_{\alpha,+} \left(\mathbf{Q}_{\alpha,+}^H \mathbf{Q}_{\alpha,+}\right)^{-1} \mathbf{Q}_{\alpha,+}^H \mathbf{d}_{\theta_s,\phi_s}}, \tag{9.57}$$

$$\mathbf{h}_{-} = \frac{\left(\mathbf{Q}_{\alpha,-}^H \mathbf{Q}_{\alpha,-}\right)^{-1} \mathbf{Q}_{\alpha,-}^H \mathbf{d}_{\theta_s,\phi_s}}{\mathbf{d}_{\theta_s,\phi_s}^H \mathbf{Q}_{\alpha,-} \left(\mathbf{Q}_{\alpha,-}^H \mathbf{Q}_{\alpha,-}\right)^{-1} \mathbf{Q}_{\alpha,-}^H \mathbf{d}_{\theta_s,\phi_s}}. \tag{9.58}$$

As a result, the binaural MWNG beamformer is

$$\mathbf{h}_{1,\mathrm{MWNG}} = \frac{\mathbf{Q}_{\alpha,+} \left(\mathbf{Q}_{\alpha,+}^H \mathbf{Q}_{\alpha,+}\right)^{-1} \mathbf{Q}_{\alpha,+}^H \mathbf{d}_{\theta_s,\phi_s}}{\mathbf{d}_{\theta_s,\phi_s}^H \mathbf{Q}_{\alpha,+} \left(\mathbf{Q}_{\alpha,+}^H \mathbf{Q}_{\alpha,+}\right)^{-1} \mathbf{Q}_{\alpha,+}^H \mathbf{d}_{\theta_s,\phi_s}}, \tag{9.59}$$

$$\mathbf{h}_{2,\mathrm{MWNG}} = \frac{\mathbf{Q}_{\alpha,-} \left(\mathbf{Q}_{\alpha,-}^H \mathbf{Q}_{\alpha,-}\right)^{-1} \mathbf{Q}_{\alpha,-}^H \mathbf{d}_{\theta_s,\phi_s}}{\mathbf{d}_{\theta_s,\phi_s}^H \mathbf{Q}_{\alpha,-} \left(\mathbf{Q}_{\alpha,-}^H \mathbf{Q}_{\alpha,-}\right)^{-1} \mathbf{Q}_{\alpha,-}^H \mathbf{d}_{\theta_s,\phi_s}}. \tag{9.60}$$

9.5 Illustrative Examples

9.5.1 *Objective Measurements*

In this subsection, we study some examples of binaural beamformers. The array that we choose to use is a ULA with $2M$ omnidirectional microphones, where the interelement spacing is 1.0 cm. We assume that the desired source signal propagates from the endfire direction of the ULA, i.e., $(\theta_s, \phi_s) = (90°, 0°)$.

We first demonstrate the performance of the binaural MDF beamformer, $\mathbf{h}_{1,\mathrm{MDF}}$ and $\mathbf{h}_{2,\mathrm{MDF}}$, given in (9.44) and (9.45), respectively. Figure 9.2 shows plots of the binaural power beampattern, $\left|\mathcal{B}_{\theta_s,\phi_s}(\mathbf{h}_1, \mathbf{h}_2)\right|^2$, of this beamformer for different numbers of sensors, $2M$, at frequency $f = 2000$ Hz. One can see that the main beam of this beamformer becomes narrower as M increases. Figure 9.3 shows plots of the DF, $\mathcal{D}(\mathbf{h}_1, \mathbf{h}_2)$, and WNG, $\mathcal{W}(\mathbf{h}_1, \mathbf{h}_2)$, of this beamformer, as a function of frequency, for different numbers of sensors, $2M$. It can be observed from the figure that the DF increases with the value of M. The beamformer has a very low WNG, particularly at low frequencies, which means that it suffers from white noise amplification. Figure 9.4 shows plots of the output ICs of (diffuse and white) noises in the output signals of this beamformer, as a function of frequency, for different numbers of sensors. As seen in Fig. 9.4a, the output ICs of the diffuse noise, $\gamma_d(\mathbf{h}_1, \mathbf{h}_2)$, of all beamformers are equal to zero, which means that the diffuse noises in the two output signals of the binaural MDF beamformer are completely incoherent, so the output signals correspond to the heterophasic case. As shown in Fig. 9.4b, the output ICs of the white noise, $\gamma_w(\mathbf{h}_1, \mathbf{h}_2)$, of all beamformers decrease toward -1 as the value of M increases at low frequencies. According to researches in psychoacoustics, the speech and white noise in the output binaural signals will be perceived from different directions/zones in space [5].

Then, we evaluate the performance of the binaural MVDR beamformer, $\mathbf{h}_{1,\mathrm{MVDR}}$ and $\mathbf{h}_{2,\mathrm{MVDR}}$, given in (9.51) and (9.52), respectively. We consider the situation

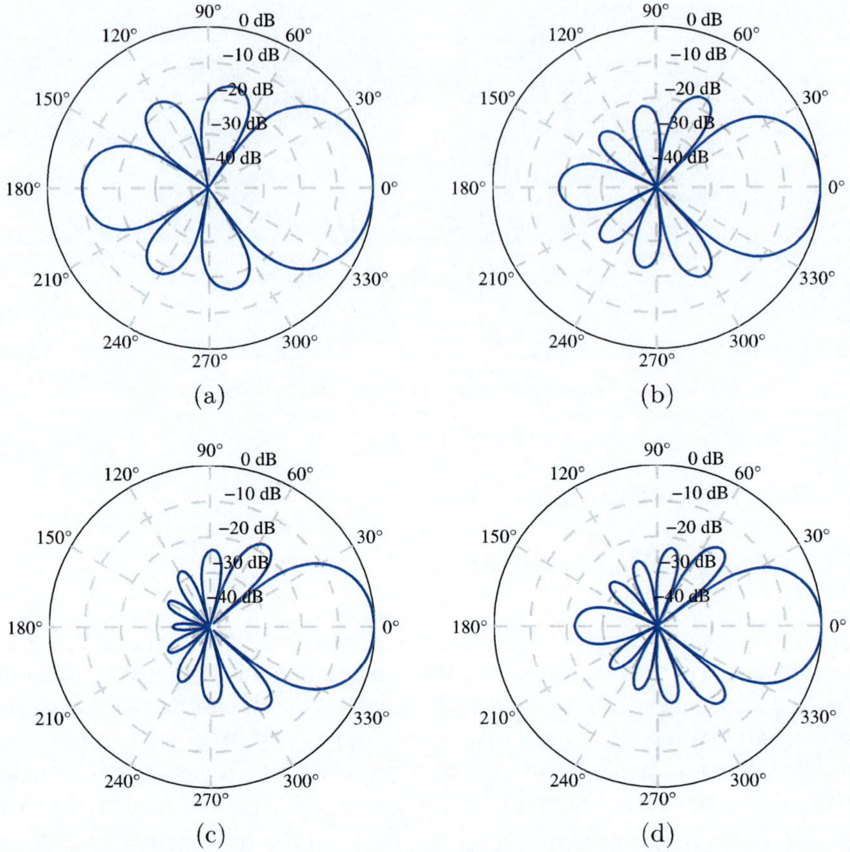

Fig. 9.2 Power patterns of the binaural MDF beamformer, $\mathbf{h}_{1,\text{MDF}}$ and $\mathbf{h}_{2,\text{MDF}}$, given in (9.44) and (9.45), respectively, for different numbers of sensors, $2M$: (**a**) $M = 4$, (**b**) $M = 6$, (**c**) $M = 8$, and (**d**) $M = 10$. Conditions: $f = 2000$ Hz, $\delta = 1.0$ cm, and $(\theta_s, \phi_s) = (90°, 0°)$

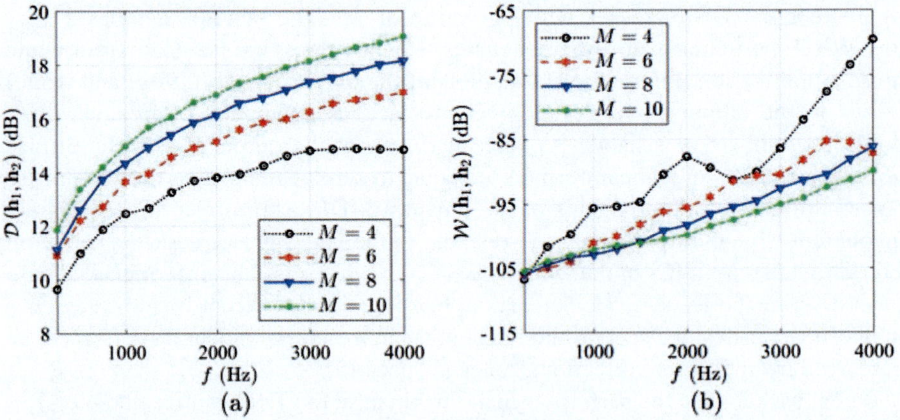

Fig. 9.3 DF and WNG of the binaural MDF beamformer, $\mathbf{h}_{1,\text{MDF}}$ and $\mathbf{h}_{2,\text{MDF}}$, given in (9.44) and (9.45), respectively, as a function of frequency, for different numbers of sensors, $2M$: (**a**) DF and (**b**) WNG. Conditions: $\delta = 1.0$ cm and $(\theta_s, \phi_s) = (90°, 0°)$

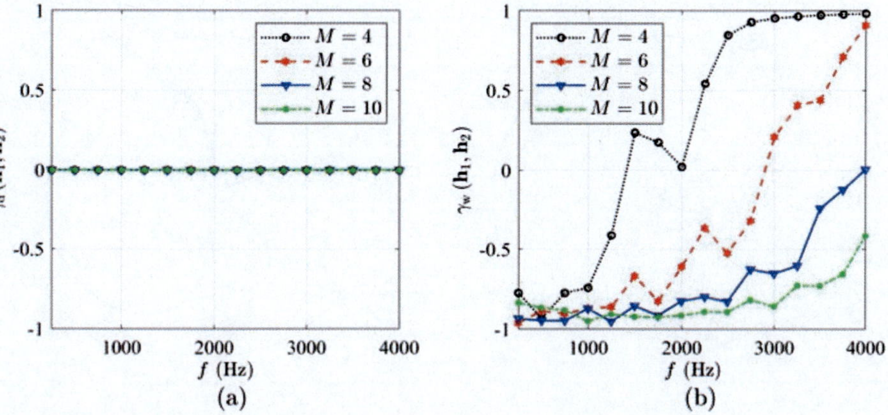

Fig. 9.4 Output ICs of noises in the output signals of the binaural MDF beamformer, $\mathbf{h}_{1,\text{MDF}}$ and $\mathbf{h}_{2,\text{MDF}}$, given in (9.44) and (9.45), respectively, as a function of frequency, for different numbers of sensors, $2M$: (**a**) diffuse noise and (**b**) white noise. Conditions: $\delta = 1.0$ cm and $(\theta_s, \phi_s) = (90°, 0°)$

in which the speech signal of interest is contaminated by both white and diffuse noises as well as noise from two independent point sources (incoherent with the desired speech signal) impinging from $(\theta_{i,1}, \phi_{i,1}) = (90°, 150°)$ and $(\theta_{i,2}, \phi_{i,2}) = (90°, 210°)$. The variances of the two interferences are the same and equal to ϕ_V. The covariance matrix of the noise signal is

$$\mathbf{\Phi}_\mathbf{v} = \phi_V \left(\mathbf{d}_{\theta_{i,1},\phi_{i,1}} \mathbf{d}^H_{\theta_{i,1},\phi_{i,1}} + \mathbf{d}_{\theta_{i,2},\phi_{i,2}} \mathbf{d}^H_{\theta_{i,2},\phi_{i,2}} \right) + \phi_{V_w} \mathbf{I}_{2M} + \phi_d \mathbf{\Gamma}_d, \tag{9.61}$$

where ϕ_{V_w} and ϕ_d are the variances of the white and diffuse noises, respectively. The variance of the desired signal is set to 1; ϕ_V, ϕ_{V_w}, and ϕ_d are computed according to the input SNRs, which are 10 dB for interferences and 20 dB for white and diffuse noises. Plots of the binaural power beampattern, $\left|\mathcal{B}_{\theta_s,\phi_s}(\mathbf{h}_1, \mathbf{h}_2)\right|^2$, of the binaural MVDR beamformer for different numbers of sensors, at frequency $f = 2000$ Hz, are shown in Fig. 9.5. One could see that the main beam of the power patterns points in the direction of the desired speech signal, while nulls are formed in the directions of the point source noises. Furthermore, the main beam becomes narrower as the value of M increases. Figure 9.6 shows plots of the binaural SNR gain, $\mathcal{G}(\mathbf{h}_1, \mathbf{h}_2)$, of the binaural MVDR beamformer, as a function of frequency, for different numbers of sensors, $2M$. It is observed that this gain increases with the value of M. Figure 9.7 shows plots of the ICs of the noise in the two output signals of the binaural MVDR beamformer, as a function of frequency, for different numbers of sensors, $2M$. As seen, the output ICs of the noises, $\gamma_{V_{rn,1}V_{rn,2}}$ in (9.20), of all beamformers are equal to zero, which means that noises in the two output signals of this beamformer are completely incoherent, so the output signals correspond to the heterophasic case.

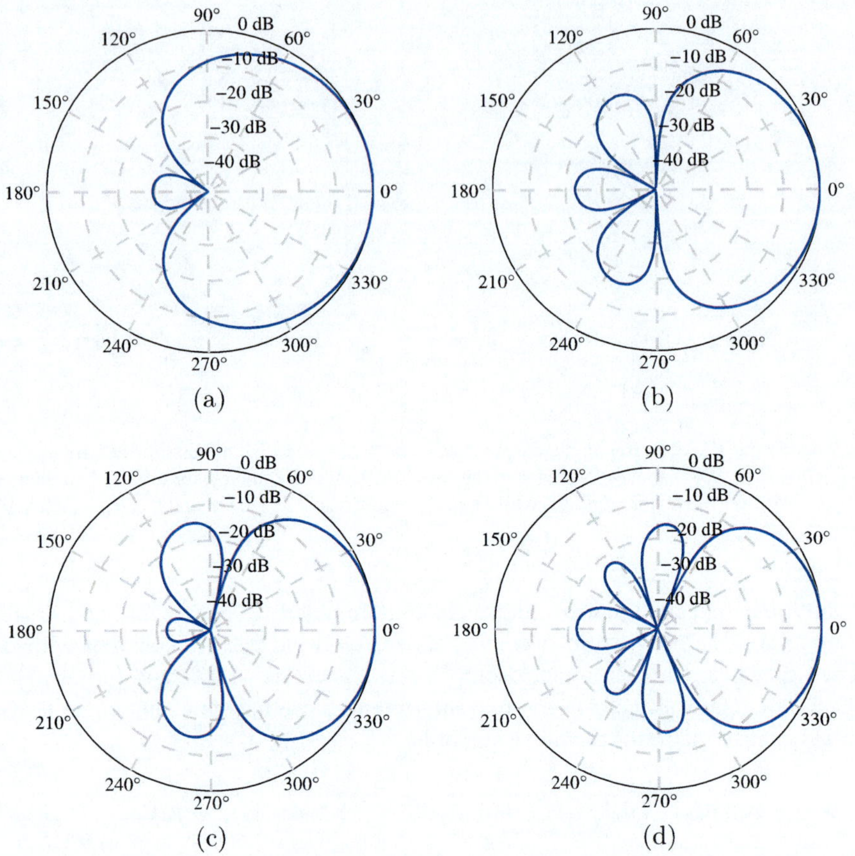

Fig. 9.5 Power patterns of the binaural MVDR beamformer, $\mathbf{h}_{1,\mathrm{MVDR}}$ and $\mathbf{h}_{2,\mathrm{MVDR}}$, given in (9.44) and (9.45), respectively, for different numbers of sensors, $2M$: **(a)** $M = 4$, **(b)** $M = 6$, **(c)** $M = 8$, and **(d)** $M = 10$. Conditions: $f = 2000$ Hz, $\delta = 1.0$ cm, and $(\theta_s, \phi_s) = (90°, 0°)$. Two point source noises propagating from $(\theta_{i,1}, \phi_{i,1}) = (90°, 150°)$ and $(\theta_{i,2}, \phi_{i,2}) = (90°, 210°)$ are added to the desired speech signal at an SNR of 10 dB. White and diffuse noises are added as an SNR of 20 dB

Finally, we study the performance of the binaural MWNG beamformer, $\mathbf{h}_{1,\mathrm{MWNG}}$ and $\mathbf{h}_{2,\mathrm{MWNG}}$, given in (9.59) and (9.60), respectively. Figure 9.8 shows plots of the binaural power beampattern, $\left|\mathcal{B}_{\theta_s,\phi_s}(\mathbf{h}_1, \mathbf{h}_2)\right|^2$, of this beamformer for different numbers of sensors, and different values of α, at frequency $f = 2000$ Hz. As seen, the main beam of the power patterns becomes narrower as the value of M increases, while it gets wider with the increase of the value of α. Figure 9.9 shows plots of the DF, $\mathcal{D}(\mathbf{h}_1, \mathbf{h}_2)$, and WNG, $\mathcal{W}(\mathbf{h}_1, \mathbf{h}_2)$, of this beamformer, as a function of frequency, for different numbers of sensors and different values of α. With the increase of the value of α, the DF decreases, while the WNG increases, which indicates that α controls the tradeoff between DF and WNG. Figure 9.10 shows plots

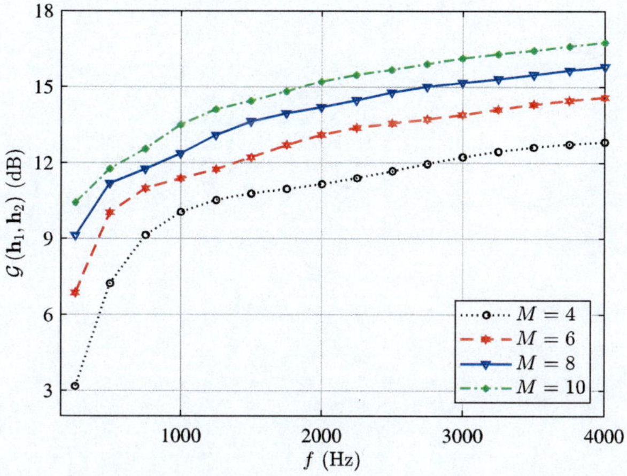

Fig. 9.6 SNR gain of the binaural MVDR beamformer, $\mathbf{h}_{1,\text{MVDR}}$ and $\mathbf{h}_{2,\text{MVDR}}$, given in (9.51) and (9.52), respectively, as a function of frequency, for different numbers of sensors, $2M$. Conditions: $\delta = 1.0\,\text{cm}$ and $(\theta_\text{s}, \phi_\text{s}) = (90°, 0°)$. Two point source noises propagating from $(\theta_{\text{i},1}, \phi_{\text{i},1}) = (90°, 150°)$ and $(\theta_{\text{i},2}, \phi_{\text{i},2}) = (90°, 210°)$ are added to the desired speech signal at an SNR of 10 dB. White and diffuse noises are added as an SNR of 20 dB

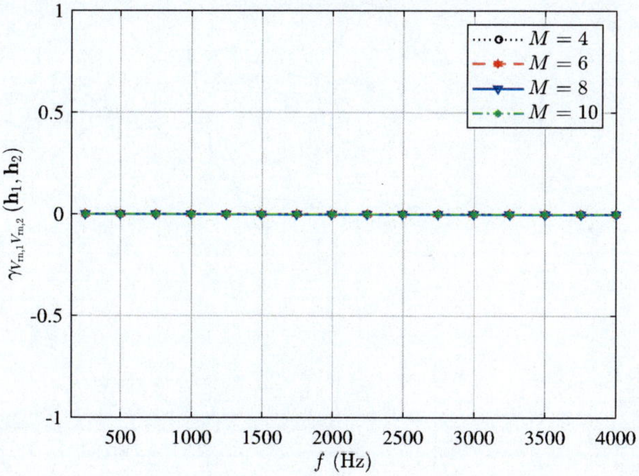

Fig. 9.7 Output ICs of noises in the output signals of the binaural MVDR beamformer, $\mathbf{h}_{1,\text{MVDR}}$ and $\mathbf{h}_{2,\text{MVDR}}$, given in (9.51) and (9.52), respectively, as a function of frequency, for different numbers of sensors, $2M$. Conditions: $\delta = 1.0\,\text{cm}$ and $(\theta_\text{s}, \phi_\text{s}) = (90°, 0°)$. Two point source noises propagating from $(\theta_{\text{i},1}, \phi_{\text{i},1}) = (90°, 150°)$ and $(\theta_{\text{i},2}, \phi_{\text{i},2}) = (90°, 210°)$ are added to the desired speech signal at an SNR of 10 dB. White and diffuse noises are added as an SNR of 20 dB

of the output ICs of noises in the output signals of the binaural MWNG beamformer, for $M = 6$, as a function of frequency, for different values of α, in diffuse, white, and diffuse-plus-white noises. As seen in Fig. 9.10a, the output ICs of the diffuse noise, $\gamma_\text{d}(\mathbf{h}_1, \mathbf{h}_2)$, of all beamformers are equal to one, which means that the diffuse

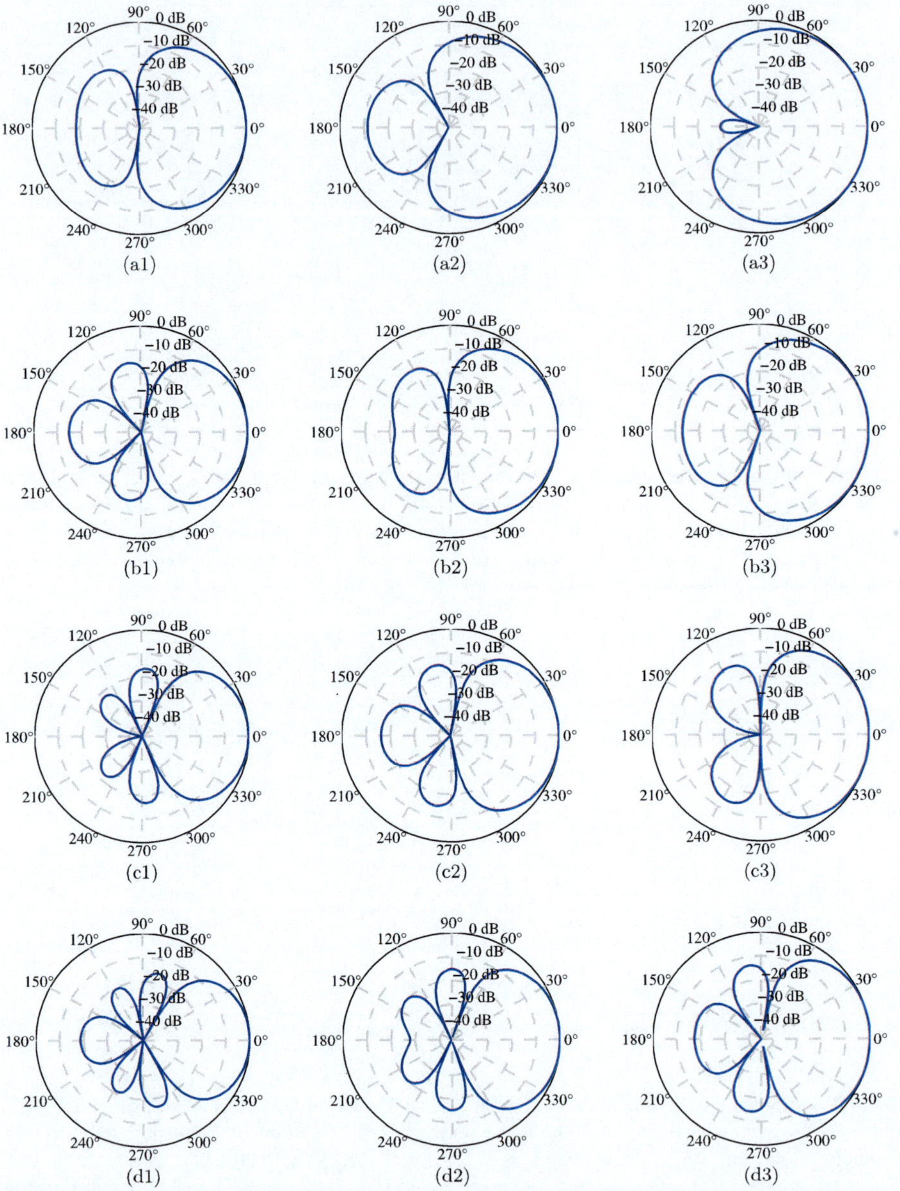

Fig. 9.8 Power patterns of the binaural MWNG beamformer, $\mathbf{h}_{1,\mathrm{MWNG}}$ and $\mathbf{h}_{2,\mathrm{MWNG}}$, given in (9.59) and (9.60), respectively, for different numbers of sensors, $2M$, and different values of α: **(a1)** $\alpha = 0.1$, $M = 4$, **(b1)** $\alpha = 0.1$, $M = 6$, **(c1)** $\alpha = 0.1$, $M = 8$, **(d1)** $\alpha = 0.1$, $M = 10$, **(a2)** $\alpha = 0.5$, $M = 4$, **(b2)** $\alpha = 0.5$, $M = 6$, **(c2)** $\alpha = 0.5$, $M = 8$, **(d2)** $\alpha = 0.5$, $M = 10$, **(a3)** $\alpha = 0.9$, $M = 4$, **(b3)** $\alpha = 0.9$, $M = 6$, **(c3)** $\alpha = 0.9$, $M = 8$, and **(d3)** $\alpha = 0.9$, $M = 10$. Conditions: $f = 2000$ Hz, $\delta = 1.0$ cm, and $(\theta_{\mathrm{s}}, \phi_{\mathrm{s}}) = (90^\circ, 0^\circ)$

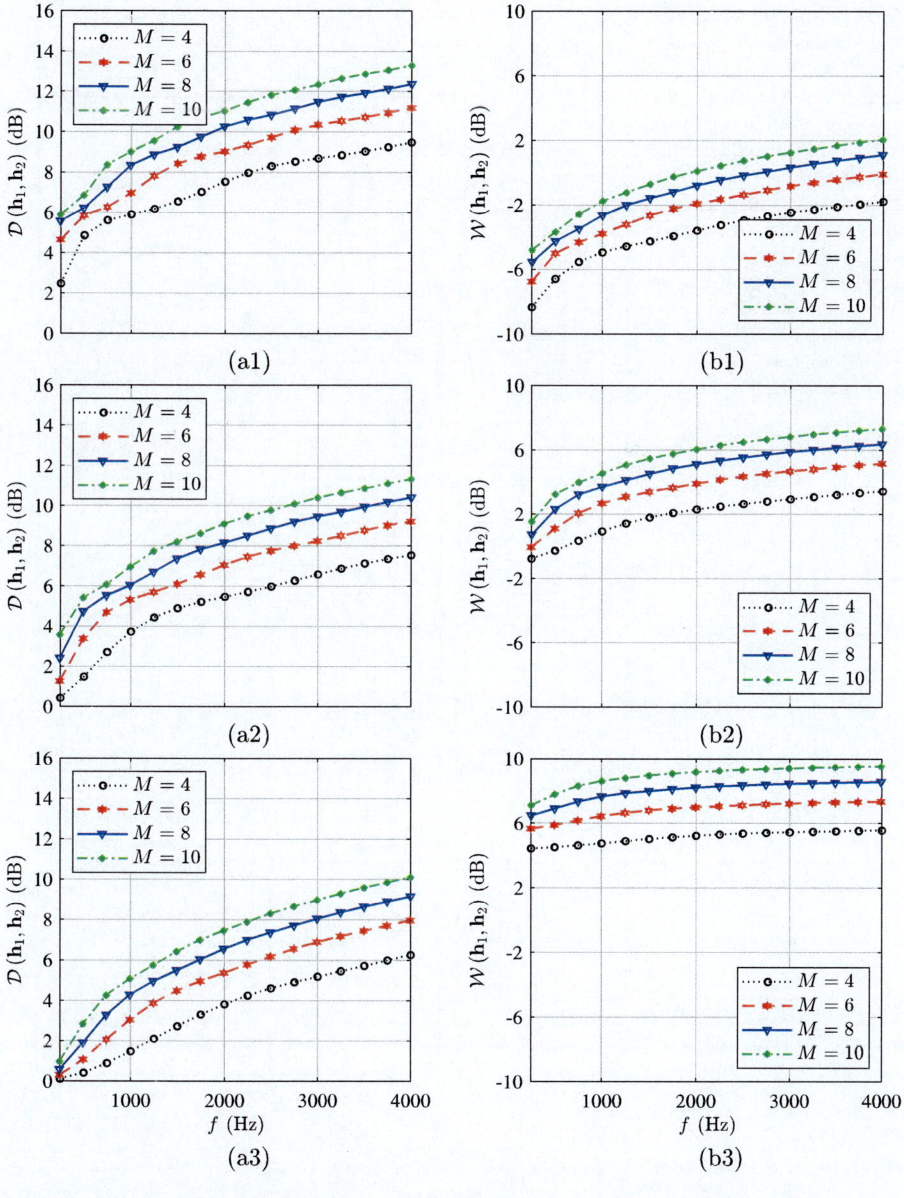

Fig. 9.9 DF and WNG of the binaural MWNG beamformer, $\mathbf{h}_{1,\text{MWNG}}$ and $\mathbf{h}_{2,\text{MWNG}}$, given in (9.59) and (9.60), respectively, as a function of frequency, for different numbers of sensors, $2M$, and different values of α: (**a1**) DF for $\alpha = 0.1$, (**b1**) WNG for $\alpha = 0.1$, (**a2**) DF for $\alpha = 0.5$, (**b2**) WNG for $\alpha = 0.5$, (**a3**) DF for $\alpha = 0.9$, and (**b3**) WNG for $\alpha = 0.9$. Conditions: $\delta = 1.0$ cm and $(\theta_s, \phi_s) = (90°, 0°)$

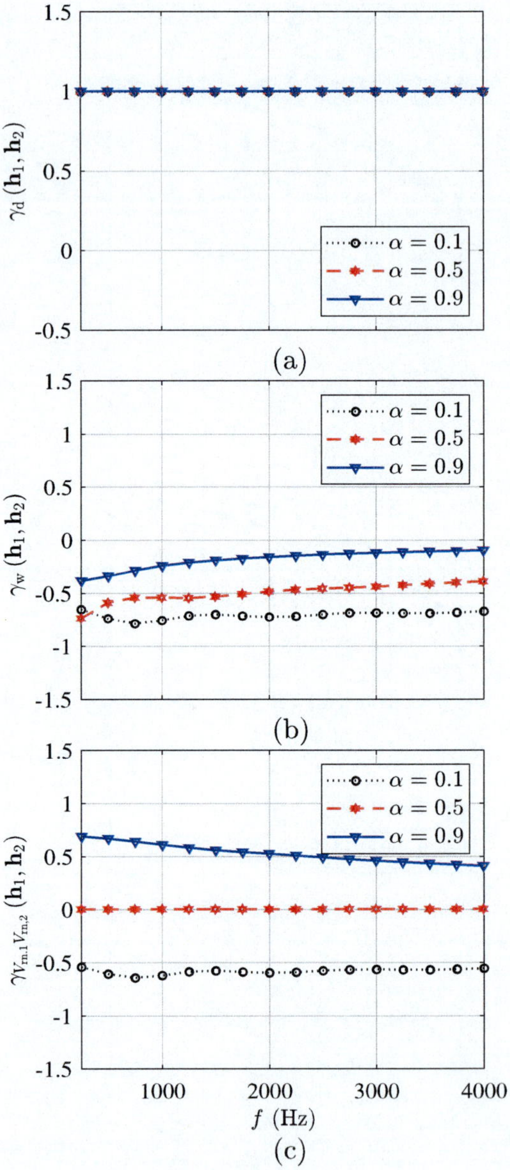

Fig. 9.10 Output ICs of noises in the output signals of the binaural MWNG beamformer, $\mathbf{h}_{1,\text{MWNG}}$ and $\mathbf{h}_{2,\text{MWNG}}$, given in (9.59) and (9.60), respectively, as a function of frequency, for different values of α: (**a**) diffuse noise, (**b**) white noise, and (**c**) diffuse noise mixed with white noise at an energy ratio of 0 dB. Conditions: $M = 6$, $\delta = 1.0$ cm, and $(\theta_s, \phi_s) = (90°, 0°)$

noises in the two output signals are completely coherent. The output ICs of the white noise, $\gamma_w(\mathbf{h}_1, \mathbf{h}_2)$, are shown in Fig. 9.10b; the IC increases as the value of α increases and is always smaller than 0, so in the output signals, white noise will be perceived from different zones from the desired signal. Then, we consider the case when diffuse noise is mixed with white noise at an energy ratio of 0 dB. Figure 9.10c shows plots of the output ICs of the noise, $\gamma_{V_{rn,1}V_{rn,2}}$ in (9.20), where one could see that the output IC increases as the value of α increases. Particularly, the output IC of the beamformer with $\alpha = 0.5$ is equal to 0, which means that noises in the two output signals of this beamformer are incoherent and the output signals correspond to the heterophasic case.

9.5.2 Listening Tests

In this subsection, we discuss listening tests that were conducted to further validate the perceptual directions/zones of the desired speech and unwanted noises. Five experienced normal hearing volunteers aged from 22 to 28 participated in the experiment. They were asked to listen to the output audio signals processed by the binaural beamformers in a quiet environment with BOSE QC35II headphones and draw out the zones of the desired speech signal and noises on the horizontal plane according to their perception. Participants were instructed to color the zones of the desired speech signal in red, while the zones of noises in blue. Results sketched by all participants are averaged as the final results.

We consider a room of size 6 m $\times$ 5 m $\times$ 3 m, of which the reflection coefficients of all walls are assumed to be frequency independent and set to be 0.9, and the corresponding reverberation time, T_{60}, is approximately 500 ms. The center of a ULA, which consists of 12 omnidirectional microphones along the x-axis and with an interelement spacing of 1.0 cm, is located at (2, 2, 1.5). The desired point source is placed at (4, 2, 1.5), where a loudspeaker playbacks some prerecorded clean speech signals (20 clean speech signals randomly selected from the TIMIT database [13]), which simulates the desired speech signal of interest. The acoustic channel impulse responses from the point source to the microphones are generated with the image model method [14]. Then, the speech signals at the sensors are generated by convolving the source signal with the corresponding impulse responses. To demonstrate the direction perception performance of the proposed binaural beamformers, different noises are added to the speech signals. All signals are sampled at 16 kHz. For each beamformer in a particular noise, 100 sketches (20 clean speech signals $\times$ 5 participants) were collected.

To demonstrate the rendering performance of the binaural MDF beamformer, $\mathbf{h}_{1,\mathrm{MDF}}$ and $\mathbf{h}_{2,\mathrm{MDF}}$, given in (9.44) and (9.45), respectively, diffuse and white noises are added to each speech signal at an SNR of 5 and 15 dB, separately. To avoid the influence of the white noise amplification problem, we use diagonal loading for regularization; specifically, $\mathbf{\Gamma}_{\mathrm{d}}$ is replaced by $\mathbf{\Gamma}_{\mathrm{d}} + \epsilon \mathbf{I}_{2M}$, where $\epsilon \geq 0$ is the regularization parameter, which is set to be 0.01 in this experiment. Figure 9.11a shows the perceived source map of this beamformer in diffuse noise. As seen, the desired signal is perceived to be in the middle front zone of the head, and the diffuse noise in the binaural outputs is perceived to locate in the respective side of the head, which corresponds to the heterophasic case. Figure 9.11b shows the perceived source map of this beamformer in white noise, where one could see that the desired signal is still perceived to be in the middle front zone of the head, and the white

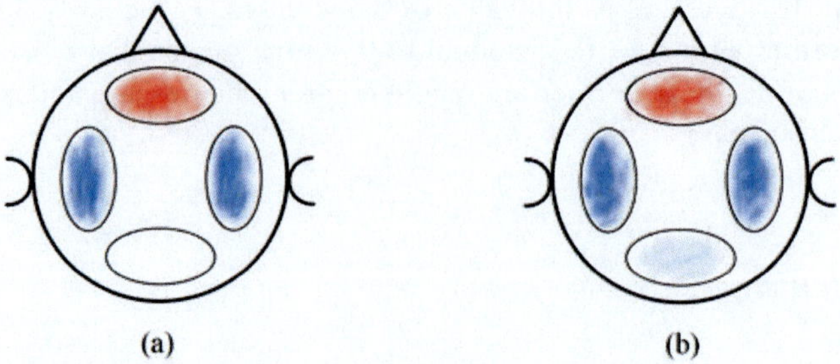

Fig. 9.11 The perceived source map on the horizontal plane averaged over 100 sketches from five listeners of the binaural MDF beamformer, $\mathbf{h}_{1,\mathrm{MDF}}$ and $\mathbf{h}_{2,\mathrm{MDF}}$, given in (9.44) and (9.45), respectively: (**a**) diffuse noise and (**b**) white noise

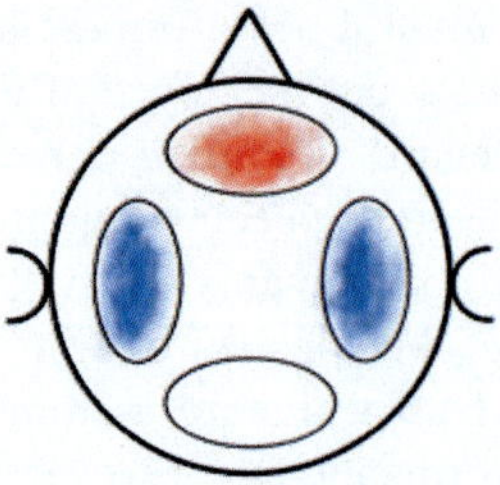

Fig. 9.12 The perceived source map on the horizontal plane averaged over 100 sketches from five listeners of the binaural MVDR beamformer, $\mathbf{h}_{1,\mathrm{MVDR}}$ and $\mathbf{h}_{2,\mathrm{MVDR}}$, given in (9.51) and (9.52), respectively

noise in the binaural outputs is perceived to be in the respective side of the head in most sketches, while in some cases, the participants perceived white noise from the middle back of the head. This phenomenon happens when the IC of the white noise in the output signals is smaller than zero as shown in Fig. 9.4; when the IC gets smaller from 0 to -1, the perceived zone of the white noise tends to move from the respective side of the head toward the middle back of the head [5].

We then study the rendering ability of the binaural MVDR beamformer, $\mathbf{h}_{1,\mathrm{MVDR}}$ and $\mathbf{h}_{2,\mathrm{MVDR}}$, given in (9.51) and (9.52), respectively. We consider the environment with two point source noises, white noise, and diffuse noise coexisting. Two loudspeakers placed at (0.27, 3, 1.5) and (0.27, 1, 1.5), which playback a clip of prerecorded babble and factory noises, respectively, are used to simulate the interference sources. Then, the interference signals at the sensors are generated by convolving the two noise source signals with the corresponding room impulse responses, which are generated by the image model method. The interferences are normalized to have the same variance. The input SNR between the desired speech signal and the sum of the two interferences is 0 dB; input SNRs of the white and diffuse noises are both set to be 20 dB. The perceived source map of this simulation is shown in Fig. 9.12, where one can see that the desired signal is perceived to be in the middle front zone of the head, and the noise in the binaural outputs is perceived to be in the respective side of the head, which corresponds to the heterophasic case.

Finally, we evaluate the rendering performance of the binaural MWNG beamformers, $\mathbf{h}_{1,\mathrm{MWNG}}$ and $\mathbf{h}_{2,\mathrm{MWNG}}$, given in (9.59) and (9.60), respectively, in diffuse,

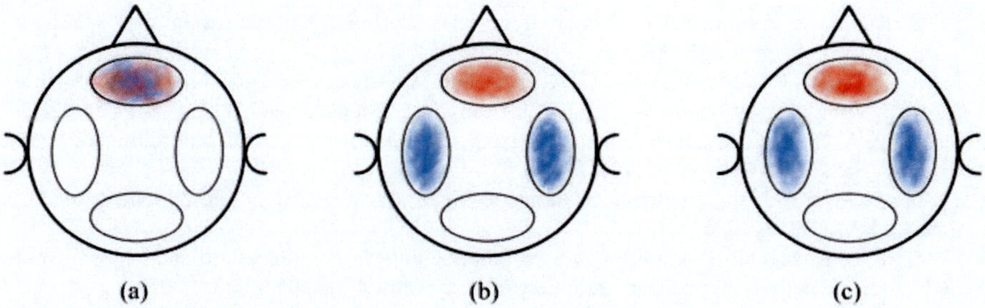

Fig. 9.13 The perceived source map on the horizontal plane averaged over 100 sketches from five listeners of the binaural MWNG beamformer, $\mathbf{h}_{1,\mathrm{MWNG}}$ and $\mathbf{h}_{2,\mathrm{MWNG}}$, given in (9.59) and (9.60), respectively: (**a**) diffuse noise, (**b**) white noise, and (**c**) diffuse noise mixed with white noise at an energy ratio of 0 dB

white, and diffuse-plus-white noises. The parameter α in (9.53) is set to be 0.5. Figure 9.13a shows the perceived source map in diffuse noise at an SNR of 5 dB. One can see that both the desired signal and diffuse noise in the binaural outputs are perceived from the median front zone of the head and are mixed together. Figure 9.13b shows the perceived source map in white noise, which is also added to the desired signal at an SNR of 5 dB. One can see that the desired signal is still perceived to be in the middle front zone of the head, while the white noise in the binaural outputs is perceived to be in the respective side of the head. To conclude, we consider the case where both diffuse and white noises coexist at an SNR of 5 dB. Figure 9.13c shows the perceived source map, where the desired signal is concentrated in the middle of the head, while the noise is perceived to be in the respective side of the head, which is consistent with the theoretical analysis as a heterophasic case.

References

1. I.J. Hirsh, I. Pollack, The role of interaural phase in loudness. J. Acoust. Soc. Am. **20**(6), 761–766 (1948)
2. I.J. Hirsh, The influence of interaural phase on interaural summation and inhibition. J. Acoust. Soc. Am. **20**(4), 592–592 (1948)
3. W.E. Kock, Binaural localization and masking. J. Acoust. Soc. Am. **22**(6), 801–804 (1950)
4. L.A. Jeffress, D.E. Robinson, Formulas for the coefficient of interaural correlation for noise. J. Acoust. Soc. Am. **34**(10), 1658–1659 (1962)
5. J. Blauert, *Spatial Hearing: The Psychophysics of Human Sound Localization* (MIT Press, Cambridge, 1996)
6. L.A. Jeffress, H.C. Blodgett, B.H. Deatherage, Effect of interaural correlation on the precision of centering a noise. J. Acoust. Soc. Am. **34**(8), 1122–1123 (1962)
7. J. Blauert, W. Lindemann, Spatial mapping of intracranial auditory events for various degrees of interaural coherence. J. Acoust. Soc. Am. **79**(3), 806–813 (1986)

8. J.C.R. Licklider, The influence of interaural phase relations upon the masking of speech by white noise. J. Acoust. Soc. Am. **20**(2), 150–159 (1948)
9. J. Jin, J. Chen, J. Benesty, Y. Wang, G. Huang, Heterophasic binaural differential beamforming for speech intelligibility improvement. IEEE Trans. Veh. Technol. **69**(11), 13497–13509 (2020)
10. Y. Wang, J. Chen, J. Benesty, J. Jin, G. Huang, Binaural heterophasic superdirective beamforming. Sensors **21**(74), 22 (2021)
11. J. Chen, Y. Wang, J. Jin, G. Huang, J. Benesty, Binaural beamforming microphone array. U.S. Patent 11,546,691 (2023)
12. U. Zimmer, E. Macaluso, High binaural coherence determines successful sound localization and increased activity in posterior auditory areas. Neuron **47**(6), 893–905 (2005)
13. J.S. Garofolo, L.F. Lamel, W.M. Fisher, J.G. Fiscus, D.S. Pallett, N.L. Dahlgren, DARPA TIMIT acoustic phonetic continuous speech corpus (1993). http://www.ldc.upenn.edu/Catalog/LDC93S1.html
14. J.B. Allen, D.A. Berkley, Image method for efficiently simulating small-room acoustics. J. Acoust. Soc. Am. **65**, 943–950 (1979)

Chapter 10
Large Array Beamforming

In this chapter, we study beamforming with very large arrays, i.e., arrays that contain a very large number of microphones. Conventional beamforming in this context, where a simple complex frequency-dependent weight is applied to each microphone, may not be very practical for obvious reasons such as high complexity, difficulty to accurately estimate statistics of the signals, and dealing with large matrices that can be very ill conditioned. Here, instead, we tackle this problem from a low-rank beamforming perspective, whose flexibility can lead to better performance.

10.1 Basic Concepts and Signal Model

We are concerned with very large arrays, which also encompass distributed arrays. Let us consider a microphone array, whose total number of sensors, M, is very large. From this array, we can form N_C clusters, which contain the same number of microphones, M_0, with $M_0 < M$, and overlap is possible, i.e., a sensor can belong to two different adjacent clusters. In each cluster, the first sensor, denoted by $\mathrm{M}_{n,1}$, $n = 1, 2, \ldots, N_C$, is the reference and $\mathrm{M}_{1,1} \neq \mathrm{M}_{2,1} \neq \cdots \neq \mathrm{M}_{N_C,1}$. After choosing $\mathrm{M}_{n,1}$, we take the $M_0 - 1$ closest microphones to $\mathrm{M}_{n,1}$, so that the cluster C_n is formed. Each cluster represents a subarray composed of M_0 microphones, and the global array, which contains all M sensors, is the union of the N_C subarrays, so that it is composed of $N_C M_0$ virtual sensors, with $N_C M_0 \geq M$. Obviously, the degree of freedoms of the global array is still equal to M. We may also assume that $N_C \leq M_0$.[1]

Now, let the desired source signal propagates from the known position (θ_s, ϕ_s) and impinges on the above global array. To be more concise, we will use the notation $\vartheta_s = (\theta_s, \phi_s)$ and, more generally, $\vartheta = (\theta, \phi)$. Then, using the conventional

[1] This assumption is not really required but it may make sense in practice.

J. Benesty et al., *Microphone Arrays*, Springer Topics in Signal Processing 22,
https://doi.org/10.1007/978-3-031-36974-2_10

array signal model [1], the observed signal vector (of length M_0) at the nth ($n = 1, 2, \ldots, N_C$) subarray is

$$\mathbf{y}_n = \mathbf{d}_{n,\vartheta_s} X + \mathbf{v}_n, \tag{10.1}$$

where

$$\mathbf{d}_{n,\vartheta_s} = \left[e^{\jmath 2\pi f \mathbf{a}_{\vartheta_s}^T \mathbf{r}_{n,1}/c} \; e^{\jmath 2\pi f \mathbf{a}_{\vartheta_s}^T \mathbf{r}_{n,2}/c} \; \cdots \; e^{\jmath 2\pi f \mathbf{a}_{\vartheta_s}^T \mathbf{r}_{n,M_0}/c} \right]^T \tag{10.2}$$

is the steering vector of the nth subarray,

$$\mathbf{a}_{\vartheta_s} = \left[\sin\theta_s \cos\phi_s \;\; \sin\theta_s \sin\phi_s \;\; \cos\theta_s \right]^T, \tag{10.3}$$

and $\mathbf{r}_{n,m}$ is the Cartesian position of the mth element of the nth subarray. We deduce that the covariance matrix of $\mathbf{y}_n$ is

$$\begin{aligned} \mathbf{\Phi}_{\mathbf{y}_n} &= E\left(\mathbf{y}_n \mathbf{y}_n^H\right) \\ &= \phi_X \mathbf{d}_{n,\vartheta_s} \mathbf{d}_{n,\vartheta_s}^H + \mathbf{\Phi}_{\mathbf{v}_n} \\ &= \phi_X \mathbf{d}_{n,\vartheta_s} \mathbf{d}_{n,\vartheta_s}^H + \phi_{\mathrm{N},n} \mathbf{\Gamma}_{\mathbf{v}_n}, \end{aligned} \tag{10.4}$$

where ϕ_X is the variance of X; $\mathbf{\Phi}_{\mathbf{v}_n} = E\left(\mathbf{v}_n \mathbf{v}_n^H\right)$ is the covariance matrix of $\mathbf{v}_n$; $\phi_{\mathrm{N},n}$ is the variance of the noise at any sensor of the nth subarray, which is assumed to be the same at all microphones within the same subarray; and $\mathbf{\Gamma}_{\mathbf{v}_n} = \mathbf{\Phi}_{\mathbf{v}_n}/\phi_{\mathrm{N},n}$ is the coherence matrix of the noise at the nth subarray. In case of spherically isotropic (diffuse) noise field, (10.4) becomes

$$\mathbf{\Phi}_{\mathbf{y}_n} = \phi_X \mathbf{d}_{n,\vartheta_s} \mathbf{d}_{n,\vartheta_s}^H + \phi_{\mathrm{DN},n} \mathbf{\Gamma}_{\mathrm{DN},n}, \tag{10.5}$$

where $\phi_{\mathrm{DN},n}$ is the variance of the diffuse noise at any sensor of the nth subarray and

$$\mathbf{\Gamma}_{\mathrm{DN},n} = \frac{1}{4\pi} \int_0^{2\pi} \int_0^{\pi} \mathbf{d}_{n,\vartheta} \mathbf{d}_{n,\vartheta}^H \sin\theta d\phi d\theta \tag{10.6}$$

is the coherence matrix of the diffuse noise, with $\mathbf{d}_{n,\vartheta}$ being the steering vector corresponding to the nth subarray for any $\vartheta = (\theta, \phi)$.

Now, for the global array, which encompasses all the N_C subarrays, we can form the observed signal matrix (of size $M_0 \times N_C$):

$$\begin{aligned} \mathbf{Y} &= \left[\mathbf{y}_1 \;\; \mathbf{y}_2 \;\; \cdots \;\; \mathbf{y}_{N_C} \right] \\ &= \mathbf{D}_{\vartheta_s} X + \mathbf{V}, \end{aligned} \tag{10.7}$$

where

$$\mathbf{D}_{\vartheta_s} = \begin{bmatrix} \mathbf{d}_{1,\vartheta_s} & \mathbf{d}_{2,\vartheta_s} & \cdots & \mathbf{d}_{N_C,\vartheta_s} \end{bmatrix} \tag{10.8}$$

is the steering matrix corresponding to the global array, and the noise signal matrix $\mathbf{V}$ is defined similarly to $\mathbf{Y}$. From (10.4) and (10.7), we find that the input SNRs of the nth subarray and global array are, respectively,

$$\text{iSNR}_n = \frac{\phi_X}{\phi_{\text{N},n}} \tag{10.9}$$

and

$$\text{iSNR} = \frac{N_C \phi_X}{\sum_{n=1}^{N_C} \phi_{\text{N},n}} = \frac{\phi_X}{\phi_{\text{N}}}. \tag{10.10}$$

Our objective in this chapter is to study beamforming for large arrays in the most objective way by exploiting the coherence among the different subarrays and quantify their contributions to the global array gain, thanks to our beamforming approach.

10.2 Beamforming

Conventional beamforming is not very practical when the number of sensors in the array is very large since complexity can be very high and large ill-conditioned matrices need to be inverted. Also, not all microphones contribute equally to the performance of an optimal beamformer; as a result, beamforming may lead to poor results, especially in an adaptive context. In this section, we propose a better decomposition based on low-rank beamforming (see Chap. 5).

Let $\mathbf{h}_l,\ l = 1, 2, \ldots, L$ be L filters of length M_0 that we apply to the left of $\mathbf{Y}$ and let $\mathbf{g}_l,\ l = 1, 2, \ldots, L$ be L filters of length N_C that we apply to the right of $\mathbf{Y}$. Then, the proposed beamforming approach is

$$Z = \sum_{l=1}^{L} \mathbf{h}_l^H \mathbf{Y} \mathbf{g}_l^*, \tag{10.11}$$

where Z is the beamformer output or the estimate of the desired signal, X, and $L \leq N_C \leq M_0$. For $L = N_C$, we exploit all available degrees of freedom [2] almost like in conventional beamforming. However, this may not be the best thing to do in practice if some subarrays are not really mutually coherent; indeed, this will only increase complexity for nothing and a huge number of observations will be required to reliably estimate the different statistics for adaptive beamforming. This method will likely lead to very poor performances. The simplest case is for $L = 1$, i.e.,

$$Z = \mathbf{h}^H \mathbf{Y} \mathbf{g}^*, \tag{10.12}$$

where $\mathbf{h}$ and $\mathbf{g}$ are filters of lengths M_0 and N_C, respectively. We see that the beamformer $\mathbf{h}$ is the same for all subarrays, which may make sense somehow in some situations. In practice, depending on the position of the subarrays as well as the position of the desired source signal, a proper value of L between 1 and N_C should be chosen.

A useful way to express (10.11) is

$$Z = \underline{\mathbf{h}}^H \left(\mathbf{I}_L \otimes \mathbf{Y} \right) \underline{\mathbf{g}}^*, \tag{10.13}$$

where

$$\underline{\mathbf{h}} = \begin{bmatrix} \mathbf{h}_1^T & \mathbf{h}_2^T & \cdots & \mathbf{h}_L^T \end{bmatrix}^T$$

is a long filter of length LM_0, $\mathbf{I}_L$ is the $L \times L$ identity matrix, and

$$\underline{\mathbf{g}} = \begin{bmatrix} \mathbf{g}_1^T & \mathbf{g}_2^T & \cdots & \mathbf{g}_L^T \end{bmatrix}^T$$

is another long filter of length LN_C. Continuing with the rewriting of Z, we also have

$$\begin{aligned} Z &= \mathrm{tr}\left[\left(\underline{\mathbf{h}}^* \underline{\mathbf{g}}^H \right)^T \left(\mathbf{I}_L \otimes \mathbf{Y} \right) \right] \\ &= \mathrm{vec}^T \left(\underline{\mathbf{h}}^* \underline{\mathbf{g}}^H \right) \mathrm{vec} \left(\mathbf{I}_L \otimes \mathbf{Y} \right) \\ &= \left(\underline{\mathbf{g}} \otimes \underline{\mathbf{h}} \right)^H \mathrm{vec} \left(\mathbf{I}_L \otimes \mathbf{Y} \right). \end{aligned} \tag{10.14}$$

One of the terms in (10.14) can conveniently be rewritten as [3]

$$\mathrm{vec} \left(\mathbf{I}_L \otimes \mathbf{Y} \right) = \left(\mathbf{I}_L \otimes \mathbf{K}_{N_C L} \otimes \mathbf{I}_{M_0} \right) \left[\mathrm{vec} \left(\mathbf{I}_L \right) \otimes \mathrm{vec} \left(\mathbf{Y} \right) \right], \tag{10.15}$$

where

$$\mathbf{K}_{N_C L} = \sum_{n=1}^{N_C} \sum_{l=1}^{L} \mathbf{U}_{nl} \otimes \mathbf{U}_{nl}^T \tag{10.16}$$

is the so-called commutation matrix of size $N_C L \times N_C L$ [3], with $\mathbf{U}_{nl}$ being an $N_C \times L$ matrix whose nlth element is 1 and whose remaining elements are 0, and $\mathbf{I}_{M_0}$ is the $M_0 \times M_0$ identity matrix. In order to simplify the presentation, we will use the following variables:

$$
\begin{aligned}
\underline{\mathbf{K}} &= \mathbf{I}_L \otimes \mathbf{K}_{N_C L} \otimes \mathbf{I}_{M_0}, \\
\text{vec}\left(\mathbf{I}_L\right) &= \underline{\mathbf{i}}, \\
\text{vec}\left(\mathbf{Y}\right) = \underline{\mathbf{y}} &= \text{vec}\left(\mathbf{D}_{\vartheta_s}\right) X + \text{vec}\left(\mathbf{V}\right) \\
&= \underline{\mathbf{d}}_{\vartheta_s} X + \underline{\mathbf{v}}.
\end{aligned}
$$

Therefore, (10.14) is

$$
Z = \left(\underline{\mathbf{g}} \otimes \underline{\mathbf{h}}\right)^H \underline{\mathbf{K}} \left(\underline{\mathbf{i}} \otimes \underline{\mathbf{y}}\right). \tag{10.17}
$$

From the following relations:

$$
\begin{aligned}
\underline{\mathbf{g}} \otimes \underline{\mathbf{h}} &= \left(\underline{\mathbf{g}} \times 1\right) \otimes \left(\mathbf{I}_{LM_0} \times \underline{\mathbf{h}}\right) \\
&= \left(\underline{\mathbf{g}} \otimes \mathbf{I}_{LM_0}\right) \underline{\mathbf{h}} \qquad (10.18)
\end{aligned}
$$

and

$$
\begin{aligned}
\underline{\mathbf{g}} \otimes \underline{\mathbf{h}} &= \left(\mathbf{I}_{LN_C} \times \underline{\mathbf{g}}\right) \otimes \left(\underline{\mathbf{h}} \times 1\right) \\
&= \left(\mathbf{I}_{LN_C} \otimes \underline{\mathbf{h}}\right) \underline{\mathbf{g}}, \qquad (10.19)
\end{aligned}
$$

where $\mathbf{I}_{LM_0}$ and $\mathbf{I}_{LN_C}$ are the $LM_0 \times LM_0$ and $LN_C \times LN_C$ identity matrices, respectively, we can express Z in two different ways:

$$
\begin{aligned}
Z &= \underline{\mathbf{h}}^H \underline{\mathbf{G}} \left(\underline{\mathbf{i}} \otimes \underline{\mathbf{y}}\right) \qquad (10.20) \\
&= \underline{\mathbf{h}}^H \underline{\mathbf{G}} \left(\underline{\mathbf{i}} \otimes \underline{\mathbf{d}}_{\vartheta_s}\right) X + \underline{\mathbf{h}}^H \underline{\mathbf{G}} \left(\underline{\mathbf{i}} \otimes \underline{\mathbf{v}}\right) \\
&= \underline{\mathbf{g}}^H \underline{\mathbf{H}} \left(\underline{\mathbf{i}} \otimes \underline{\mathbf{y}}\right) \qquad (10.21) \\
&= \underline{\mathbf{g}}^H \underline{\mathbf{H}} \left(\underline{\mathbf{i}} \otimes \underline{\mathbf{d}}_{\vartheta_s}\right) X + \underline{\mathbf{g}}^H \underline{\mathbf{H}} \left(\underline{\mathbf{i}} \otimes \underline{\mathbf{v}}\right),
\end{aligned}
$$

where

$$
\begin{aligned}
\underline{\mathbf{G}} &= \left(\underline{\mathbf{g}} \otimes \mathbf{I}_{LM_0}\right)^H \underline{\mathbf{K}}, \\
\underline{\mathbf{H}} &= \left(\mathbf{I}_{LN_C} \otimes \underline{\mathbf{h}}\right)^H \underline{\mathbf{K}}.
\end{aligned}
$$

From (10.20) and (10.21), we see that the distortionless constraint is

$$1 = \underline{\mathbf{h}}^H \underline{\mathbf{G}} \left(\underline{\mathbf{i}} \otimes \underline{\mathbf{d}}_{\vartheta_s} \right) \tag{10.22}$$

$$= \underline{\mathbf{g}}^H \underline{\mathbf{H}} \left(\underline{\mathbf{i}} \otimes \underline{\mathbf{d}}_{\vartheta_s} \right). \tag{10.23}$$

Then, the variance of Z is

$$\phi_Z = \underline{\mathbf{h}}^H \underline{\mathbf{G}} \left[\left(\underline{\mathbf{i}}\, \underline{\mathbf{i}}^T \right) \otimes \mathbf{\Phi}_{\underline{\mathbf{y}}} \right] \underline{\mathbf{G}}^H \underline{\mathbf{h}} \tag{10.24}$$

$$= \underline{\mathbf{g}}^H \underline{\mathbf{H}} \left[\left(\underline{\mathbf{i}}\, \underline{\mathbf{i}}^T \right) \otimes \mathbf{\Phi}_{\underline{\mathbf{y}}} \right] \underline{\mathbf{H}}^H \underline{\mathbf{g}}, \tag{10.25}$$

where $\mathbf{\Phi}_{\underline{\mathbf{y}}} = E\left(\underline{\mathbf{y}}\, \underline{\mathbf{y}}^H \right)$. Since the degree of freedoms is M and in order to make sure that all covariance matrices are invertible, we must always choose L in such a way that $LM_0 \leq M$.

10.3 Performance Measures

In this section, we briefly discuss the most important performance measures in our context of large arrays.

The first measure is the beampattern, which is given by

$$\mathcal{B}_\vartheta \left(\underline{\mathbf{h}}, \underline{\mathbf{g}} \right) = \left(\underline{\mathbf{g}} \otimes \underline{\mathbf{h}} \right)^H \underline{\mathbf{K}} \left(\underline{\mathbf{i}} \otimes \underline{\mathbf{d}}_\vartheta \right)$$

$$= \underline{\mathbf{h}}^H \underline{\mathbf{G}} \left(\underline{\mathbf{i}} \otimes \underline{\mathbf{d}}_\vartheta \right) \tag{10.26}$$

$$= \underline{\mathbf{g}}^H \underline{\mathbf{H}} \left(\underline{\mathbf{i}} \otimes \underline{\mathbf{d}}_\vartheta \right). \tag{10.27}$$

Now, all next performance measures are defined twice by assuming that one of the two beamforming filters $\underline{\mathbf{h}}$ and $\underline{\mathbf{g}}$ is fixed. This will facilitate the derivation of optimal beamformers in the next section.

From ϕ_Z and assuming that $\underline{\mathbf{g}}$ is fixed, we can express the output SNR as

$$\text{oSNR}\left(\underline{\mathbf{h}} | \underline{\mathbf{g}} \right) = \frac{\phi_X \left| \underline{\mathbf{h}}^H \underline{\mathbf{G}} \left(\underline{\mathbf{i}} \otimes \underline{\mathbf{d}}_{\vartheta_s} \right) \right|^2}{\underline{\mathbf{h}}^H \underline{\mathbf{G}} \left[\left(\underline{\mathbf{i}}\, \underline{\mathbf{i}}^T \right) \otimes \mathbf{\Phi}_{\underline{\mathbf{v}}} \right] \underline{\mathbf{G}}^H \underline{\mathbf{h}}}, \tag{10.28}$$

where $\mathbf{\Phi}_{\underline{\mathbf{v}}} = E\left(\underline{\mathbf{v}}\, \underline{\mathbf{v}}^H \right)$. In the same way, by assuming that $\underline{\mathbf{h}}$ is fixed, we can define the output SNR as

$$\text{oSNR}\left(\underline{\mathbf{g}} | \underline{\mathbf{h}} \right) = \frac{\phi_X \left| \underline{\mathbf{g}}^H \underline{\mathbf{H}} \left(\underline{\mathbf{i}} \otimes \underline{\mathbf{d}}_{\vartheta_s} \right) \right|^2}{\underline{\mathbf{g}}^H \underline{\mathbf{H}} \left[\left(\underline{\mathbf{i}}\, \underline{\mathbf{i}}^T \right) \otimes \mathbf{\Phi}_{\underline{\mathbf{v}}} \right] \underline{\mathbf{H}}^H \underline{\mathbf{g}}}. \tag{10.29}$$

Then, the SNR gains when $\underline{\mathbf{g}}$ is fixed and when $\underline{\mathbf{h}}$ is fixed are, respectively,

$$\begin{aligned}\mathcal{G}\left(\underline{\mathbf{h}}|\underline{\mathbf{g}}\right) &= \frac{\text{oSNR}\left(\underline{\mathbf{h}}|\underline{\mathbf{g}}\right)}{\text{iSNR}} \\ &= \frac{\phi_{\text{N}}\left|\underline{\mathbf{h}}^H\underline{\mathbf{G}}\left(\underline{\mathbf{i}}\otimes\underline{\mathbf{d}}_{\vartheta_s}\right)\right|^2}{\underline{\mathbf{h}}^H\underline{\mathbf{G}}\left[\left(\underline{\mathbf{i}}\,\underline{\mathbf{i}}^T\right)\otimes\mathbf{\Phi}_{\underline{\mathbf{v}}}\right]\underline{\mathbf{G}}^H\underline{\mathbf{h}}}\end{aligned} \tag{10.30}$$

and

$$\begin{aligned}\mathcal{G}\left(\underline{\mathbf{g}}|\underline{\mathbf{h}}\right) &= \frac{\text{oSNR}\left(\underline{\mathbf{g}}|\underline{\mathbf{h}}\right)}{\text{iSNR}} \\ &= \frac{\phi_{\text{N}}\left|\underline{\mathbf{g}}^H\underline{\mathbf{H}}\left(\underline{\mathbf{i}}\otimes\underline{\mathbf{d}}_{\vartheta_s}\right)\right|^2}{\underline{\mathbf{g}}^H\underline{\mathbf{H}}\left[\left(\underline{\mathbf{i}}\,\underline{\mathbf{i}}^T\right)\otimes\mathbf{\Phi}_{\underline{\mathbf{v}}}\right]\underline{\mathbf{H}}^H\underline{\mathbf{g}}}.\end{aligned} \tag{10.31}$$

To be able to properly define the WNG and DF, we will assume that $\phi_{\text{N},1} \approx \cdots \approx \phi_{\text{N},N}$.

We deduce that the WNGs when $\underline{\mathbf{g}}$ is fixed and when $\underline{\mathbf{h}}$ is fixed are, respectively,

$$\mathcal{W}\left(\underline{\mathbf{h}}|\underline{\mathbf{g}}\right) = \frac{\left|\underline{\mathbf{h}}^H\underline{\mathbf{G}}\left(\underline{\mathbf{i}}\otimes\underline{\mathbf{d}}_{\vartheta_s}\right)\right|^2}{\underline{\mathbf{h}}^H\underline{\mathbf{G}}\left[\left(\underline{\mathbf{i}}\,\underline{\mathbf{i}}^T\right)\otimes\underline{\mathbf{I}}_{M_0N_\text{C}}\right]\underline{\mathbf{G}}^H\underline{\mathbf{h}}} \tag{10.32}$$

and

$$\mathcal{W}\left(\underline{\mathbf{g}}|\underline{\mathbf{h}}\right) = \frac{\left|\underline{\mathbf{g}}^H\underline{\mathbf{H}}\left(\underline{\mathbf{i}}\otimes\underline{\mathbf{d}}_{\vartheta_s}\right)\right|^2}{\underline{\mathbf{g}}^H\underline{\mathbf{H}}\left[\left(\underline{\mathbf{i}}\,\underline{\mathbf{i}}^T\right)\otimes\underline{\mathbf{I}}_{M_0N_\text{C}}\right]\underline{\mathbf{H}}^H\underline{\mathbf{g}}}, \tag{10.33}$$

where $\underline{\mathbf{I}}_{M_0N_\text{C}}$ is mostly the $M_0N_\text{C} \times M_0N_\text{C}$ identity matrix, except for some of its off-diagonal elements, where sensors overlap in different clusters; in these cases, they are equal to 1.

We also find that the DFs when $\underline{\mathbf{g}}$ is fixed and when $\underline{\mathbf{h}}$ is fixed are, respectively,

$$\mathcal{D}\left(\underline{\mathbf{h}}|\underline{\mathbf{g}}\right) = \frac{\left|\underline{\mathbf{h}}^H\underline{\mathbf{G}}\left(\underline{\mathbf{i}}\otimes\underline{\mathbf{d}}_{\vartheta_s}\right)\right|^2}{\underline{\mathbf{h}}^H\underline{\mathbf{G}}\left[\left(\underline{\mathbf{i}}\,\underline{\mathbf{i}}^T\right)\otimes\underline{\mathbf{\Gamma}}_{\text{DN}}\right]\underline{\mathbf{G}}^H\underline{\mathbf{h}}} \tag{10.34}$$

and

$$\mathcal{D}\left(\underline{\mathbf{g}}|\underline{\mathbf{h}}\right) = \frac{\left|\underline{\mathbf{g}}^H\underline{\mathbf{H}}\left(\underline{\mathbf{i}}\otimes\underline{\mathbf{d}}_{\vartheta_s}\right)\right|^2}{\underline{\mathbf{g}}^H\underline{\mathbf{H}}\left[\left(\underline{\mathbf{i}}\,\underline{\mathbf{i}}^T\right)\otimes\underline{\mathbf{\Gamma}}_{\text{DN}}\right]\underline{\mathbf{H}}^H\underline{\mathbf{g}}}, \tag{10.35}$$

where

$$\underline{\mathbf{\Gamma}}_{\mathrm{DN}} = \frac{1}{4\pi} \int_0^{2\pi} \int_0^{\pi} \underline{\mathbf{d}}_{\vartheta} \underline{\mathbf{d}}_{\vartheta}^H \sin\theta d\phi d\theta. \tag{10.36}$$

Finally, the last measure of interest is the MSE criterion. When $\underline{\mathbf{g}}$ is fixed, the MSE criterion is

$$\begin{aligned}\mathcal{J}\left(\underline{\mathbf{h}}|\underline{\mathbf{g}}\right) &= E\left[\left|X - \underline{\mathbf{h}}^H \underline{\mathbf{G}}\left(\underline{\mathbf{i}} \otimes \underline{\mathbf{y}}\right)\right|^2\right] \\ &= \phi_X + \underline{\mathbf{h}}^H \underline{\mathbf{G}}\left[\left(\underline{\mathbf{i}}\,\underline{\mathbf{i}}^T\right) \otimes \mathbf{\Phi}_{\underline{\mathbf{y}}}\right] \underline{\mathbf{G}}^H \underline{\mathbf{h}} - \phi_X \underline{\mathbf{h}}^H \underline{\mathbf{G}}\left(\underline{\mathbf{i}} \otimes \underline{\mathbf{d}}_{\vartheta_{\mathrm{s}}}\right) \\ &\quad - \phi_X \left(\underline{\mathbf{i}} \otimes \underline{\mathbf{d}}_{\vartheta_{\mathrm{s}}}\right)^H \underline{\mathbf{G}}^H \underline{\mathbf{h}}\end{aligned} \tag{10.37}$$

and when $\underline{\mathbf{h}}$ is fixed, the MSE criterion is

$$\begin{aligned}\mathcal{J}\left(\underline{\mathbf{g}}|\underline{\mathbf{h}}\right) &= E\left[\left|X - \underline{\mathbf{g}}^H \underline{\mathbf{H}}\left(\underline{\mathbf{i}} \otimes \underline{\mathbf{y}}\right)\right|^2\right] \\ &= \phi_X + \underline{\mathbf{g}}^H \underline{\mathbf{H}}\left[\left(\underline{\mathbf{i}}\,\underline{\mathbf{i}}^T\right) \otimes \mathbf{\Phi}_{\underline{\mathbf{y}}}\right] \underline{\mathbf{H}}^H \underline{\mathbf{g}} - \phi_X \underline{\mathbf{g}}^H \underline{\mathbf{H}}\left(\underline{\mathbf{i}} \otimes \underline{\mathbf{d}}_{\vartheta_{\mathrm{s}}}\right) \\ &\quad - \phi_X \left(\underline{\mathbf{i}} \otimes \underline{\mathbf{d}}_{\vartheta_{\mathrm{s}}}\right)^H \underline{\mathbf{H}}^H \underline{\mathbf{g}}.\end{aligned} \tag{10.38}$$

10.4 Examples of Optimal Beamformers

In this section, we give some examples of fixed and adaptive beamformers, which are based on the approach developed in the previous section.

The maximum WNG (MWNG) filters are obtained from the maximization of the WNGs. We get

$$\underline{\mathbf{h}}_{\mathrm{MWNG}} = \frac{\left\{\underline{\mathbf{G}}\left[\left(\underline{\mathbf{i}}\,\underline{\mathbf{i}}^T\right) \otimes \underline{\mathbf{I}}_{M_0 N_{\mathrm{C}}}\right] \underline{\mathbf{G}}^H\right\}^{-1} \underline{\mathbf{G}}\,\underline{\mathbf{s}}_{\vartheta_{\mathrm{s}}}}{\underline{\mathbf{s}}_{\vartheta_{\mathrm{s}}}^H \underline{\mathbf{G}}^H \left\{\underline{\mathbf{G}}\left[\left(\underline{\mathbf{i}}\,\underline{\mathbf{i}}^T\right) \otimes \underline{\mathbf{I}}_{M_0 N_{\mathrm{C}}}\right] \underline{\mathbf{G}}^H\right\}^{-1} \underline{\mathbf{G}}\,\underline{\mathbf{s}}_{\vartheta_{\mathrm{s}}}}, \tag{10.39}$$

$$\underline{\mathbf{g}}_{\mathrm{MWNG}} = \frac{\left\{\underline{\mathbf{H}}\left[\left(\underline{\mathbf{i}}\,\underline{\mathbf{i}}^T\right) \otimes \underline{\mathbf{I}}_{M_0 N_{\mathrm{C}}}\right] \underline{\mathbf{H}}^H\right\}^{-1} \underline{\mathbf{H}}\,\underline{\mathbf{s}}_{\vartheta_{\mathrm{s}}}}{\underline{\mathbf{s}}_{\vartheta_{\mathrm{s}}}^H \underline{\mathbf{H}}^H \left\{\underline{\mathbf{H}}\left[\left(\underline{\mathbf{i}}\,\underline{\mathbf{i}}^T\right) \otimes \underline{\mathbf{I}}_{M_0 N_{\mathrm{C}}}\right] \underline{\mathbf{H}}^H\right\}^{-1} \underline{\mathbf{H}}\,\underline{\mathbf{s}}_{\vartheta_{\mathrm{s}}}}, \tag{10.40}$$

where

$$\underline{\mathbf{s}}_{\vartheta_{\mathrm{s}}} = \underline{\mathbf{i}} \otimes \underline{\mathbf{d}}_{\vartheta_{\mathrm{s}}}.$$

Using the ALS algorithm, i.e., by alternatively iterating between $\underline{\mathbf{h}}_{\text{MWNG}}$ and $\underline{\mathbf{g}}_{\text{MWNG}}$ as in [2, 4, 5], these two optimal filters will converge to the MWNG beamformers just after a couple of iterations.

Analogously, the maximum DF (MDF) filters are obtained from the maximization of the DFs. We get

$$\underline{\mathbf{h}}_{\text{MDF}} = \frac{\left\{\underline{\mathbf{G}}\left[\left(\underline{\mathbf{i}}\,\underline{\mathbf{i}}^T\right) \otimes \underline{\boldsymbol{\Gamma}}_{\text{DN}}\right]\underline{\mathbf{G}}^H\right\}^{-1} \underline{\mathbf{G}}\,\underline{\mathbf{s}}_{\vartheta_s}}{\underline{\mathbf{s}}_{\vartheta_s}^H \underline{\mathbf{G}}^H \left\{\underline{\mathbf{G}}\left[\left(\underline{\mathbf{i}}\,\underline{\mathbf{i}}^T\right) \otimes \underline{\boldsymbol{\Gamma}}_{\text{DN}}\right]\underline{\mathbf{G}}^H\right\}^{-1} \underline{\mathbf{G}}\,\underline{\mathbf{s}}_{\vartheta_s}}, \tag{10.41}$$

$$\underline{\mathbf{g}}_{\text{MDF}} = \frac{\left\{\underline{\mathbf{H}}\left[\left(\underline{\mathbf{i}}\,\underline{\mathbf{i}}^T\right) \otimes \underline{\boldsymbol{\Gamma}}_{\text{DN}}\right]\underline{\mathbf{H}}^H\right\}^{-1} \underline{\mathbf{H}}\,\underline{\mathbf{s}}_{\vartheta_s}}{\underline{\mathbf{s}}_{\vartheta_s}^H \underline{\mathbf{H}}^H \left\{\underline{\mathbf{H}}\left[\left(\underline{\mathbf{i}}\,\underline{\mathbf{i}}^T\right) \otimes \underline{\boldsymbol{\Gamma}}_{\text{DN}}\right]\underline{\mathbf{H}}^H\right\}^{-1} \underline{\mathbf{H}}\,\underline{\mathbf{s}}_{\vartheta_s}}. \tag{10.42}$$

Thanks to the ALS algorithm, we get the MDF beamformers.

In order to compromise between WNG and DF, we could alternate between $\underline{\mathbf{h}}_{\text{MDF}}$ and $\underline{\mathbf{g}}_{\text{MWNG}}$, or between $\underline{\mathbf{h}}_{\text{MWNG}}$ and $\underline{\mathbf{g}}_{\text{MDF}}$. The two approaches may give very different results.

The above beamformers are fixed. Our first proposed adaptive beamformer is the MVDR. Indeed, the two MVDR filters are obtained from the maximization of the output SNRs. We get

$$\begin{aligned}
\underline{\mathbf{h}}_{\text{MVDR}} &= \frac{\left\{\underline{\mathbf{G}}\left[\left(\underline{\mathbf{i}}\,\underline{\mathbf{i}}^T\right) \otimes \boldsymbol{\Phi}_{\underline{\mathbf{y}}}\right]\underline{\mathbf{G}}^H\right\}^{-1} \underline{\mathbf{G}}\,\underline{\mathbf{s}}_{\vartheta_s}}{\underline{\mathbf{s}}_{\vartheta_s}^H \underline{\mathbf{G}}^H \left\{\underline{\mathbf{G}}\left[\left(\underline{\mathbf{i}}\,\underline{\mathbf{i}}^T\right) \otimes \boldsymbol{\Phi}_{\underline{\mathbf{y}}}\right]\underline{\mathbf{G}}^H\right\}^{-1} \underline{\mathbf{G}}\,\underline{\mathbf{s}}_{\vartheta_s}} \\
&= \frac{\left\{\underline{\mathbf{G}}\left[\left(\underline{\mathbf{i}}\,\underline{\mathbf{i}}^T\right) \otimes \boldsymbol{\Phi}_{\underline{\mathbf{y}}}\right]\underline{\mathbf{G}}^H\right\}^{-1} \underline{\mathbf{G}}\,\underline{\mathbf{s}}_{\vartheta_s}}{\underline{\mathbf{s}}_{\vartheta_s}^H \underline{\mathbf{G}}^H \left\{\underline{\mathbf{G}}\left[\left(\underline{\mathbf{i}}\,\underline{\mathbf{i}}^T\right) \otimes \boldsymbol{\Phi}_{\underline{\mathbf{y}}}\right]\underline{\mathbf{G}}^H\right\}^{-1} \underline{\mathbf{G}}\,\underline{\mathbf{s}}_{\vartheta_s}},
\end{aligned} \tag{10.43}$$

$$\begin{aligned}
\underline{\mathbf{g}}_{\text{MVDR}} &= \frac{\left\{\underline{\mathbf{H}}\left[\left(\underline{\mathbf{i}}\,\underline{\mathbf{i}}^T\right) \otimes \boldsymbol{\Phi}_{\underline{\mathbf{y}}}\right]\underline{\mathbf{H}}^H\right\}^{-1} \underline{\mathbf{H}}\,\underline{\mathbf{s}}_{\vartheta_s}}{\underline{\mathbf{s}}_{\vartheta_s}^H \underline{\mathbf{H}}^H \left\{\underline{\mathbf{H}}\left[\left(\underline{\mathbf{i}}\,\underline{\mathbf{i}}^T\right) \otimes \boldsymbol{\Phi}_{\underline{\mathbf{y}}}\right]\underline{\mathbf{H}}^H\right\}^{-1} \underline{\mathbf{H}}\,\underline{\mathbf{s}}_{\vartheta_s}} \\
&= \frac{\left\{\underline{\mathbf{H}}\left[\left(\underline{\mathbf{i}}\,\underline{\mathbf{i}}^T\right) \otimes \boldsymbol{\Phi}_{\underline{\mathbf{y}}}\right]\underline{\mathbf{H}}^H\right\}^{-1} \underline{\mathbf{H}}\,\underline{\mathbf{s}}_{\vartheta_s}}{\underline{\mathbf{s}}_{\vartheta_s}^H \underline{\mathbf{H}}^H \left\{\underline{\mathbf{H}}\left[\left(\underline{\mathbf{i}}\,\underline{\mathbf{i}}^T\right) \otimes \boldsymbol{\Phi}_{\underline{\mathbf{y}}}\right]\underline{\mathbf{H}}^H\right\}^{-1} \underline{\mathbf{H}}\,\underline{\mathbf{s}}_{\vartheta_s}}.
\end{aligned} \tag{10.44}$$

From the ALS algorithm, we find the MVDR beamformers.

From the minimization of the MSE criteria, we find the two Wiener filters:

$$\begin{aligned}
\underline{\mathbf{h}}_{\text{W}} &= \phi_X \left\{\underline{\mathbf{G}}\left[\left(\underline{\mathbf{i}}\,\underline{\mathbf{i}}^T\right) \otimes \boldsymbol{\Phi}_{\underline{\mathbf{y}}}\right]\underline{\mathbf{G}}^H\right\}^{-1} \underline{\mathbf{G}}\,\underline{\mathbf{s}}_{\vartheta_s} \\
&= \frac{\phi_X \left\{\underline{\mathbf{G}}\left[\left(\underline{\mathbf{i}}\,\underline{\mathbf{i}}^T\right) \otimes \boldsymbol{\Phi}_{\underline{\mathbf{y}}}\right]\underline{\mathbf{G}}^H\right\}^{-1} \underline{\mathbf{G}}\,\underline{\mathbf{s}}_{\vartheta_s}}{1 + \phi_X \underline{\mathbf{s}}_{\vartheta_s}^H \underline{\mathbf{G}}^H \left\{\underline{\mathbf{G}}\left[\left(\underline{\mathbf{i}}\,\underline{\mathbf{i}}^T\right) \otimes \boldsymbol{\Phi}_{\underline{\mathbf{y}}}\right]\underline{\mathbf{G}}^H\right\}^{-1} \underline{\mathbf{G}}\,\underline{\mathbf{s}}_{\vartheta_s}},
\end{aligned} \tag{10.45}$$

$$\underline{\mathbf{g}}_{\mathrm{W}} = \phi_X \left\{ \underline{\mathbf{H}} \left[\left(\underline{\mathbf{i}}\, \underline{\mathbf{i}}^T \right) \otimes \mathbf{\Phi}_{\underline{\mathbf{y}}} \right] \underline{\mathbf{H}}^H \right\}^{-1} \underline{\mathbf{H}}\, \underline{\mathbf{s}}_{\vartheta_s}$$
$$= \frac{\phi_X \left\{ \underline{\mathbf{H}} \left[\left(\underline{\mathbf{i}}\, \underline{\mathbf{i}}^T \right) \otimes \mathbf{\Phi}_{\underline{\mathbf{v}}} \right] \underline{\mathbf{H}}^H \right\}^{-1} \underline{\mathbf{H}}\, \underline{\mathbf{s}}_{\vartheta_s}}{1 + \phi_X \underline{\mathbf{s}}_{\vartheta_s}^H \underline{\mathbf{H}}^H \left\{ \underline{\mathbf{H}} \left[\left(\underline{\mathbf{i}}\, \underline{\mathbf{i}}^T \right) \otimes \mathbf{\Phi}_{\underline{\mathbf{v}}} \right] \underline{\mathbf{H}}^H \right\}^{-1} \underline{\mathbf{H}}\, \underline{\mathbf{s}}_{\vartheta_s}}. \quad (10.46)$$

By alternatively iterating between $\underline{\mathbf{h}}_{\mathrm{W}}$ and $\underline{\mathbf{g}}_{\mathrm{W}}$, we find the optimal Wiener beamformers.

Very often in practice, it is desirable to be able to compromise between noise reduction and speech distortion. For this purpose, we can derive the tradeoff filters:

$$\underline{\mathbf{h}}_{\mathrm{T},\mu} = \frac{\phi_X \left\{ \underline{\mathbf{G}} \left[\left(\underline{\mathbf{i}}\, \underline{\mathbf{i}}^T \right) \otimes \mathbf{\Phi}_{\underline{\mathbf{v}}} \right] \underline{\mathbf{G}}^H \right\}^{-1} \underline{\mathbf{G}}\, \underline{\mathbf{s}}_{\vartheta_s}}{\mu + \phi_X \underline{\mathbf{s}}_{\vartheta_s}^H \underline{\mathbf{G}}^H \left\{ \underline{\mathbf{G}} \left[\left(\underline{\mathbf{i}}\, \underline{\mathbf{i}}^T \right) \otimes \mathbf{\Phi}_{\underline{\mathbf{v}}} \right] \underline{\mathbf{G}}^H \right\}^{-1} \underline{\mathbf{G}}\, \underline{\mathbf{s}}_{\vartheta_s}}, \quad (10.47)$$

$$\underline{\mathbf{g}}_{\mathrm{T},\mu} = \frac{\phi_X \left\{ \underline{\mathbf{H}} \left[\left(\underline{\mathbf{i}}\, \underline{\mathbf{i}}^T \right) \otimes \mathbf{\Phi}_{\underline{\mathbf{v}}} \right] \underline{\mathbf{H}}^H \right\}^{-1} \underline{\mathbf{H}}\, \underline{\mathbf{s}}_{\vartheta_s}}{\mu + \phi_X \underline{\mathbf{s}}_{\vartheta_s}^H \underline{\mathbf{H}}^H \left\{ \underline{\mathbf{H}} \left[\left(\underline{\mathbf{i}}\, \underline{\mathbf{i}}^T \right) \otimes \mathbf{\Phi}_{\underline{\mathbf{v}}} \right] \underline{\mathbf{H}}^H \right\}^{-1} \underline{\mathbf{H}}\, \underline{\mathbf{s}}_{\vartheta_s}}, \quad (10.48)$$

where $\mu \geq 0$ is a tuning parameter. From the ALS algorithm, we find the tradeoff beamformers. One can see that

- $\mu = 1$ gives the Wiener beamformers;
- $\mu = 0$ leads to the MVDR beamformers;
- $\mu > 1$ results in beamformers with low residual noise (from a broadband perspective) at the expense of high desired signal distortion (as compared to Wiener); and
- $\mu < 1$ results in beamformers with high residual noise and low desired signal distortion (as compared to Wiener).

Another interesting way to compromise between noise reduction and speech distortion is to alternate between $\underline{\mathbf{h}}_{\mathrm{MVDR}}$ and $\underline{\mathbf{g}}_{\mathrm{W}}$, or between $\underline{\mathbf{h}}_{\mathrm{W}}$ and $\underline{\mathbf{g}}_{\mathrm{MVDR}}$.

10.5 Illustrative Examples

Let us study some examples with a large array that consists of N_{C} subarrays, where each subarray is a uniform circular array (UCA) composed of M_0 omnidirectional microphones with radius r, as shown in Fig. 10.1, and with $(\theta_s, \phi_s) = (90°, 0°)$. We assume that there is no overlap between different subarrays. The two beamforming filters $\underline{\mathbf{h}}$ and $\underline{\mathbf{g}}$ are optimized by using the ALS algorithm, where the filter $\underline{\mathbf{g}}$ is initialized in a similar way to the low-rank beamformer in Chap. 5. In the implementation of a matrix inversion, a small regularization parameter, 10^{-6}, is added to the diagonal elements of the matrix.

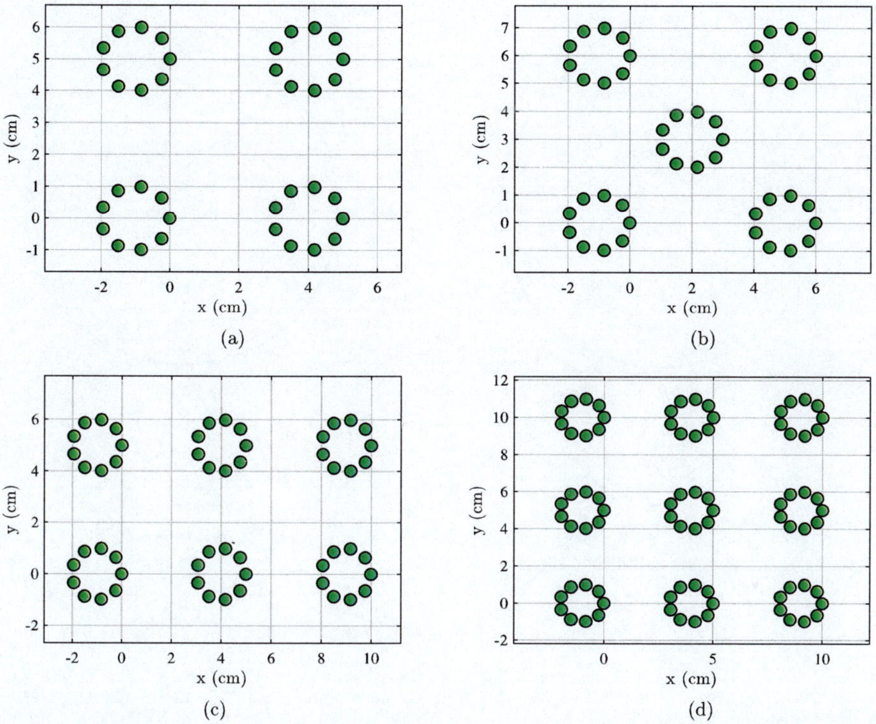

Fig. 10.1 Illustration of a large array that consists of N_C subarrays: (**a**) $N_C = 4$, the center of the four subarrays are $(-1, 0)$ cm, $(4, 0)$ cm, $(-1, 5)$ cm, and $(4, 5)$ cm, respectively; (**b**) $N_C = 5$, the center of the five subarrays are $(-1, 0)$ cm, $(5, 0)$ cm, $(2, 3)$ cm, $(-1, 6)$ cm, and $(5, 6)$ cm, respectively; (**c**) $N_C = 6$, the center of the six subarrays are $(-1, 0)$ cm, $(4, 0)$ cm, $(9, 0)$ cm, $(-1, 5)$ cm, $(4, 5)$ cm, and $(9, 5)$ cm, respectively; and (**d**) $N_C = 9$, the center of the nine subarrays are $(-1, 0)$ cm, $(4, 0)$ cm, $(9, 0)$ cm, $(-1, 5)$ cm, $(4, 5)$ cm, $(9, 5)$ cm, $(-1, 10)$ cm, $(4, 10)$ cm, and $(9, 10)$ cm, respectively. Each subarray is a UCA composed of M_0 omnidirectional microphones with radius r

To demonstrate the performance of the MWNG beamformers, $\underline{\mathbf{h}}_{\text{MWNG}}$ and $\underline{\mathbf{g}}_{\text{MWNG}}$, given in (10.39) and (10.40), respectively, we choose $M_0 = 9$ and $L = 1$. Figure 10.2 shows plots of the power patterns, $\left|\mathcal{B}_\vartheta\left(\underline{\mathbf{h}}, \underline{\mathbf{g}}\right)\right|^2$, of the MWNG beamformers versus the azimuth angle for $f = 1000$ Hz, and different values of N_C. As seen, the power patterns have a wide main beam and the main beam of the power patterns points in the desired direction. It can also be observed that their main beam becomes narrower when the value of N_C increases. Figure 10.3 shows plots of the DF when $\underline{\mathbf{g}}$ is fixed, $\mathcal{D}\left(\underline{\mathbf{h}}|\underline{\mathbf{g}}\right)$; DF when $\underline{\mathbf{h}}$ is fixed, $\mathcal{D}\left(\underline{\mathbf{g}}|\underline{\mathbf{h}}\right)$; WNG when $\underline{\mathbf{g}}$ is fixed, $\mathcal{W}\left(\underline{\mathbf{h}}|\underline{\mathbf{g}}\right)$; and WNG when $\underline{\mathbf{h}}$ is fixed, $\mathcal{W}\left(\underline{\mathbf{g}}|\underline{\mathbf{h}}\right)$, of the MWNG filters as a function of frequency for different values of N_C. It can be observed that both the WNG and DF increase when N_C increases. It can also be observed that the two filters have the

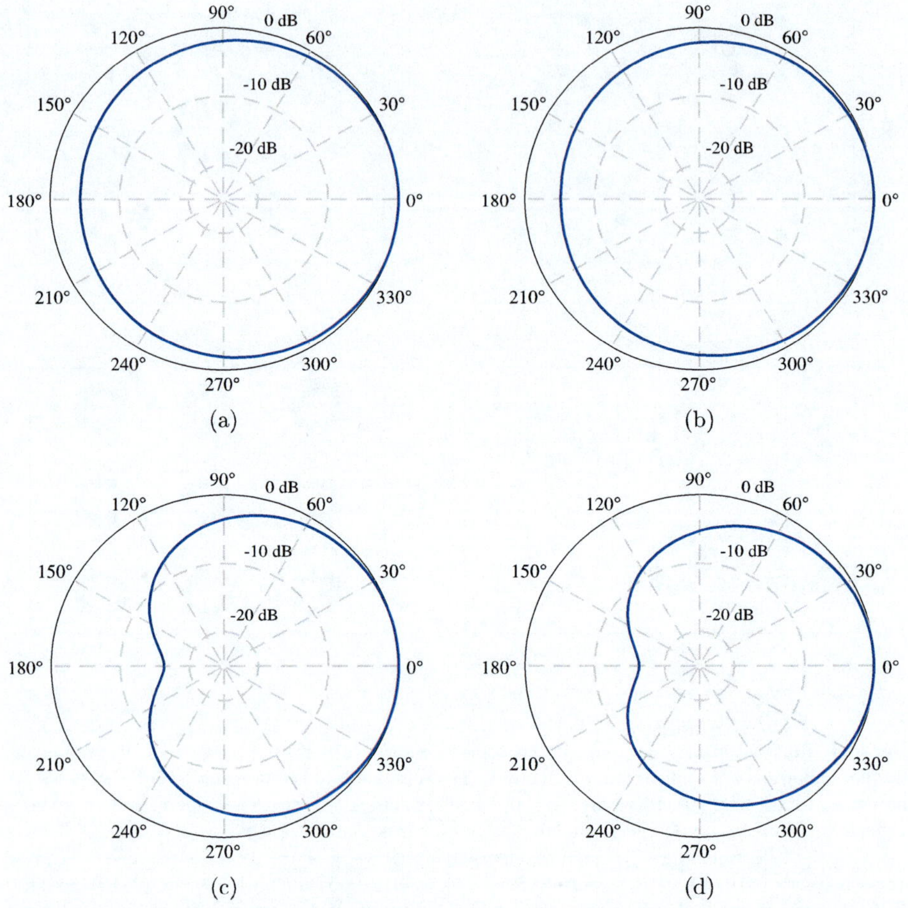

Fig. 10.2 Power patterns of the MWNG beamformers, $\underline{\mathbf{h}}_{\text{MWNG}}$ and $\underline{\mathbf{g}}_{\text{MWNG}}$, versus the azimuth angle, for different values of N_{C}: **(a)** $N_{\text{C}} = 4$, **(b)** $N_{\text{C}} = 5$, **(c)** $N_{\text{C}} = 6$, and **(d)** $N_{\text{C}} = 9$. Conditions: $M_0 = 9$, $f = 1000\,\text{Hz}$, $L = 1$, $(\theta_{\text{s}}, \phi_{\text{s}}) = (90°, 0°)$, and each subarray is a UCA with $r = 1.0\,\text{cm}$

same DF and WNG. This should not come as a surprise since, after convergence, the two filters will have the same performance, i.e., $\mathcal{D}\left(\underline{\mathbf{h}}|\underline{\mathbf{g}}\right) = \mathcal{D}\left(\underline{\mathbf{g}}|\underline{\mathbf{h}}\right)$ and $\mathcal{W}\left(\underline{\mathbf{h}}|\underline{\mathbf{g}}\right) = \mathcal{W}\left(\underline{\mathbf{g}}|\underline{\mathbf{h}}\right)$. For the same reason, we will have $\mathcal{G}\left(\underline{\mathbf{h}}|\underline{\mathbf{g}}\right) = \mathcal{G}\left(\underline{\mathbf{g}}|\underline{\mathbf{h}}\right)$. So, in the following examples, we will only show the performance of the beamformers when $\underline{\mathbf{g}}$ is fixed, i.e., the DF when $\underline{\mathbf{g}}$ is fixed, $\mathcal{D}\left(\underline{\mathbf{h}}|\underline{\mathbf{g}}\right)$; the WNG when $\underline{\mathbf{g}}$ is fixed, $\mathcal{W}\left(\underline{\mathbf{h}}|\underline{\mathbf{g}}\right)$; and the SNR gain when $\underline{\mathbf{g}}$ is fixed, $\mathcal{G}\left(\underline{\mathbf{h}}|\underline{\mathbf{g}}\right)$.

To demonstrate the performance of the MDF filters, $\underline{\mathbf{h}}_{\text{MDF}}$ and $\underline{\mathbf{g}}_{\text{MDF}}$, given in (10.41) and (10.42), respectively, we choose $M_0 = 5$ and $N_{\text{C}} = 5$. Figure 10.4

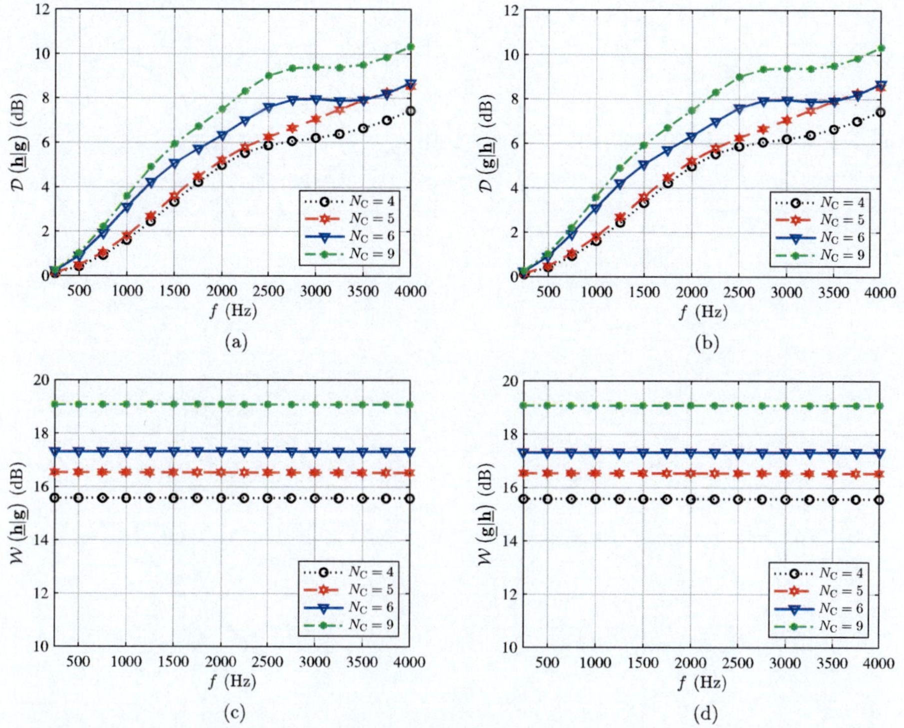

Fig. 10.3 DF and WNG of the MWNG beamformers, $\underline{\mathbf{h}}_{\text{MWNG}}$ and $\underline{\mathbf{g}}_{\text{MWNG}}$, as a function of frequency, for different values of N_C: **(a)** $\mathcal{D}\left(\underline{\mathbf{h}}|\underline{\mathbf{g}}\right)$, **(b)** $\mathcal{D}\left(\underline{\mathbf{g}}|\underline{\mathbf{h}}\right)$, **(c)** $\mathcal{W}\left(\underline{\mathbf{h}}|\underline{\mathbf{g}}\right)$, and **(d)** $\mathcal{W}\left(\underline{\mathbf{g}}|\underline{\mathbf{h}}\right)$. Conditions: $M_0 = 9$, $L = 1$, $(\theta_s, \phi_s) = (90°, 0°)$, and each subarray is a UCA with $r = 1.0$ cm

shows plots of the power patterns, $\left|\mathcal{B}_\vartheta\left(\underline{\mathbf{h}}, \underline{\mathbf{g}}\right)\right|^2$, of the MDF beamformer versus the azimuth angle for $f = 1000$ Hz and different values of L. As seen, the power patterns have much narrower main beams as compared to the ones with the MWNG beamformer. It can also be observed that the power patterns achieve narrower main beams when the value of L increases. Figure 10.5 shows plots of the DF, $\mathcal{D}\left(\underline{\mathbf{h}}|\underline{\mathbf{g}}\right)$, and WNG, $\mathcal{W}\left(\underline{\mathbf{h}}|\underline{\mathbf{g}}\right)$, of this beamformer as a function of frequency for different values of L. It is observed that the DF increases, while the WNG decreases when L increases.

Now, let us study the performance of the adaptive beamformers. We consider an environment with two point source noises, diffuse noise, and white noise, where the two statistically independent interferences impinge on the global array from two different directions $\left(\theta_{i,1}, \phi_{i,1}\right)$ and $\left(\theta_{i,2}, \phi_{i,2}\right)$. The two interferences are incoherent with the desired signal. The covariance matrix of the noise signal, $\mathbf{\Phi}_\mathbf{v}$, is given in (4.95). We choose $(\theta_s, \phi_s) = (90°, 0°)$, $\left(\theta_{i,1}, \phi_{i,1}\right) = (90°, 150°)$, and

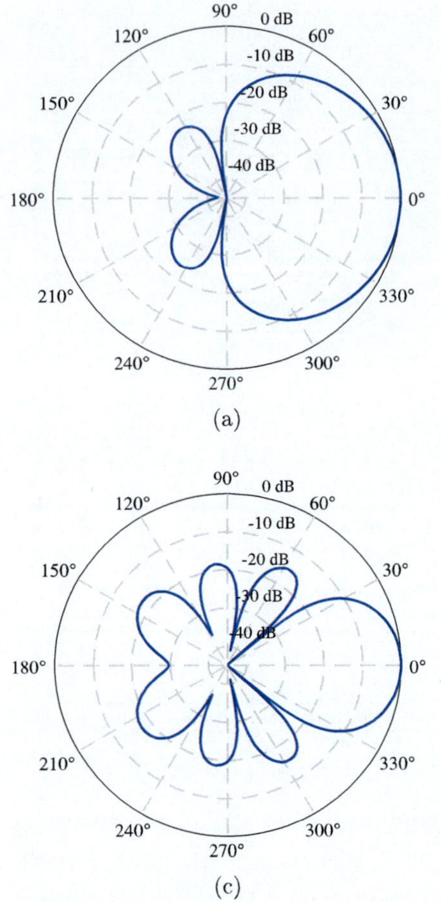

Fig. 10.4 Power patterns of the MDF beamformers, $\underline{\mathbf{h}}_{\mathrm{MDF}}$ and $\underline{\mathbf{g}}_{\mathrm{MDF}}$, versus the azimuth angle, for different values of L: **(a)** $L = 1$, **(b)** $L = 2$, **(c)** $L = 3$, and **(d)** $L = 4$. Conditions: $M_0 = 5$, $N_{\mathrm{C}} = 5$, $f = 1000\,\mathrm{Hz}$, $(\theta_{\mathrm{s}}, \phi_{\mathrm{s}}) = (90°, 0°)$, and each subarray is a UCA with $r = 1.0\,\mathrm{cm}$

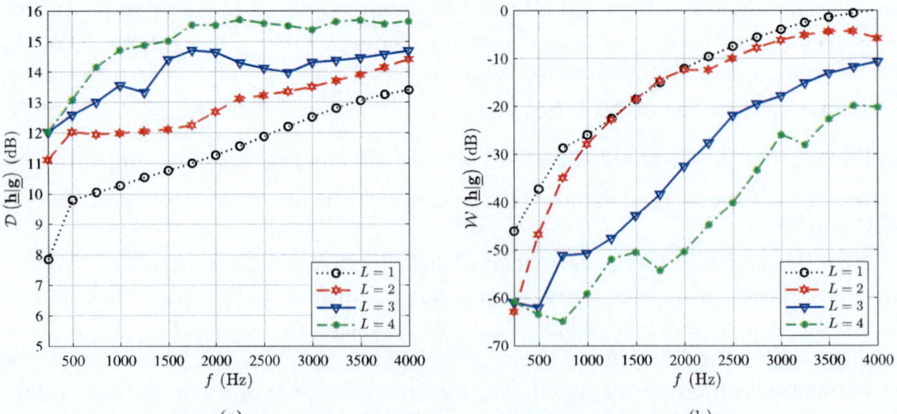

Fig. 10.5 DF and WNG of the MDF beamformers, $\underline{\mathbf{h}}_{\mathrm{MDF}}$ and $\underline{\mathbf{g}}_{\mathrm{MDF}}$, as a function of frequency, for different values of L: **(a)** DF and **(b)** WNG. Conditions: $M_0 = 5$, $N_{\mathrm{C}} = 5$, $(\theta_{\mathrm{s}}, \phi_{\mathrm{s}}) = (90°, 0°)$, and each subarray is a UCA with $r = 1.0\,\mathrm{cm}$

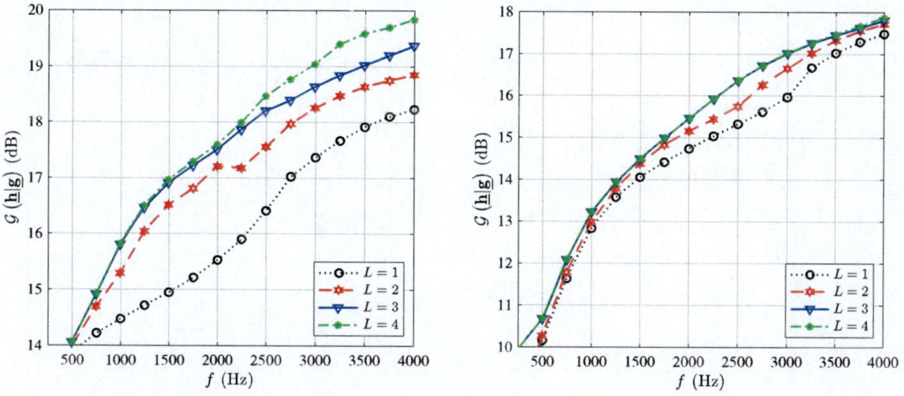

(a) (b)

Fig. 10.6 SNR gain of the MVDR beamformers, $\underline{\mathbf{h}}_{\mathrm{MVDR}}$ and $\underline{\mathbf{g}}_{\mathrm{MVDR}}$, as a function of frequency, for different values of α and different values of L: **(a)** $\alpha = 10^{-4}$ and **(b)** $\alpha = 10^{-2}$. Conditions: $M_0 = 5$, $N_C = 5$, $(\theta_s, \phi_s) = (90°, 0°)$, iSNR = 0 dB, and each subarray is a UCA with $r = 1.0$ cm

$(\theta_{i,2}, \phi_{i,2}) = (90°, 210°)$. The variance of the desired signal is set to 1 and the variances of the noises are computed according to the specified input SNR and values of α ($\alpha > 0$ is related to the ratio between the powers of the white and diffuse noises).

To show the performance of the MVDR beamformers, $\underline{\mathbf{h}}_{\mathrm{MVDR}}$ and $\underline{\mathbf{g}}_{\mathrm{MVDR}}$, given in (10.43) and (10.44), respectively, we choose $M_0 = 5$, $N_C = 5$, and iSNR = 0 dB. Figure 10.6 shows plots of the SNR gain, $\mathcal{G}\left(\underline{\mathbf{h}}|\underline{\mathbf{g}}\right)$, of these beamformers as a function of frequency for different values of L and different values of α. As seen, the SNR gain increases when the value of L increases. It is observed that the SNR gain decreases when the value of α increases since the noise signals at the microphone arrays are less coherent. It can also be observed that the improvement in SNR gain becomes less obvious with L when the value of α increases. This explains the fact that when the noise signals at different sensors are not coherent enough, a good beamforming performance can be achieved with a small value of L.

To demonstrate the performance of the Wiener beamformers, $\underline{\mathbf{h}}_{\mathrm{W}}$ and $\underline{\mathbf{g}}_{\mathrm{W}}$, given in (10.45) and (10.46), respectively, we choose $M_0 = 5$, $N_C = 5$, and iSNR = 0 dB. Figure 10.7 shows plots of the SNR gain, $\mathcal{G}\left(\underline{\mathbf{h}}|\underline{\mathbf{g}}\right)$, of the Wiener beamformers as a function of frequency for different values of L and different values of α. Similarly, the SNR gain increases as the value of L increases, but it decreases as the value of α increases. As seen, the Wiener and MVDR beamformers have the same performance in terms of the SNR gain. It is not surprising since these beamformers are identical up to a scaling factor and are special cases of the tradeoff beamformers, $\underline{\mathbf{h}}_{\mathrm{T},\mu}$ and $\underline{\mathbf{g}}_{\mathrm{T},\mu}$, i.e., $\mu = 0$ leads to the MVDR beamformers, while $\mu = 1$ gives the Wiener beamformers. It is easily observed from (10.47) and (10.48) that the SNR gains of

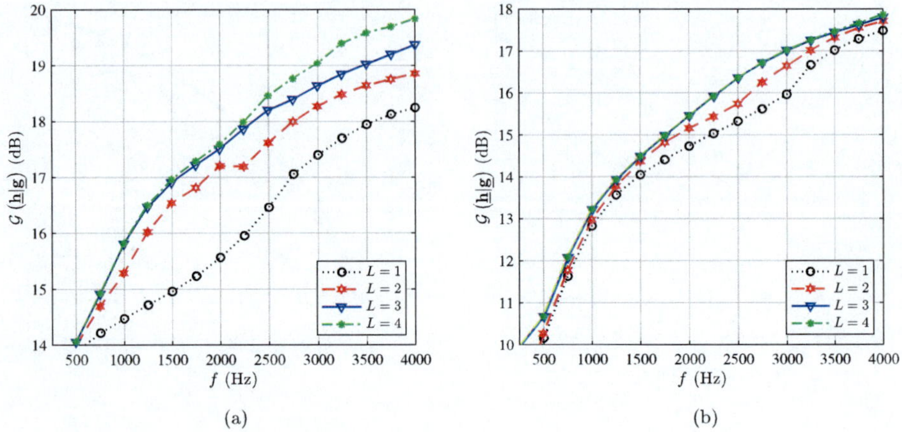

Fig. 10.7 SNR gain of the Wiener beamformers, $\underline{\mathbf{h}}_\mathrm{W}$ and $\underline{\mathbf{g}}_\mathrm{W}$, as a function of frequency, for different values of α and different values of L: (**a**) $\alpha = 10^{-4}$ and (**b**) $\alpha = 10^{-2}$. Conditions: $M_0 = 5$, $N_\mathrm{C} = 5$, $(\theta_\mathrm{s}, \phi_\mathrm{s}) = (90°, 0°)$, iSNR = 0 dB, and each subarray is a UCA with $r = 1.0$ cm

the tradeoff beamformers are independent of the parameter μ from a narrowband perspective.

To show the performance of the tradeoff beamformers, $\underline{\mathbf{h}}_{\mathrm{T},\mu}$ and $\underline{\mathbf{g}}_{\mathrm{T},\mu}$, given in (10.47) and (10.48), respectively, we choose $M_0 = 5$, $N_\mathrm{C} = 5$, $\alpha = 10^{-2}$, and $L = 3$. We use broadband performance metrics to better evaluate the performance of these beamformers. We define the broadband SNR gain, noise reduction factor, desired signal reduction factor, and desired signal distortion index in a similar way to (3.78), (3.79), (3.80), and (3.81). Figure 10.8 shows plots of the broadband SNR gain, $\overline{\mathcal{G}\left(\underline{\mathbf{h}}|\underline{\mathbf{g}}\right)}$; broadband noise reduction factor, $\overline{\xi\left(\underline{\mathbf{h}}|\underline{\mathbf{g}}\right)}$; broadband desired signal reduction factor, $\overline{\xi_\mathrm{d}\left(\underline{\mathbf{h}}|\underline{\mathbf{g}}\right)}$; and broadband desired signal distortion index, $\overline{\upsilon\left(\underline{\mathbf{h}}|\underline{\mathbf{g}}\right)}$, of the tradeoff beamformer as a function of the input SNR for different values of μ. It can be observed that the SNR gain, noise reduction factor, desired noise reduction factor, and desired signal distortion index all increase as the value of μ increases, which indicates that the low-rank tradeoff beamformer is able to achieve different compromises between noise reduction and desired signal distortion from a broadband perspective.

In the previous examples, we assumed that all subarrays that form the global array are the same. Let us now consider a large array with different subarrays. We consider an array that is composed of N_C subarrays, where each subarray is still a UCA composed of M_0 omnidirectional microphones, as shown in Fig. 10.1, but with a different radius. We choose $M_0 = 5$ and $N_\mathrm{C} = 5$, and the radii of the five subarrays are 0.8 cm, 0.9 cm, 1.0 cm, 1.1 cm, and 1.2 cm, respectively. Figure 10.9 shows plots of the broadband SNR gain, $\overline{\mathcal{G}\left(\underline{\mathbf{h}}|\underline{\mathbf{g}}\right)}$; broadband noise reduction factor,

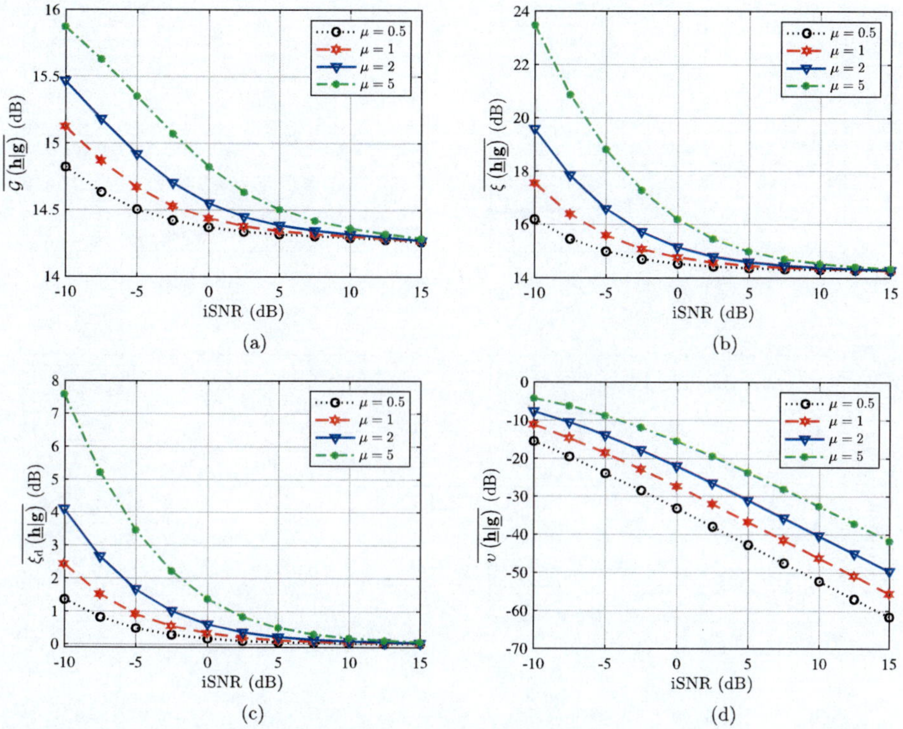

Fig. 10.8 Performance of the tradeoff beamformers, $\underline{\mathbf{h}}_{T,\mu}$ and $\underline{\mathbf{g}}_{T,\mu}$, as a function of the input SNR, for different values of μ: (**a**) broadband SNR gain, (**b**) broadband noise reduction factor, (**c**) broadband desired signal reduction factor, and (**d**) broadband desired signal distortion index. Conditions: $M_0 = 5$, $N_C = 5$, $\alpha = 10^{-2}$, $L = 3$, $(\theta_s, \phi_s) = (90°, 0°)$, and each subarray is a UCA with $r = 1.0$ cm

$\overline{\xi\left(\underline{\mathbf{h}}|\underline{\mathbf{g}}\right)}$; broadband desired signal reduction factor, $\overline{\xi_d\left(\underline{\mathbf{h}}|\underline{\mathbf{g}}\right)}$; and broadband desired signal distortion index, $\overline{\upsilon\left(\underline{\mathbf{h}}|\underline{\mathbf{g}}\right)}$, of the tradeoff beamformer as a function of the input SNR for $\alpha = 10^{-2}$, $L = 3$, and different values of μ. Similarly, the SNR gain, noise reduction factor, desired noise reduction factor, and desired signal distortion index all increase when μ increases.

We also show the performance of the MDF beamformer in this case. Figure 10.10 shows plots of the DF, $\mathcal{D}\left(\underline{\mathbf{h}}|\underline{\mathbf{g}}\right)$, and WNG, $\mathcal{W}\left(\underline{\mathbf{h}}|\underline{\mathbf{g}}\right)$, of this beamformer as a function of frequency for different values of L. It is observed that the performance is similar to that in Fig. 10.5, where the DF increases, while the WNG decreases as the value of L increases. As one can see, the MDF beamformer has a very small value of WNG at low frequencies, indicating that this beamformer suffers from serious white noise amplification. To help improve the WNG, a small regularization parameter, 10^{-2}, is added to the diagonal elements of the matrix $\underline{\mathbf{\Gamma}}_{DN}$. Figure 10.11 shows

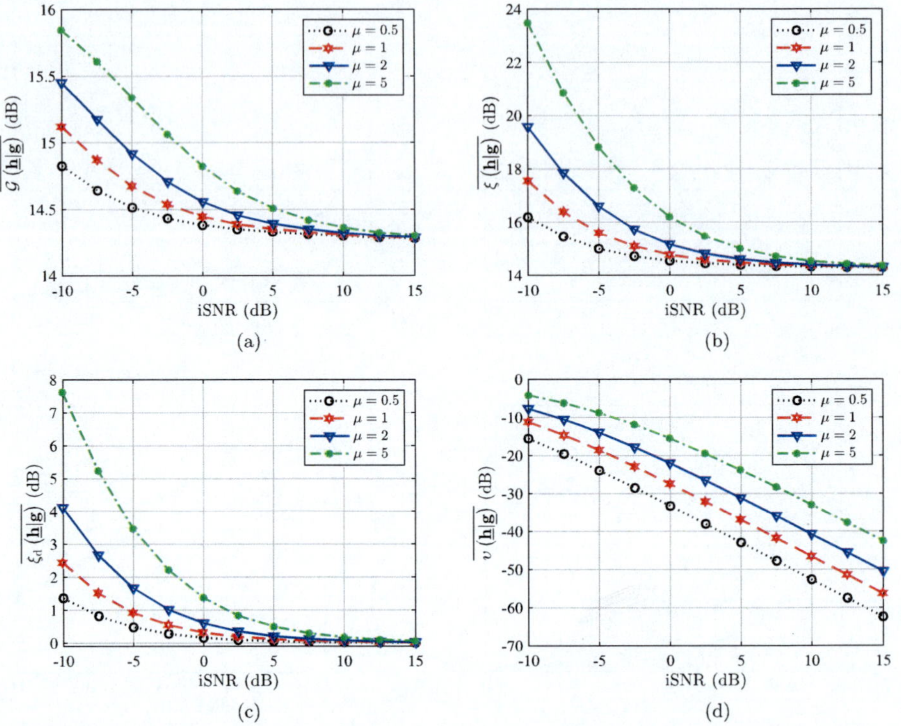

Fig. 10.9 Performance of the tradeoff beamformers, $\underline{\mathbf{h}}_{\mathrm{T},\mu}$ and $\underline{\mathbf{g}}_{\mathrm{T},\mu}$, as a function of the input SNR, for different values of μ: (**a**) broadband SNR gain, (**b**) broadband noise reduction factor, (**c**) broadband desired signal reduction factor, and (**d**) broadband desired signal distortion index. Conditions: $M_0 = 5$, $N_\mathrm{C} = 5$, $\alpha = 10^{-2}$, $L = 3$, $(\theta_\mathrm{s}, \phi_\mathrm{s}) = (90°, 0°)$, and the radii of the five subarrays are 0.8 cm, 0.9 cm, 1.0 cm, 1.1 cm, and 1.2 cm, respectively

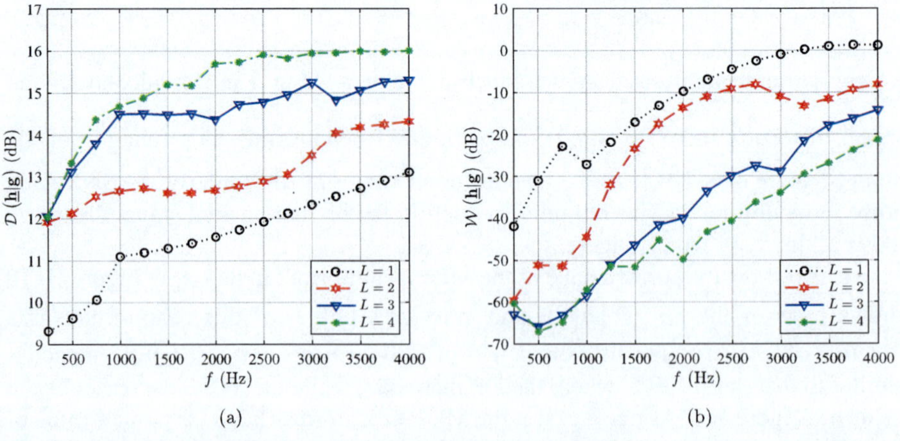

Fig. 10.10 DF and WNG of the MDF beamformers, $\underline{\mathbf{h}}_{\mathrm{MDF}}$ and $\underline{\mathbf{g}}_{\mathrm{MDF}}$, as a function of frequency, for different values of L: (**a**) DF and (**b**) WNG. Conditions: $M_0 = 5$, $N_\mathrm{C} = 5$, $(\theta_\mathrm{s}, \phi_\mathrm{s}) = (90°, 0°)$, and the radii of the five subarrays are 0.8 cm, 0.9 cm, 1.0 cm, 1.1 cm, and 1.2 cm, respectively

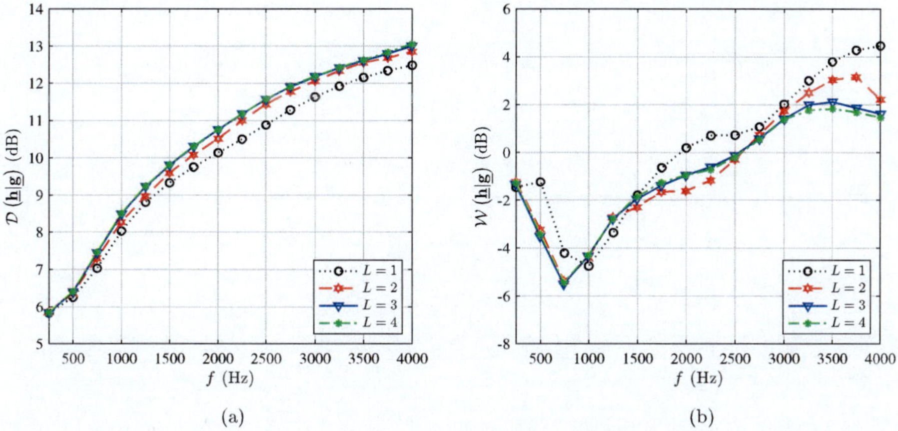

Fig. 10.11 DF and WNG of the MDF beamformers, $\underline{\mathbf{h}}_{\mathrm{MDF}}$ and $\underline{\mathbf{g}}_{\mathrm{MDF}}$ with more regularization, as a function of frequency, for different values of L: (**a**) DF and (**b**) WNG. Conditions: $M_0 = 5$, $N_{\mathrm{C}} = 5$, $(\theta_{\mathrm{s}}, \phi_{\mathrm{s}}) = (90°, 0°)$, and the radii of the five subarrays are 0.8 cm, 0.9 cm, 1.0 cm, 1.1 cm, and 1.2 cm, respectively

plots of the DF, $\mathcal{D}\left(\underline{\mathbf{h}}|\underline{\mathbf{g}}\right)$, and WNG, $\mathcal{W}\left(\underline{\mathbf{h}}|\underline{\mathbf{g}}\right)$, of this regularized beamformer as a function of frequency for different values of L, under the same conditions. As seen, the value of L does not affect much the DF and WNG, which indicates that we can use a small value of L to design a robust MDF beamformer.

References

1. J. Benesty, J. Chen, Y. Huang, *Microphone Array Signal Processing* (Springer, Berlin, 2008)
2. J. Benesty, C. Paleologu, L.-M. Dogariu, S. Ciochină, Identification of linear and bilinear systems: a unified study. Electronics **10**(15), 33 (2021)
3. D.A. Harville, *Matrix Algebra from a Statistician's Perspective* (Springer, New York, 1997)
4. J. Benesty, C. Paleologu, S. Ciochină, On the identification of bilinear forms with the Wiener filter. IEEE Signal Process. Lett. **24**, 653–657 (2017)
5. C. Paleologu, J. Benesty, S. Ciochină, Linear system identification based on a Kronecker product decomposition. IEEE/ACM Trans. Audio Speech Language Process. **26**, 1793–1808 (2018)

Index

J. Benesty et al., *Microphone Arrays*, Springer Topics in Signal Processing 22,
https://doi.org/10.1007/978-3-031-36974-2